Compartmental Models and Their Application

Compartmental Models and Their Application

Keith Godfrey

Department of Engineering,

University of Warwick, UK

1983

Academic Press

A Subsidiary of Harcourt Brace Jovanovich, Publishers

London New York

Paris San Diego San Francisco

São Paulo Sydney Tokyo Toronto

ACADEMIC PRESS INC. (LONDON) LTD
24/28 Oval Road,
London NW1 7DX

United States Edition published by
ACADEMIC PRESS INC.
111 Fifth Avenue,
New York, New York 10003

British Library Cataloguing in Publication Data

Godfrey, K.
Compartmental models and their applications.
1. Compartmental analysis (Biology)
I. Title
574.1′92 QP519.9.C/

ISBN 0-12-286970-2

Filmset by Eta Services (Typesetters) Ltd, Beccles, Suffolk
Printed in Great Britain by Galliard (Printers) Ltd, Great Yarmouth

83 010478

Foreword

The successful application of systems methods and control techniques is critically dependent on the generation and analysis of accurate models for specific classes of system. It is refreshing and heartening to welcome a text which concentrates on just such a topic. Dr Godfrey's thorough and comprehensive treatment of compartmental models and their application is a most welcome and timely contribution to the development of applied systems analysis methods. It is welcome because of the careful and lucid way in which he has organized and explained his material. It is timely on two counts. First, because of the need to digest and present in an accessible form the widespread and proliferating literature on compartmental models. Second, because the recent sudden increase in the availability of cheap computing equipment will make these techniques readily accessible to a wide group of potential users.

The careful treatment of the necessary basic dynamical theory required for a sound understanding of compartmental model behaviour which is given here, coupled with a clear treatment of many specific examples, should make this book particularly useful to workers in the fields of biomedicine, pharmacokinetics and ecology. The interesting properties of this specific class of dynamical model should however give it a much wider appeal, and many engineering specialists in control and systems analysis would profit from a study of the many detailed examples investigated here. A particularly welcome feature is the breadth of treatment given—the nonlinear, time-varying, and stochastic aspects of this class of model are not neglected, and attention is also given to the numerical aspects of the computations involved. There is a very comprehensive set of references.

I personally found this an instructive and interesting text, and I am pleased to welcome a most useful addition to the technical literature. I hope that it finds the wide readership which it deserves.

November 1982

A. G. J. MacFarlane
Cambridge

Preface

Compartmental models are widely used for the modelling of systems in biomedicine, pharmacokinetics (and anaesthesia) and ecology, and the number of applications reported in these areas runs into many hundreds. A small number of applications in engineering have appeared recently. A good deal of theoretical work on compartmental models has also been reported over the last few years, in particular, on obtaining the parameters of linear, time-invariant models, on the analysis and identification of nonlinear models, on the stability of linear but time-varying models and on stochastic models.

In view of this widespread interest, it is perhaps rather surprising that, to date, there have only been two texts aiming to cover both the theory and application of compartmental models; these are the monograph of Atkins (1969) and the book by Jacquez (1972). There are a number of books and special issues of journals giving selected papers from conferences and symposia on compartmental modelling, for example Hart (ed.) (1963), Bergner and Lushbaugh (eds.) (1967) and Matis, Patten and White (eds.) (1979), while several books have chapters giving introductions to compartmental analysis, for example Rescigno and Beck (1972) and Llaurado (1973). Certain aspects of compartmental modelling have been discussed in books primarily on other topics, for instance Shipley and Clark (1972) on tracer methods in physiology and Gibaldi and Perrier (1975) on pharmacokinetics.

Developments have been thick and fast since Jacquez's book was published in 1972 and a new book seems timely. The present book is intended for anyone interested in compartmental modelling and its application. For those new to the field, the mathematical treatment is developed informally and heuristically (there is an absence of theorems, lemmas and proofs), although it is necessary to understand something of Laplace transforms and matrices to follow the theory properly. Many examples are included throughout the book. Those already familiar with the field will, I hope, find something of interest in the later chapters of the book, which will also illustrate to those working in one of the areas of biomedicine, pharmacokinetics and ecosystem modelling that there is interesting work on compartmental modelling in progress in the other two areas.

The book starts with an introductory chapter outlining the history of the subject and the main application areas at present. The main part of the book is divided into four parts, describing the analysis of compartmental models with linear, time-invariant dynamics; the identifiability of such models; nonlinear models, linear but time-varying models and stochastic models; and some applications selected to emphasize different aspects of modelling. The book concludes with a short chapter detailing some unresolved problems in compartmental modelling and two Appendices concerned with Laplace transforms and elementary matrix algebra.

The first part of the book (Chapters 2 to 4) deals with the analysis of linear, time-invariant models. By "analysis" is meant the determination of the form of the compartmental responses to external perturbation of one or more compartments, given the rate constants. Such analysis is a prerequisite for tackling the important problem of identifiability—that is, whether, given some or all of the responses, the unknown rate constants can be found. Although most of the analysis material is well known, it is somewhat scattered through the literature. With extensive use of examples, it is hoped that this part of the book will give anyone new to the field a reasonably comprehensive introduction. Chapter 2 gives the essentials of the analysis of linear systems in general (introductory material to Laplace transforms and matrices being given in Appendices 1 and 2). Chapter 3 deals with the analysis of linear, time-invariant compartmental systems with one or two compartments, giving an exhaustive analysis of all possible two-compartment models. Chapter 4 analyses a number of systems with three or more compartments. For systems with more than two compartments, the roots (eigenvalues) of the system's characteristic equation are no longer necessarily real, and in Chapter 4, a particular form of compartmental system with complex eigenvalues is analysed.

In the second part of the book (Chapters 5 to 8), identifiability of linear, time-invariant compartmental systems is considered. Jacquez (1972) noted that "... there is still much work to be done on the inverse problem, that is, the problem of system identification and parameter estimation. This is the important problem for the working scientist, for he has to go from experimental data to a plausible description of the system which he is studying." Since 1972, a good deal of effort has gone into this problem so that now, several methods are available for the analysis of the deterministic identifiability (i.e. identifiability with perfect data) of linear, time-invariant compartmental models, although there is still much scope for work on the numerical identifiability problem, that is, how imperfections in the observations affect the accuracy of the estimated parameters. The whole problem of identifiability is discussed in Chapter 5, where reference is made to a number of approaches to deterministic identifiability analysis. The most familiar of these, based on the Laplace transforms of the observations, is described in Chapter 6 and is applied to a number of models of practical interest. The effects of perturbing more than one compartment and of observing more than one compartment are illustrated. Compartmental equations are constrained by the need for all flow rates to be non-negative, and these constraints place bounds on the rate constants in unidentifiable models. The bounds are evaluated for some unidentifiable models of practical interest in Chapter 7, and the second part of the book concludes with a discussion in Chapter 8 of the extremely important, but hitherto rarely discussed, problem of numerical identifiability.

The third part of the book discusses nonlinear compartmental models in Chapter 9 and linear but time-varying models and stochastic models in Chapter 10. The Laplace transform and most other methods of analysing deterministic identifiability are not applicable to nonlinear or time-varying systems, but an alternative approach is available, based on Taylor series expansion of the observations. This method is applied to a number of examples of nonlinear systems (some previously reported and some new) in Chapter 9 and to some examples of periodically varying systems in Chapter 10. One particular type of nonlinear system, with capacity-limited (Michaelis–Menten) elimination, is discussed in some detail in Chapter 9. Relatively little has appeared on time-varying systems (other than some stability analyses) probably due to the difficulty of obtaining analytical solutions, and what there has been is discussed in Chapter 10. In the same chapter, the substantial body of theory on stochastic compartmental models is discussed.

In the fourth part of the book (Chapter 11) some applications are described. Rather than include brief descriptions of a large number of applications, somewhat longer descriptions are given of nine applications, selected to illustrate particular aspects of modelling. The examples are drawn from biomedicine, pharmacokinetics, ecology and engineering.

Within the stipulated maximum length of the book, a certain amount of rather specialist material has had to be omitted. The selection of what to omit has been personal and as far as possible, I have given adequate reference to a fuller account of the material. Two particular topics only covered briefly are chemical reaction kinetics and stochastic compartmental models, both well covered elsewhere.

My own interest in this field was fostered through discussions with Professor Ludvik Finkelstein and Dr Ewart Carson of The City University, London and through two grants made available by the UK Science and Engineering Research Council, one concerned with linear models and the other (still in progress at the time of writing) with nonlinear models. Thanks are due to Alan Hulme, Manager of the S.E.R.C. Prime Computing Facility at the University of Warwick for considerable help with the computing of the responses presented in Chapter 3, 4, 8, 9 and 10.

Special thanks are due to Dr John Norton of the University of

Birmingham who has given much encouragement and advice in the preparation of this book. Since coming to the University of Warwick for six months' study leave in 1978 (from the University of Tasmania, where he was then on the staff), John has made several distinguished contributions on the identifiability of compartmental models. I am most grateful to him for reading through the original draft of the book and for making many suggestions for improvements.

November 1982 Keith Godfrey

Contents

1 Introduction to Compartmental Modelling

1.1 What is a compartmental system?

Compartmental systems consist of a finite number of homogeneous, well-mixed, lumped subsystems, called compartments, which exchange with each other and with the environment so that the quantity or concentration of material within each compartment may be described by a first-order differential equation. A compartmental system may be used to model either the kinetics of one substance, in which case the compartments occupy different spaces and the inter-compartment transfers represent flow of material from one location to another, or the kinetics of two or more substances (such as a drug and its metabolites) in which case different compartments may occupy the same space and some of the inter-compartment transfers represent transformation from one substance to another.

In the literature, the words "compartment" and "pool" have, to a considerable extent, come to be interchangeable, although Atkins (1969) observes that, strictly speaking, there is a distinction between the two. He notes that Sprinson and Rittenberg (1949) defined a metabolic pool of an animal (or organ or cell) as "that mixture of compounds derived either from the diet or from the breakdown of the tissues, which the animal (or organ or cell) employs for the synthesis of tissue constituents." On the other hand, Atkins defines a compartment—a term first used in this context by Sheppard (1948)—as "a quantity of a substance which has a uniform and distinguishable kinetics of transformation or transport." With this distinction, a pool will often be distributed among many compartments, although when referring to plasma, the pool and compartment will usually be the same. In this book, the term compartment will be used, as defined above.

The most general form of compartmental equations (Sandberg, 1978) for a system with p compartments is:

$$\frac{\mathrm{d}x_i}{\mathrm{d}t} = f_{i0} + \sum_{\substack{j=1 \\ j \neq i}}^{p} f_{ij} - \sum_{\substack{j=1 \\ j \neq i}}^{p} f_{ji} - f_{0i}, \qquad i = 1, 2, \ldots, p \tag{1.1}$$

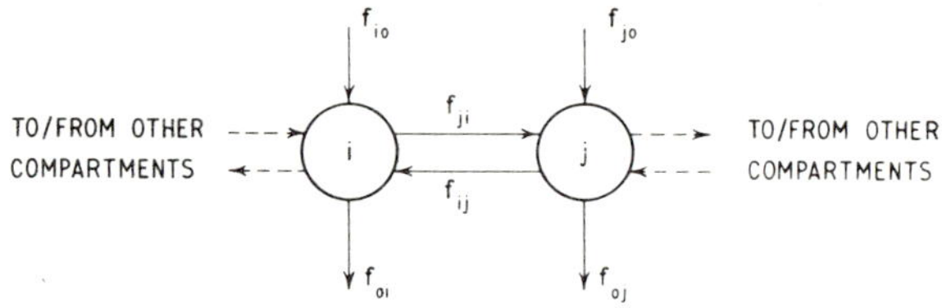

Figure 1.1 Two compartments of a general compartmental model, as described by eqn (1.1). Arrowed quantities are flow rates.

where x_i is the amount of material in compartment i (x_i can also be taken to represent concentration, but in this book it will be used for quantity); f_{ij} is the flow rate *to* compartment i *from* compartment j (note that in the pharmacokinetics literature, the order of the subscripts is reversed); and subscript 0 denotes the environment.

If the flow rates from all compartments to the environment are zero ($f_{0i} = 0, i = 1, 2, \ldots, p$), the system is said to be closed—otherwise it is open. Equation (1.1) is illustrated for two of the p compartments in Fig. 1.1.

We see that compartmental models are simply sets of constrained first-order differential equations, the constraints being the physical requirement that flow rates are non-negative. It is difficult to make progress with equations of as general a form as eqn (1.1) and most of the work reported to date has assumed a more specific form.

By far the biggest body of theory and applications has been concerned with linear, time-invariant compartmental models, for which the flow rates are directly proportional to the quantity in the donor compartment, the constant of proportionality being referred to as a rate constant. Equation (1.1) then simplifies to:

$$\frac{\mathrm{d}x_i}{\mathrm{d}t} = \sum_{\substack{j=1 \\ j \neq i}}^{p} k_{ij}x_j - \sum_{\substack{j=1 \\ j \neq i}}^{p} k_{ji}x_i - k_{0i}x_i + u_i(t), \qquad i = 1, 2, \ldots, p \qquad (1.2)$$

where the input flow rate f_{i0} from the environment has been written as $u_i(t)$ to conform with the usual notation of linear systems theory. Note that in this model, the flow rate to compartment i from compartment j is $k_{ij}x_j$, i.e. directly proportional to the quantity (x_j) in the donor compartment but independent of the quantity (x_i) in the receptor compartment. Equation (1.2) is illustrated for two of the p compartments in Fig. 1.2.

Such a model is a gross simplification of the actual system being modelled, but it does seem to provide a good description of the responses of many systems when a small perturbation is made to a system previously in a steady state, for example to describe the kinetics of a labelled tracer added to a

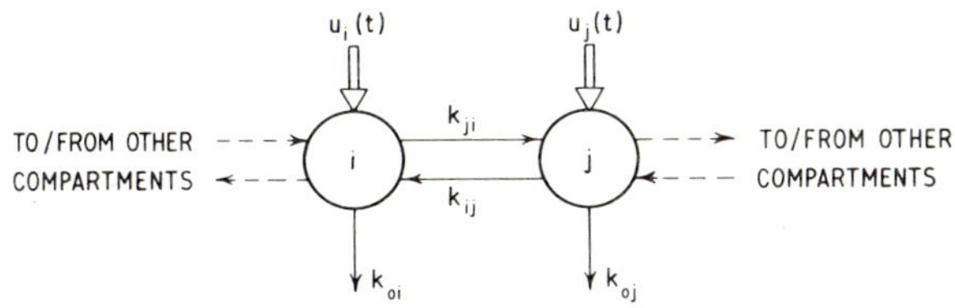

Figure 1.2 Two compartments of a linear, time-invariant compartmental model, as described by eqn (1.2). The k's are rate constants, so that flow rates are $k_{ij}x_j$, etc.; the inputs $u_i(t)$ and $u_j(t)$ are flow rates.

system. We will see in Chapters 3 and 4 that the responses of a linear, time-invariant compartmental system to an impulsive perturbation consist of sums of exponentials (usually with as many exponentials as compartments). It is interesting to note how often tracer response data from human beings and other animals approximate closely to this form, indicating that, over the range of values covered by the responses, a linear-dynamics model is reasonable.

To date, much less work has appeared on nonlinear systems, and it has concentrated on two forms of simplification of eqn (1.1). The first simplification is for systems where inter-compartment transfers, though nonlinear, are still donor-controlled. Such a system is still described by eqn (1.2), but some (or all) of the k_{ij} and k_{ji} are no longer constants, and so are referred to as rate coefficients. One frequently-occurring example describes capacity-limited (Michaelis–Menten) excretion of a drug from a compartment (i, say), when

$$\text{Excretion flow rate } f_{0i} = \frac{V_m}{K_m + x_i(t)}\, x_i(t) \tag{1.3}$$

where V_m is a constant having dimensions of $(\text{time})^{-1}$ and K_m is a constant having the same units as $x_i(t)$. (This is an adaptation of the Michaelis–Menten form of enzyme kinetics; a good introduction will be found in Riggs (1963, Chapter 11).) For small values of $x_i(t)$ such that $K_m \gg x_i(t)$,

$$f_{0i} \simeq \frac{V_m}{K_m}\, x_i(t) \tag{1.4}$$

which is of first-order linear form, with $\dfrac{V_m}{K_m}$ replacing the rate constant k_{0i} in eqn (1.2). For large values of $x_i(t)$ such that $K_m \ll x_i(t)$,

$$f_{0i} \simeq V_m \tag{1.5}$$

which is constant. The elimination does not now depend on $x_i(t)$ and is usually referred to as zero-order elimination. The model is no longer linear; doubling the input will more than double the responses of the compartmental quantities.

The second form of simplification of eqn (1.1) is for systems where some of the inter-compartmental transfers are functions of two or more states (quantities or concentrations) of the system:

$$k_{ij} = k_{ij}(x_1, x_2, \ldots, x_p). \tag{1.6}$$

Such transfer occurs in chemical reaction kinetics (see Section 1.2). One particular example occurs when the rate coefficient is a function of the state of the receptor compartment, as well as that of the donor compartment:

$$k_{ij} = k_{ij}(x_i, x_j). \tag{1.7}$$

An example is discussed in Application Example 3 in Chapter 11.

Some work has also been reported recently on two further departures from the linear, time-invariant compartmental model of eqn (1.2), the first concerning time-varying models and the second, stochastic models. Most work reported to date on time-varying models has assumed what to the control engineer is a linear, time-varying model, that is, a model where the rate coefficients in eqn (1.2) vary with time, but are not functions of the states:

$$k_{ij} = k_{ij}(t). \tag{1.8}$$

In the biochemistry literature, such a model is often called nonlinear. Of particular interest are systems where some of the rate coefficients are periodic functions of time, for example

$$k_{ij}(t) = k_{ij}(1 + \gamma \sin \omega t) \tag{1.9}$$

where γ is a constant which must be less than one to preserve the non-negativity of the rate coefficient.

Stochastic models by contrast have some feature (or features) which vary *randomly* with time, and a substantial body of theory has been built up over the past few years concerning such systems. The theory may be classified into approaches where the rate coefficients are assumed stochastic and those where the number of particles of interest within a compartment is assumed stochastic (the rate coefficients being constant); some recent work has combined the two approaches. Time-varying models and stochastic models are discussed in Chapter 10.

1.2 Some processes described by first-order differential equations

Many processes can be described by first-order differential equations, and many such equations are compartmental in form, although often not described as such. A familiar example is the equation for radioactive decay:

$$\frac{\mathrm{d}N}{\mathrm{d}t} = -\lambda N \tag{1.10}$$

where $N(t)$ is the number of radioactive particles at time t and λ is a constant. This is compartmental since N is non-negative and λN is non-negative, $\frac{\mathrm{d}N}{\mathrm{d}t}$ being non-positive.

The diffusion of material across a membrane according to Fick's Law is also a first-order process. The material transfer takes place from regions of high concentration to those of low concentration along the line of the concentration gradient, and the flow of substance per unit area is proportional to the gradient. If the concentrations in two constant-volume compartments (volumes V_1 and V_2) are C_1 and C_2, then the mass transfer equations according to Fick's Law are:

$$V_1 \frac{\mathrm{d}C_1}{\mathrm{d}t} = k \cdot \frac{S}{X} (C_2 - C_1) \tag{1.11a}$$

$$V_2 \frac{\mathrm{d}C_2}{\mathrm{d}t} = k \cdot \frac{S}{X} (C_1 - C_2) \tag{1.11b}$$

where k is the coefficient of diffusion, S is the effective surface area between the compartments, and X is the effective separation between the compartments. If k, S and X are constants, eqns (1.11) are linear and time-invariant.

Another process in which first-order differential equations arise is a simple irreversible, single-stage chemical reaction whereby chemicals A and B react to give products. The change of mass, x_A, of A is described by the equation

$$\frac{\mathrm{d}x_A}{\mathrm{d}t} = -k_{\alpha+\beta} C_A^{\alpha} C_B^{\beta} \tag{1.12}$$

where C_A is the concentration of A; C_B is the concentration of B; α is the order of the reaction with respect to A; β is the order of the reaction with respect to B; $\alpha + \beta$ is the order of the reaction; $k_{\alpha+\beta}$ is a constant with dimensions dependent on $\alpha + \beta$.

For a first-order reaction ($\alpha + \beta = 1$) with A (only) reacting to give products

$$\frac{dx_A}{dt} = -k_1 C_A, \tag{1.13}$$

and if the reaction takes place in a vessel of constant volume,

$$\frac{dx_A}{dt} = -kx_A. \tag{1.14}$$

For a reversible first-order reaction

$$A \underset{k_2}{\overset{k_1}{\rightleftharpoons}} B$$

$$\frac{dx_A}{dt} = -k_1 x_A + k_2 x_B. \tag{1.15}$$

For a reaction of the form

$$A + B \underset{k_2}{\overset{k_1}{\rightleftharpoons}} C$$

the forward reaction is at rate $k_1 C_A C_B$ and the backward reaction is at rate $k_2 C_c$, so that

$$\frac{dC_A}{dt} = -k_1 C_A C_B + k_2 C_c \tag{1.16}$$

which is of nonlinear compartmental form, with the first term on the right-hand side dependent on two states. Many equations involving conservation of mass or energy are of compartmental form.

1.3 The origins of compartmental modelling

The fact that many first-order differential equations are compartmental without necessarily being described as such makes tracing the exact origins of compartmental models rather difficult. In the biological area, Atkins (1969) notes that, while radioactive tracers were first applied to biological systems as long ago as 1923, the majority of work on such systems up to the early 1940s was of a qualitative nature. During the 1940s, there was a trend towards treating data in a more quantitative manner, so that the use of differential equations (and, consequently, compartments) to describe the behaviour of tracers became increasingly important. The first good quantitative treatment was probably that of Zilversmit, Entenman and Fishler (1943), but the term "compartment" was first introduced by Sheppard in 1948. A further significant early paper was that of Sheppard and Householder (1951) discussing the mathematical basis for interpreting tracer experiments.

The origins of pharmacokinetics are usually attributed to Teorell who published a systematic study of the kinetics of drugs introduced into the mammalian body via extravascular routes (1937a) and via intravascular routes (1937b). He gave four differential equations representing the transfer of material from the subcutaneous depot to the blood, elimination of drug from the blood to the urine, the exchange between blood and tissue, and tissue elimination. Teorell derived expressions for the quantity of drug in the subcutaneous depot (for the extravascular administration), in the blood and in the tissue and found that the last two were tri-exponential.

In a review of linear compartmental analysis in ecosystem modelling, O'Neill (1979) notes that linear, time-invariant compartmental modelling has played an important role in the development of systems ecology since its inception in the late 1950s, but models closely resembling compartmental models had appeared much earlier than this—certainly as early as 1935 in the atmospheric sciences.

These, then, seem to be the beginning of compartmental modelling in the three main applications areas. In Section 1.4, the present-day role of compartmental modelling will be assessed.

1.4 The role of compartmental modelling

Compartmental modelling has been and is being used widely in biomedicine (particularly in the modelling of metabolic processes), in pharmacokinetics and in ecosystem modelling.

In modelling metabolic system dynamics, compartmental modelling is the most widely-used approach, although it is not universally applicable. Probably the best guide on when to consider using such a model is to return to the definition given at the beginning of this chapter and for experimenters to ask themselves whether the concept of homogeneous, well-stirred tanks is appropriate to their particular application. Two recent reviews have given guidelines on when it is appropriate to model metabolic systems using compartments (Carson, Cobelli and Finkelstein, 1981; Carson, Godfrey and Reeve, 1982). In some instances, it is appropriate to model only part of a system with compartments; for example, models of the role of insulin in the regulation of blood glucose level (Carson, Godfrey and Reeve, 1982, Section 1.4.3) have equations of compartmental form in the forward path, with a non-compartmental feedback control mechanism incorporated into the overall model. Carson, Cobelli and Finkelstein (1981) note that where a compartmental structure is not well-defined, or where experimental data are insufficient for its identification, it may be more appropriate to adopt a non-compartmental approach, focusing on overall input–output relationships, especially at the level of the whole organism; several examples are cited.

Compartmental models have also been used elsewhere in biomedicine, for example, to model gas exchange in lungs, where, with modern measuring devices, it is now possible to verify such models by externally scanning radioactivity in the lungs. Examples of such modelling are given by Bajzer and Nosil (1977) and Bache, Gray and Murray-Smith (1981). External scanning has also featured in the modelling of metabolic systems—see Application Example 2 in Chapter 11.

In pharmacokinetics, it is rare for compartments to be assigned a physiological label. For linear modelling, the model is tailored to the observations, the number of compartments of an open model being made equal to the number of observed exponentials. This approach is often referred to as "black-box" modelling. If the drug in plasma concentration-time curve has two exponentials, the two compartments are called the central compartment and the peripheral compartment. If there are three exponentials, a form of model is assumed with a central compartment exchanging with a shallow peripheral compartment and a deep peripheral compartment; the peripheral compartments do not exchange directly with each other. The adjectives "shallow" and "deep" reflect fast and slow exchange with the central compartment and the only physiology in the model is that the central compartment contains the systemic plasma.

With no assumption of physiological significance for the compartments, the purpose of a linear compartmental model describing the time course of drug in plasma (or urine) becomes difficult to define. The lack of relevance of such models has been commented upon by experts in the pharmacokinetics field (see, for example, Wagner, 1975). As noted above, an approach focusing on overall input–output relationships seems preferable (provided, of course, that the resulting model is as economical a description as the compartmental alternative). Such a relationship in this case would be the fitting of exponentials (or the corresponding Laplace transfer function) to the responses of compartmental quantities to impulse or step perturbation. Going further, some pharmacokinetic quantities of interest, such as the area under the curve (AUC) and, following oral administration, peak concentration (C_{max}) and time to peak (T_{max}) do not even require the fitting of exponentials and can be found from a plot of concentration v. time.

Linear compartmental models are useful in pharmacokinetics in the simultaneous description of the kinetics of a drug and its metabolites. A metabolite is a substance formed from a drug by biotransformation (usually mainly in the liver) and is of considerable importance in that, while it may occasionally have the same therapeutic effect as the drug, it quite often produces side-effects which can limit the quantity of drug administered, particularly orally. A multi-species example, with two compartments each for the drug and its metabolites (Boxenbaum and Riegelman, 1976) is discussed in Application Example 4 in Chapter 11.

In most pharmacokinetics trials, drugs are administered in therapeutic quantities rather than minute (trace) quantities; so that some nonlinear effects might be expected (Wagner, 1973b). Although, as noted above, linear compartmental models do not seem necessary for describing the time course of drugs alone, modelling nonlinear drug kinetics within a compartmental framework proves particularly useful. Much of the work hitherto reported has been on systems with Michaelis–Menten elimination kinetics (Wagner, 1973a). Wagner (1973b) has given examples from the literature to suggest that the elimination pathways (both hepatic and renal) for many drugs are very liable to become saturated. The importance of recognizing capacity-limited elimination cannot be overemphasized. Pharmacokinetics measured following a single administration may appear linear but if a multiple-dosing regimen is then designed on the basis of linear (first-order) elimination, it can lead to a considerable build-up of drug in the body if the elimination pathway becomes saturated at higher drug levels. This has been graphically illustrated in simulation results by Wagner (1978), while Garrettson and Jusko (1975) quote an example of children overdosed with diphenylhydantoin which led these authors to fit Michaelis–Menten kinetics to the subsequent decline in drug concentration in the children.

The first application describing fitting Michaelis–Menten elimination kinetics to a single concentration-time curve seems to be that of Lundquist and Wolthers (1958) concerning alcohol elimination. The recent impetus towards nonlinear modelling in pharmacokinetics has been fostered by the publication of a single journal (*Journal of Pharmacokinetics and Biopharmaceutics*) which contains a high proportion of the good-quality papers on mathematical modelling in pharmacokinetics.

A further recent development involving compartmental models in the pharmacokinetics area concerns the simultaneous measurement of drug concentration in venous plasma (pharmacokinetics) and the pharmacological response (pharmacodynamics). The pharmacodynamics are traditionally modelled as a nonlinear function of the drug concentration at the receptor site, but this gives rise to two problems. Firstly exactly what form of nonlinear function should be used, and secondly, how should the concentration at the receptor site be related to the (measured) concentration in the central compartment and/or the (unmeasured) concentration in the peripheral compartment of the model describing the pharmacokinetics? Recent work in this area will be discussed in Application Example 7 in Chapter 11.

In the area of anaesthetics, interest in compartmental modelling has generally mirrored that in the pharmacokinetics literature, but with an emphasis on modelling pharmacological response. A good survey is given in two companion review papers, one discussing the pharmacokinetics of competitive muscle relaxants (Miller, 1982) and the other dealing with the pharmacodynamics of neuromuscular blocking agents (Hull, 1982).

Turning now to the role of compartmental models in ecology, a recent text (Matis *et al.*, 1979) presents sixteen conference papers selected to show the state of the art in this field. The editors in their preface note that ecosystem modelling began in the late 1950s with the use of compartmental models to mimic storages and flows of radiotracers, materials and energy in food chains and food webs, while during the late 1960s and early 1970s, large-scale computer models, mostly compartment-oriented, were developed. These modelled the steady state, with appropriate inputs from and outputs to the environment. The ease of computer simulation in recent years has allowed extensive perturbation experiments on the simulated systems and in some examples this has allowed the extrapolation of a trend, the beginning of which had been noticed in the real system, to its conclusion well in advance of actuality. Such extrapolation requires confidence in the validity of the model. Some of the dangers of drawing conclusions from models fitted using a limited length of data are discussed in Chapter 8; problems could arise from other sources, such as using a linear model to describe an essentially nonlinear process.

A recent review paper by O'Neill (1979) on *linear* compartmental models in ecology contains over a hundred references on the application of such models. Many ecosystem models are nonlinear, and some authors (e.g. Bledsoe (1976)) have been critical of the use of linear models. O'Neill himself notes that many of the perturbations imposed on ecosystems, particularly by human society, are not minor and cannot be considered to remain within the neighbourhood of a system equilibrium. Nevertheless, the considerable mathematical tools which can be brought to bear on a linear problem mean that linear compartmental models will continue to be used either as preliminary models or as final models, subject to constraints on size of perturbation.

The most recent developments in ecosystem modelling have been concerned with stochastic modelling and statistical inference within a compartmental framework.

The areas above are those where compartmental models have been most widely used. In engineering, equations of compartmental form occur in chemical reaction kinetics; they are usually nonlinear and the solutions for many such sets of equations will be found in the chemical engineering literature, for example Gavalas (1968). In electrical engineering, many sets of equations, for example lumped equivalent models of transmission lines, equations for filters and ladder networks, are compartmental in form but are not recognized explicitly as such. In the process industries, models of systems are usually either low-order, largely black-box models used for on-line prediction and control, or high-order models, based on the physics and chemistry of the process and used particularly in the plant design stage. In the latter type of model, many equations reflect conservation of mass or energy and these are usually of compartmental form. Typical examples are the mass balance equations of a grinding process (Kortela and Niemi, 1978) and the energy balance equations of a distillation column (Espana and Landau, 1978); the latter is discussed further as Application Example 9 in Chapter 11.

The problem with viewing engineering system models in terms of compartments is that while most equations are of compartmental form, many systems will have some which are not. For example, equations concerned with conservation of momentum are often non-compartmental, while if the model includes equations for closed-loop control, these will rarely be compartmental. The compartmental approach is then only of limited relevance.

2 The Dynamics of Linear, Time-invariant Systems

Before considering the dynamics of compartmental systems in particular in Section 2.4, we will review the input–output relationships for systems with linear, time-invariant dynamics in general in Sections 2.1 to 2.3.

2.1 Single-input, single-output (SISO) systems

Consider a dynamic system with input $u(t)$ and response to that input $x(t)$, as shown in Fig. 2.1. Assume the following:

(i). The system is lumped, rather than distributed; that is, its behaviour can be described over a finite number of points in space. The differential equations describing the behaviour are ordinary rather than partial.

(ii). The system is linear so that the principle of superposition applies; that is, the response to an input $u_1(t) + u_2(t)$ is the sum of responses to $u_1(t)$ and $u_2(t)$ applied separately.

(iii). The system is in steady state prior to the application of the input at time $t = 0$. The quantities referred to in the remainder of this chapter are deviations from the (steady) values $u(0^-)$ and $x(0^-)$ immediately prior to the input perturbation. Higher-order derivatives are zero at $t = 0^-$.

The input and output are then related by the *convolution integral*:

$$x(t) = \int_0^t g(t - \lambda)u(\lambda)\,\mathrm{d}\lambda \tag{2.1}$$

where λ is a time variable ranging over the interval up to the present time t during which the input has been applied and $g(t)$ is the *weighting function* of the system, which weights *past* values of the input to give the *present* value of the output.

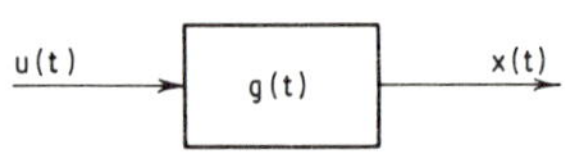

Figure 2.1 Dynamic system with weighting function $g(t)$.

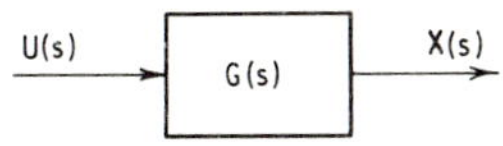

Figure 2.2 Dynamic system with transfer function $G(s)$.

We have further assumed that the system is time-invariant, that is, its dynamics do not change with time. The weighting function is then the same as the response of the system to a unit impulse.

Except for simple forms of input, the convolution integral of eqn (2.1) is difficult to evaluate analytically and use is then made of the corresponding Laplace transform relationship:

$$X(s) = G(s)U(s) \tag{2.2}$$

where $X(s)$ and $U(s)$ are the Laplace transforms of $x(t)$ and $u(t)$ respectively and $G(s)$, the Laplace transform of $g(t)$, is called the (Laplace) *transfer function* of the system (Fig. 2.2).

See Appendix 1 for an exposition of Laplace transforms.

2.1.1 *First-order systems*

For a first-order, linear, time-invariant system, the input and output are related by the ordinary differential equation:

$$T \cdot \dot{x}(t) + x(t) = K \cdot u(t) \tag{2.3}$$

where K is the *steady-state gain* of the system (i.e. the proportional factor between x and u in the steady state), T is the *time constant* of the system and $\dot{x}(t) \equiv \frac{\mathrm{d}x}{\mathrm{d}t}$. The general solution of eqn (2.3) is given by the sum of the complementary function and particular integral:

$$x(t) = \mathrm{e}^{-t/T}x(0^-) + \int_0^t \mathrm{e}^{-(t-\lambda)/T}\frac{K}{T}u(\lambda)\,\mathrm{d}\lambda. \tag{2.4}$$

The Laplace transform of eqn (2.3) is

$$(sT + 1)X(s) - T \cdot x(0^-) = K \cdot U(s) \tag{2.5}$$

which, on rearrangement, gives

$$X(s) = \frac{T \cdot x(0^-)}{sT + 1} + \frac{K \cdot U(s)}{sT + 1}. \tag{2.6}$$

Assuming that $x(0^-)$ is zero, or has been subtracted from all subsequent values of $x(t)$, eqns (2.4) and (2.6) become

$$x(t) = \int_0^t \frac{K}{T} e^{-(t-\lambda)/T} u(\lambda)\, d\lambda \tag{2.7}$$

$$X(s) = \frac{K \cdot U(s)}{sT + 1}. \tag{2.8}$$

Comparing these with eqns (2.1) and (2.3), it may be seen that the weighting function is given by

$$g(t) = \frac{K}{T} e^{-t/T} \tag{2.9}$$

and that the transfer function is given by

$$G(s) = \frac{X(s)}{U(s)} = \frac{K}{sT + 1}. \tag{2.10}$$

If a unit impulse is applied to the system at $t = 0$, then $U(s) = 1$, $X(s) = G(s)$ and $x(t) = g(t)$. The unit impulse response of a first-order system is shown in Fig. 2.3. At time $t = T$, the value of the response has fallen to

$$x(T) = \frac{K}{T} e^{-1} \simeq 0.368 \frac{K}{T} \tag{2.11}$$

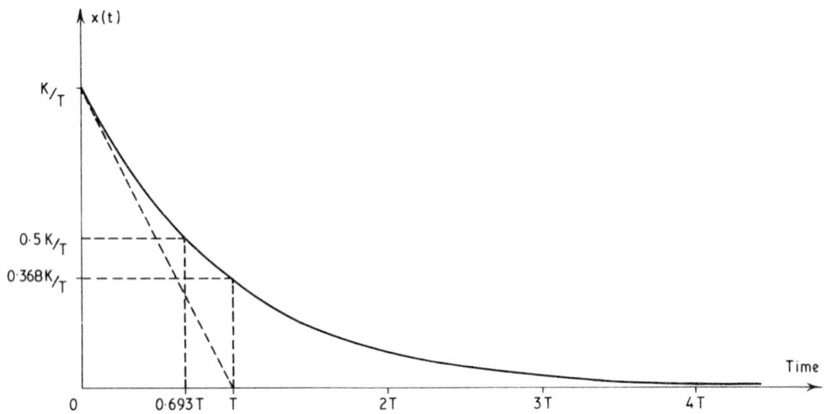

Figure 2.3 Unit impulse response of first-order system described by eqn (2.3).

so that, in one time constant, the response has fallen to 36.8% of its maximum value (at $t = 0$). The response falls to 1% of its maximum value when

$$0.01 = e^{-t/T}$$

i.e.

$$t_{0.01} = T \log_e 100 \simeq 4.605T. \tag{2.12}$$

In the medical and biological literature, the half-life, $t_{1/2}$, is more often quoted than the time constant. The half-life is the time taken for the response to fall to 50% of its maximum value and so is related to the time constant by

$$0.5 = e^{-t_{1/2}/T}$$

whence

$$t_{1/2} = T \cdot \log_e 2 \simeq 0.693T. \tag{2.13}$$

The unit step is the integral of the unit impulse and the unit-step response may be found by integrating the unit-impulse response of eqn (2.9):

$$x(t) = \int_0^t \frac{K}{T} e^{-t/T} \, dt = K(1 - e^{-t/T}) \tag{2.14}$$

which is illustrated in Fig. 2.4. The unit-step response may also be found from the transfer function using eqn (2.10). For a unit step, $U(s) = \frac{1}{s}$ so that

$$X(s) = G(s)U(s) = \frac{K}{s(sT + 1)}. \tag{2.15}$$

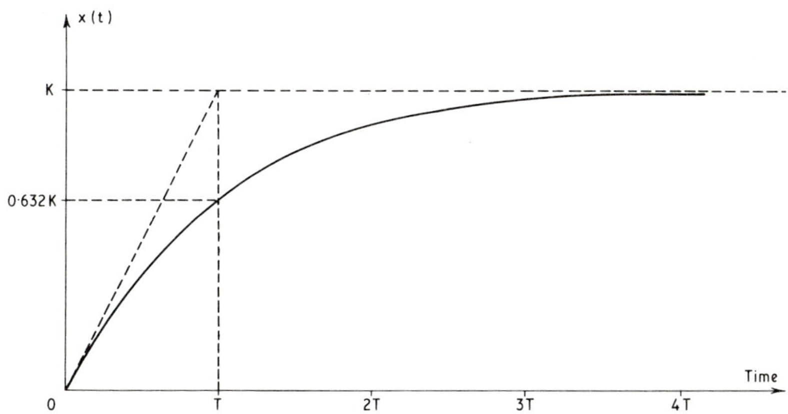

Figure 2.4 Unit step response of first-order system described by eqn (2.3).

Taking partial fractions (see Appendix 1),

$$X(s) = \frac{K}{s} - \frac{KT}{sT + 1} \tag{2.16}$$

and using the Table of Laplace transforms in Appendix 1,

$$x(t) = K - K\,e^{-t/T}, \qquad t \geqslant 0 \tag{2.17}$$

which is the same as eqn (2.14).

Example 2.1. *A resistor R and inductor L are connected in series as shown in Fig. 2.5. The current is zero when, at time $t = 0$, the switch is closed. Find the current $i(t)$ for $t \geqslant 0$.*

Solution. The voltage drops v_R and v_L across the resistor and inductor are:

$$v_R = i \cdot R \quad \text{and} \quad v_L = L\frac{di}{dt}.$$

Applying Kirchhoff's Voltage Law (that the sum of voltage drops around a circuit equals the applied voltage),

$$L\frac{di}{dt} + iR = e(t) = E, \qquad t \geqslant 0.$$

Comparing this with eqn (2.3), it may be seen that $T = \frac{L}{R}$ and $K = \frac{1}{R}$ so that substituting into eqn (2.14),

$$i(t) = \frac{E}{R}(1 - e^{-Rt/L}).$$

Alternatively, Laplace transformation of the differential equation above gives

$$(sL + R)I(s) = \frac{E}{s} \quad \text{so that} \quad I(s) = \frac{E}{s(sL + R)}.$$

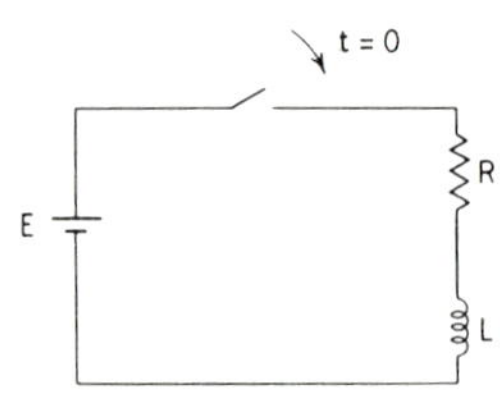

Figure 2.5 Series RL circuit of Example 2.1.

Taking partial fractions (see Appendix 1),

$$I(s) = \frac{E}{R}\cdot\frac{1}{s} - \frac{EL}{R}\cdot\frac{1}{sL+R}$$

giving $i(t) = \frac{E}{R}(1 - e^{-Rt/L})$, on inverse Laplace transformation.

The question of units for an impulsive input sometimes gives rise to confusion. The Dirac delta function is defined by

$$\delta(t) = 0 \qquad t \neq 0 \tag{2.18a}$$

$$\int_{-\infty}^{\infty} \delta(t)\,dt = 1. \tag{2.18b}$$

Consider the pulse function $f(t)$ shown in Fig. 2.6, of height $\frac{1}{\varepsilon}$, width ε and unit area. The delta function is given by

$$\delta(t) = \lim_{\varepsilon \to 0} f(t) \tag{2.19}$$

so that the delta function is a function of infinitely large height acting for an infinitely short time. Such an idealized function cannot be realized in practice, but its shape is approximated in a number of practical instances. For example, an impulsive blow of magnitude P can be regarded as a *force* of $P \cdot \delta(t)$ while a rapidly-administered dose of size D can be regarded as an input *rate* of $D \cdot \delta(t)$. Note the use of "rate", because $\delta(t)$ has units of $(\text{time})^{-1}$. The delta function is clearly the derivative of the unit step function $H(t)$, i.e.

$$\delta(t) = \frac{d}{dt} H(t). \tag{2.20}$$

To complete the discussion of the use of the delta function, let us return to the circuit of Fig. 2.5, but suppose that, instead of a step E being applied at

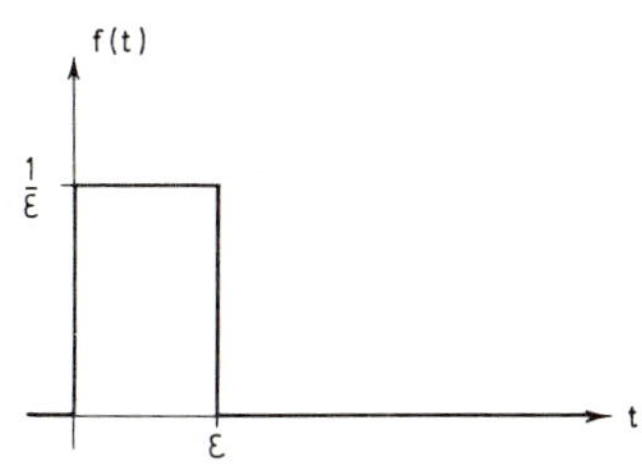

Figure 2.6 A pulse function of unit area.

$t = 0$, an impulsive voltage $E_0\,\delta(t)$ is applied. The impulsive voltage is one whose time integral is E_0 and since the units of $E_0\,\delta(t)$ are volts and those of $\delta(t)$ are $(\text{time})^{-1}$, the units of E_0 are (volts) × (time). The differential equation is thus

$$L\frac{\mathrm{d}i}{\mathrm{d}t} + iR = E_0\,\delta(t).$$

With $i(0^-)$ zero,

$$(sL + R)\cdot I(s) = E_0$$

$$I(s) = \frac{E_0}{sL + R}$$

giving, on inverse Laplace transformation,

$$i(t) = \frac{E_0}{L}\,\mathrm{e}^{-Rt/L}, \qquad t \geqslant 0.$$

2.1.2 *Second-order systems*

The dynamics of a second-order, linear, time-invariant system may be described by the differential equation

$$\frac{1}{\omega_0^2}\ddot{x} + \frac{2\zeta}{\omega_0}\dot{x} + x = u \tag{2.21}$$

where ζ, the *damping ratio* defines the degree of damping in the system, being a measure of the presence and extent of oscillation in the system response and ω_0, the *undamped natural frequency* is the frequency of oscillation (radians per second) of the system in the absence of damping, i.e. with $\zeta = 0$.

The transfer function of the system is

$$G(s) \triangleq \frac{X(s)}{U(s)} = \frac{\omega_0^2}{s^2 + 2\zeta\omega_0 s + \omega_0^2}. \tag{2.22}$$

Unit step responses for various values of ζ are shown in Fig. 2.7; corresponding impulse responses may be found by differentiation. The unit step responses are given by

$$x(t) = \mathscr{L}^{-1}\left(\frac{G(s)}{s}\right)$$

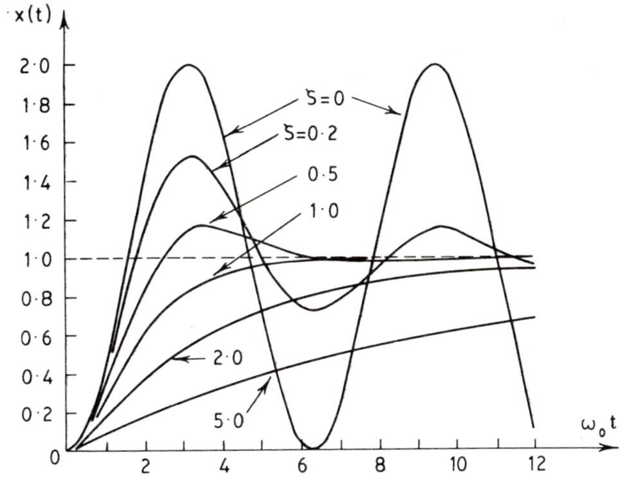

Figure 2.7 Unit step responses (for various values of ζ) of second-order system described by eqn (2.21).

and to find the inverse Laplace transform using the table given in Appendix 1, it is necessary to complete the square in the denominator of $G(s)$, by writing

$$s^2 + 2\zeta\omega_0 s + \omega_0^2 \equiv (s + \zeta\omega_0)^2 + \omega_0^2(1 - \zeta^2). \tag{2.23}$$

The shape of the response depends on whether $\omega_d^2 \triangleq \omega_0^2(1 - \zeta^2)$ is positive, zero or negative, and we will consider these three cases in turn.

(*a*) $0 \leqslant \zeta < 1$. Since ω_0 is real, $\omega_0^2(1 - \zeta^2)$ is positive and the system is said to be *underdamped*. The unit step response is the inverse transform of

$$X(s) = \frac{1}{s} - \frac{s + \zeta\omega_0}{(s + \zeta\omega_0)^2 + (1 - \zeta^2)\omega_0^2} - \frac{\zeta\omega_0}{(s + \zeta\omega_0)^2 + (1 - \zeta^2)\omega_0^2}$$

i.e. $$x(t) = 1 - e^{-\zeta\omega_0 t}\left(\cos\omega_d t + \frac{\zeta}{(1 - \zeta^2)^{1/2}}\sin\omega_d t\right). \tag{2.24}$$

If $\zeta = 0$, the oscillations are undamped, but if $\zeta > 0$, the oscillations are damped by the $e^{-\zeta\omega_0 t}$ factor (damped because ζ and ω_0 are non-negative). The roots of eqn (2.23) are a complex conjugate pair:

$$s = -\alpha \pm j\beta \tag{2.25}$$

where $\alpha = \zeta\omega_0$ is real and non-negative and $\beta = \omega_d = \omega_0(1 - \zeta^2)^{1/2}$ is real.

(*b*) $\zeta = 1$. The term $\omega_0^2(1 - \zeta^2)$ is zero and the system is said to be *critically damped*. The unit step response is the inverse transform of

$$X(s) = \frac{\omega_0^2}{s(s + \omega_0)^2} = \frac{1}{s} - \frac{s + 2\omega_0}{(s + \omega_0)^2}$$

i.e.

$$x(t) = 1 - e^{-\omega_0 t}(1 + \omega_0 t). \tag{2.26}$$

The denominator of eqn (2.22) consists of a repeated real root ($s = -\omega_0$). The step response is no longer oscillatory (see Fig. 2.7).

(*c*) $\zeta > 1$. The term $\omega_0^2(1 - \zeta^2)$ is negative and the system is said to be *overdamped*. The unit step response is the inverse transform of:

$$X(s) = \frac{\omega_0^2}{s(s + \zeta\omega_0 + \omega_0(\zeta^2 - 1)^{1/2})(s + \zeta\omega_0 - \omega_0(\zeta^2 - 1)^{1/2})}$$

i.e.

$$x(t) = 1 + \frac{\omega_0}{2(\zeta^2 - 1)^{1/2}} \times \left\{\frac{\exp[-(\zeta + (\zeta^2 - 1)^{1/2})\omega_0 t]}{\zeta + (\zeta^2 - 1)^{1/2}} - \frac{\exp[-(\zeta - (\zeta^2 - 1)^{1/2}\omega_0 t]]}{\zeta - (\zeta^2 - 1)^{1/2}}\right\}. \tag{2.27}$$

The roots of eqn (2.22) are real and distinct and it may be seen from Fig. 2.7 that the unit step response is again non-oscillatory, and gets progressively slower as ζ gets larger.

Example 2.2. *A resistor R, inductor L and capacitor C are connected in series as shown in Fig. 2.8. The current i and capacitor charge q are zero when, at* $t = 0$, *the switch is closed. Find the current* $i(t)$ *for* $t \geqslant 0$.

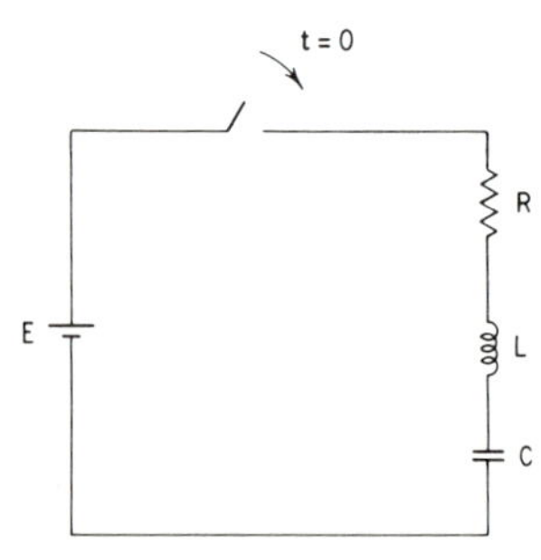

Figure 2.8 Series RLC circuit of Example 2.2.

Solution. There are two storage elements (the inductor and the capacitor) so a second-order differential equation is expected. The voltage drops across the elements of the circuit are:

$$v_L = L\frac{\mathrm{d}i}{\mathrm{d}t} = L\frac{\mathrm{d}^2q}{\mathrm{d}t^2}$$

$$v_R = Ri = R\frac{\mathrm{d}q}{\mathrm{d}t}$$

$$v_C = \frac{1}{C}\int_{0^-}^{t} i\,\mathrm{d}t = \frac{q}{C}$$

so that, applying Kirchhoff's Voltage Law,

$$L\frac{\mathrm{d}^2q}{\mathrm{d}t^2} + R\frac{\mathrm{d}q}{\mathrm{d}t} + \frac{q}{C} = E, \qquad t \geqslant 0.$$

Comparing this with eqn (2.21), it may be seen that

$$\frac{1}{\omega_0^2} = LC \quad \text{so} \quad \omega_0 = \frac{1}{(LC)^{1/2}}$$

$$\frac{2\zeta}{\omega_0} = RC \quad \text{so} \quad \zeta = \frac{R}{2}\left(\frac{C}{L}\right)^{1/2}.$$

Laplace transformation of the differential equation gives, with $i(0^-) = q(0^-) = 0$,

$$(s^2 + (R/L)s + 1/LC)Q(s) = \frac{E/L}{s}$$

whence
$$Q(s) = \frac{E}{L}\frac{1}{s(s^2 + (R/L)s + 1/LC)}.$$

But $i = \dfrac{\mathrm{d}q}{\mathrm{d}t}$ so that, with $q(0^-) = 0$, $I(s) = s \cdot Q(s)$, giving

$$I(s) = \frac{E}{L}\cdot\frac{1}{s^2 + (R/L)s + 1/LC}.$$

Let us consider one case only, with circuit values so that $\dfrac{R}{2}\left(\dfrac{C}{L}\right)^{1/2} < 1$, which, from the foregoing, will yield an underdamped system, since

$0 \leqslant \zeta < 1$. From eqns (2.22) and (2.24),

$$i(t) = \frac{E}{\omega_0^2 L}\frac{\omega_0}{(1-\zeta^2)^{1/2}}\,\mathrm{e}^{-\zeta\omega_0 t}\sin(\omega_0(1-\zeta^2)^{1/2}t)$$

$$= \frac{E}{L}\,\mathrm{e}^{-\zeta\omega_0 t}\,\frac{\sin(\omega_0(1-\zeta^2)^{1/2}t)}{\omega_0(1-\zeta^2)^{1/2}}$$

$$= \frac{E}{L}\,\mathrm{e}^{-\alpha t}\,\frac{\sin\omega t}{\omega}$$

where
$$\alpha = \zeta\omega_0 = \frac{R}{2L}$$

and
$$\omega^2 = \omega_0^2(1-\zeta^2) = \frac{1}{LC}\left(1 - \frac{R^2 C}{4L}\right) = \frac{1}{LC} - \frac{R^2}{4L^2}.$$

2.2 The state variables of a dynamic system

Much of the recent literature on time-domain analysis of a system utilizes the concept of the *state* of a system. The state is a set of numbers such that the knowledge of these numbers and the input functions will, together with the equations describing the dynamics, give the future state and output of the system. The *state variable* representation of a multi-input, multi-output system is:

$$\dot{\mathbf{x}} = \mathbf{A}\mathbf{x} + \mathbf{B}\mathbf{u}. \tag{2.28}$$

In this *state equation*, the vector $\mathbf{x}$ represents the state of the system. In some applications, it is difficult to choose variables of direct physical significance (and the state variable set is not necessarily unique) but in compartmental modelling this problem does not exist because all the differential equations are first-order and the state variables are compartmental quantities or concentrations.

Equation (2.28) indicates that the first derivative of state variable x_i is a linear function of the p state variables and m inputs:

$$\dot{x}_i = a_{i1}x_1 + a_{i2}x_2 + \ldots + a_{ip}x_p + b_{i1}u_1 + b_{i2}u_2 + \ldots + b_{im}u_m \tag{2.29}$$

where the a's and b's are constants for a time-invariant system. Equations of the form of (2.29) may be written for all the states:

$$\begin{bmatrix} \dot{x}_1 \\ \vdots \\ \dot{x}_p \end{bmatrix} = \begin{bmatrix} a_{11} & a_{12} & \ldots & a_{1p} \\ \vdots & & & \\ a_{p1} & a_{p2} & \ldots & a_{pp} \end{bmatrix} \begin{bmatrix} x_1 \\ \vdots \\ x_p \end{bmatrix} + \begin{bmatrix} b_{11} & b_{12} & \ldots & b_{1m} \\ \vdots & & & \\ b_{p1} & b_{p2} & \ldots & b_{pm} \end{bmatrix} \begin{bmatrix} u_1 \\ \vdots \\ u_m \end{bmatrix} \tag{2.30}$$

as in eqn (2.28). The outputs (or observations) are not necessarily the state variables themselves, but may be linear combinations of them:

$$y_i = c_{i1}x_1 + c_{i2}x_2 + \ldots + c_{ip}x_p \tag{2.31}$$

that is

$$\mathbf{y} = \mathbf{C}\mathbf{x}. \tag{2.32}$$

Note that the *observation equation* (2.32) is a non-dynamic relationship, being an algebraic, rather than differential, equation.

If there are p states, m inputs and n outputs (observations), $\mathbf{A}$ is a $p \times p$ matrix, $\mathbf{B}$ is a $p \times m$ matrix and $\mathbf{C}$ is a $n \times p$ matrix.

Before proceeding further, let us see how the first-order and second-order SISO systems considered earlier can be formulated in state variable form. For the first-order system, eqn (2.3) can be rearranged as

$$\dot{x}(t) = -\frac{1}{T}x(t) + \frac{K}{T}u(t) \tag{2.33}$$

which is to be compared with the state variable form

$$\dot{x}(t) = Ax(t) + Bu(t). \tag{2.34}$$

Direct comparison shows that $A = -\dfrac{1}{T}$ and $B = \dfrac{K}{T}$.

The second-order differential eqn (2.21) can be expressed as two simultaneous first-order differential equations. Define $x_1 \triangleq x$, $x_2 \triangleq \dot{x}$ so that $\dot{x} \equiv \dot{x}_1 \equiv x_2$. Then from eqn (2.21),

$$\begin{aligned}\ddot{x} \equiv \dot{x}_2 &= -2\zeta\omega_0\dot{x}_1 - \omega_0^2 x_1 + \omega_0^2 u \\ &= -2\zeta\omega_0 x_2 - \omega_0^2 x_1 + \omega_0^2 u\end{aligned}$$

which may be rearranged to give

$$\begin{aligned}\dot{x}_1 &= \qquad\qquad x_2 \\ \dot{x}_2 &= -\omega_0^2 x_1 - 2\zeta\omega_0 x_2 + \omega_0^2 u.\end{aligned}$$

In vector-matrix form,

$$\begin{bmatrix}\dot{x}_1 \\ \dot{x}_2\end{bmatrix} = \begin{bmatrix}0 & 1 \\ -\omega_0^2 & -2\zeta\omega_0\end{bmatrix}\begin{bmatrix}x_1 \\ x_2\end{bmatrix} + \begin{bmatrix}0 \\ \omega_0^2\end{bmatrix}u$$

i.e.

$$\dot{\mathbf{x}} = \mathbf{A}\mathbf{x} + \mathbf{b}u. \tag{2.35}$$

The approach is readily extended to higher-order equations.

2.3 The dynamics of multivariable systems

2.3.1 *The transfer function matrix*

The transfer function matrix $\mathbf{G}(s)$ relates all the inputs making up $\mathbf{U}(s)$ to the outputs $\mathbf{Y}(s)$. Taking Laplace transforms of eqns (2.28) and (2.32), with $\mathbf{x}(0^-)$ assumed zero,

$$s \cdot \mathbf{X}(s) = \mathbf{AX}(s) + \mathbf{BU}(s) \tag{2.36}$$

$$\mathbf{Y}(s) = \mathbf{CX}(s). \tag{2.37}$$

Rearranging eqn (2.36),

$$(s\mathbf{I} - \mathbf{A})\mathbf{X}(s) = \mathbf{BU}(s)$$

whence

$$\mathbf{X}(s) = (s\mathbf{I} - \mathbf{A})^{-1}\mathbf{BU}(s). \tag{2.38}$$

Substituting into eqn (2.37),

$$\mathbf{Y}(s) = \mathbf{C}(s\mathbf{I} - \mathbf{A})^{-1}\mathbf{BU}(s) \tag{2.39}$$

so that the transfer function matrix $\mathbf{G}(s)$ is given by

$$\mathbf{G}(s) = \mathbf{C}(s\mathbf{I} - \mathbf{A})^{-1}\mathbf{B}. \tag{2.40}$$

Example 2.3. *Establish the transfer function matrix for the system whose state equations are:*

$$\begin{aligned}\dot{x}_1 &= -2x_1 + x_2 + u_1\\ \dot{x}_2 &= 2x_1 - 3x_2 + u_2\\ y_1 &= x_1 \quad \text{and} \quad y_2 = x_2.\end{aligned}$$

Solution. For this system,

$$\mathbf{C} = \begin{bmatrix} 1 & 0 \\ 0 & 1 \end{bmatrix} = \mathbf{I}, \quad \mathbf{B} = \begin{bmatrix} 1 & 0 \\ 0 & 1 \end{bmatrix} = \mathbf{I} \quad \text{and} \quad \mathbf{A} = \begin{bmatrix} -2 & 1 \\ 2 & -3 \end{bmatrix}$$

$$s\mathbf{I} - \mathbf{A} = \begin{bmatrix} s+2 & -1 \\ -2 & s+3 \end{bmatrix}$$

so

$$(s\mathbf{I} - \mathbf{A})^{-1} = \frac{1}{\Delta(s)}\begin{bmatrix} s+3 & 1 \\ 2 & s+2 \end{bmatrix}$$

where

$$\Delta(s) = \begin{vmatrix} s+2 & -1 \\ -2 & s+3 \end{vmatrix} = s^2 + 5s + 4 = (s+4)(s+1).$$

Thus
$$\mathbf{G}(s) = \begin{bmatrix} \dfrac{s+3}{(s+4)(s+1)} & \dfrac{1}{(s+4)(s+1)} \\ \dfrac{2}{(s+4)(s+1)} & \dfrac{s+2}{(s+4)(s+1)} \end{bmatrix}$$

and
$$\mathbf{Y}(s) = \mathbf{G}(s)\mathbf{U}(s).$$

If there are m inputs and n outputs, $\mathbf{G}(s)$ is an $n \times m$ matrix. Term $G_{ij}(s)$ within the matrix is the transfer function between the jth input and the ith output.

If only a few of the $G_{ij}(s)$ terms are required, for example because only a few of the states may be perturbed and only a few observed, it is usually easier to use Cramer's rule rather than inverting $(s\mathbf{I} - \mathbf{A})$—see Appendix 2. Thus, in Example 2.3, if it were possible only to perturb x_1 and only to observe x_1 (i.e. $u_2 = 0$, $y_2 = 0$), and it were required to find only $G_{11}(s)$, this could be more easily done as follows:

Transforming the two scalar-state equations,

$$(s+2)X_1(s) - X_2(s) = U_1(s)$$
$$-2X_1(s) + (s+3)X_2(s) = 0$$

so
$$X_1(s) = \frac{\begin{vmatrix} U_1(s) & -1 \\ 0 & s+3 \end{vmatrix}}{\begin{vmatrix} s+2 & -1 \\ -2 & s+3 \end{vmatrix}} = \frac{(s+3)U_1(s)}{(s+4)(s+1)}$$

and since $Y_1(s) = X_1(s)$,

$$G_{11}(s) = \frac{Y_1(s)}{U_1(s)} = \frac{s+3}{(s+4)(s+1)}.$$

2.3.2 *The transition matrix*

The solution of eqn (2.28) is given by

$$\mathbf{x}(t) = \mathrm{e}^{\mathbf{A}t}\,\mathbf{x}(0^-) + \int_{0^-}^{t} \mathrm{e}^{\mathbf{A}(t-\lambda)}\,\mathbf{Bu}(\lambda)\,\mathrm{d}\lambda \tag{2.41}$$

(cf. the solution of eqn (2.4) for the first-order system described by eqn (2.3)).

The matrix $e^{\mathbf{A}t}$ is called the *transition matrix* and is sometimes written as $\mathbf{\Phi}(t)$. It is the sum of the infinite series

$$e^{\mathbf{A}t} = \mathbf{I} + \mathbf{A}t + \mathbf{A}^2\frac{t^2}{2!} + \mathbf{A}^3\frac{t^3}{3!} + \ldots \tag{2.42}$$

which is usually an inconvenient expression to evaluate in practice because of the uncertainty as to how many terms must be computed for acceptable accuracy and the need to compute the powers of **A**. It is often easier to use Laplace transforms. Transforming eqn (2.28),

$$s \cdot \mathbf{X}(s) - \mathbf{x}(0^-) = \mathbf{AX}(s) + \mathbf{BU}(s) \tag{2.43}$$

whence

$$(s\mathbf{I} - \mathbf{A})\mathbf{X}(s) = \mathbf{x}(0^-) + \mathbf{BU}(s)$$

giving

$$\mathbf{X}(s) = (s\mathbf{I} - \mathbf{A})^{-1}\mathbf{x}(0^-) + (s\mathbf{I} - \mathbf{A})^{-1}\mathbf{BU}(s). \tag{2.44}$$

Inverse transformation gives

$$\mathbf{x}(t) = \mathscr{L}^{-1}[(s\mathbf{I} - \mathbf{A})^{-1}\mathbf{x}(0^-)] + \mathscr{L}^{-1}[(s\mathbf{I} - \mathbf{A})^{-1}\mathbf{BU}(s)]. \tag{2.45}$$

Comparing eqns (2.41) and (2.45), it may be seen that

$$e^{\mathbf{A}t} = \mathscr{L}^{-1}[(s\mathbf{I} - \mathbf{A})^{-1}]. \tag{2.46}$$

For a system with zero initial conditions, we have, from eqns (2.32) and (2.45),

$$\mathbf{y}(t) = \mathbf{Cx}(t) = \mathbf{C}\mathscr{L}^{-1}[(s\mathbf{I} - \mathbf{A})^{-1}\mathbf{BU}(s)] = \mathscr{L}^{-1}[\mathbf{G}(s)\mathbf{U}(s)],$$

as expected.

Example 2.4. *Find the transition matrix for the system of Example 2.3.*

Solution. For that system,

$$(s\mathbf{I} - \mathbf{A})^{-1} = \begin{bmatrix} \dfrac{s+3}{(s+4)(s+1)} & \dfrac{1}{(s+4)(s+1)} \\[2ex] \dfrac{2}{(s+4)(s+1)} & \dfrac{s+2}{(s+4)(s+1)} \end{bmatrix}$$

$$= \begin{bmatrix} \dfrac{1/3}{s+4} + \dfrac{2/3}{s+1} & \dfrac{-1/3}{s+4} + \dfrac{1/3}{s+1} \\[2ex] \dfrac{-2/3}{s+4} + \dfrac{2/3}{s+1} & \dfrac{2/3}{s+4} + \dfrac{1/3}{s+1} \end{bmatrix}$$

so the transition matrix $= \mathscr{L}^{-1}[(s\mathbf{I} - \mathbf{A})^{-1}]$ is given by

$$\mathrm{e}^{\mathbf{A}t} = \mathbf{\Phi}(t) = \begin{bmatrix} \frac{1}{3}\mathrm{e}^{-4t} + \frac{2}{3}\mathrm{e}^{-t} & -\frac{1}{3}\mathrm{e}^{-4t} + \frac{1}{3}\mathrm{e}^{-t} \\ -\frac{2}{3}\mathrm{e}^{-4t} + \frac{2}{3}\mathrm{e}^{-t} & \frac{2}{3}\mathrm{e}^{-4t} + \frac{1}{3}\mathrm{e}^{-t} \end{bmatrix}.$$

2.3.3 *Eigenvalues and eigenvectors*

The *eigenvalues* λ_i of a $p \times p$ matrix $\mathbf{A}$ are the solutions of the equation:

$$|\lambda\mathbf{I} - \mathbf{A}| = 0. \tag{2.47}$$

As we have seen in Section 2.3.2, the transition matrix $\mathrm{e}^{\mathbf{A}t}$ is given by

$$\begin{aligned} \mathrm{e}^{\mathbf{A}t} &= \mathscr{L}^{-1}[(s\mathbf{I} - \mathbf{A})^{-1}] \\ &= \mathscr{L}^{-1}\left[\frac{\operatorname{adj}(s\mathbf{I} - \mathbf{A})}{|s\mathbf{I} - \mathbf{A}|}\right] \end{aligned}$$

(see the material on matrix inversion in Appendix 2). Writing the individual elements of the transformed matrix as $\alpha_{ij}(s)$, so that

$$\mathrm{e}^{\mathbf{A}t} = \mathscr{L}^{-1}\begin{bmatrix} \alpha_{11}(s) & \alpha_{12}(s) & \ldots & \alpha_{1p}(s) \\ \vdots & & & \\ \alpha_{p1}(s) & \alpha_{p2}(s) & \ldots & \alpha_{pp}(s) \end{bmatrix} \tag{2.48}$$

we note that $|s\mathbf{I} - \mathbf{A}|$ appears in the denominator of all the elements $\alpha_{ij}(s)$.

From the definition of the eigenvalues, $|s\mathbf{I} - \mathbf{A}|$ has factors $s - \lambda_k$, $k = 1, 2, \ldots, p$ corresponding to the roots λ_k of eqn (2.47). Thus

$$\alpha_{ij}(s) = \frac{a_{ij}(s)}{(s - \lambda_1)(s - \lambda_2)\ldots(s - \lambda_p)}$$

where $a_{ij}(s)$ is a polynomial in s of degree $(p - 1)$ or less. By taking partial fractions, assuming the eigenvalues are all different,

$$\alpha_{ij}(s) = \frac{\beta_1}{s - \lambda_1} + \frac{\beta_2}{s - \lambda_2} + \ldots + \frac{\beta_p}{s - \lambda_p} \tag{2.49}$$

where $\beta_1 \ldots \beta_p$ are constants. Thus each element of the transition matrix $\mathrm{e}^{\mathbf{A}t}$ will have the form $\beta_1\,\mathrm{e}^{\lambda_1 t} + \beta_2\,\mathrm{e}^{\lambda_2 t} + \ldots + \beta_p\,\mathrm{e}^{\lambda_p t}$.

If we consider solution of the homogeneous equation $\dot{\mathbf{x}} = \mathbf{A}\mathbf{x}$, the foregoing suggests that we can find solutions of the form $\mathbf{x} = \mathrm{e}^{\lambda t}\mathbf{m}$, where $\mathbf{m}$ is a non-zero constant vector. From the system equation,

$$\dot{\mathbf{x}} = \lambda\,\mathrm{e}^{\lambda t}\,\mathbf{m} = \mathbf{A}\,\mathrm{e}^{\lambda t}\,\mathbf{m}$$

whence
$$\lambda \mathbf{m} = \mathbf{A}\mathbf{m}$$
i.e.
$$(\lambda \mathbf{I} - \mathbf{A})\mathbf{m} = \mathbf{0}. \tag{2.50}$$

The vector **m** is called an *eigenvector* (of **A**), and there is an eigenvector associated with each eigenvalue. The p eigenvectors are linearly independent, and the general solution of $\dot{\mathbf{x}} = \mathbf{A}\mathbf{x}$ may be written as

$$\mathbf{x} = c_1\,\mathrm{e}^{\lambda_1 t}\mathbf{m}_1 + c_2\,\mathrm{e}^{\lambda_2 t}\mathbf{m}_2 + \ldots + c_p\,\mathrm{e}^{\lambda_p t}\mathbf{m}_p \tag{2.51}$$

where $c_1, c_2, \ldots, c_p$ are arbitrary constants which depend on the initial conditions. The eigenvectors form the columns of the *modal matrix* **M**, which is used in a number of approaches to the identifiability problem (see Chapter 5). For the purposes of this book, the above discussion of eigenvectors is sufficient, and for a more detailed description of the use of the modal matrix in system dynamic analysis, the reader is referred to Chapter 2 of Power and Simpson (1978) or Chapter 1 of Owens (1981).

Example 2.5. *For the system of Example 2.3, find the eigenvalues of* **A** *and find the responses* $y_1(t)$ *and* $y_2(t)$ *to perturbation* $u_1(t) = \delta(t)$, $u_2(t) = 0$. *Determine the eigenvectors of* **A**.

Solution. The eigenvalues are given by:

$$\begin{vmatrix} \lambda + 2 & -1 \\ -2 & \lambda + 3 \end{vmatrix} = 0$$

giving $\lambda^2 + 5\lambda + 4 = 0$ whence $\lambda_1 = -4, \lambda_2 = -1$.

$$\begin{aligned}
\mathbf{Y}(s) &= \mathbf{C}(s\mathbf{I} - \mathbf{A})^{-1}\mathbf{B}\mathbf{U}(s) \\
&= \begin{bmatrix} \dfrac{s+3}{(s+4)(s+1)} & \dfrac{1}{(s+4)(s+1)} \\ \dfrac{2}{(s+4)(s+1)} & \dfrac{s+2}{(s+4)(s+1)} \end{bmatrix} \begin{bmatrix} 1 \\ 0 \end{bmatrix} \\
&= \begin{bmatrix} \dfrac{s+3}{(s+4)(s+1)} \\ \dfrac{2}{(s+4)(s+1)} \end{bmatrix}
\end{aligned}$$

whence
$$y_1(t) = \tfrac{1}{3}\,\mathrm{e}^{-4t} + \tfrac{2}{3}\,\mathrm{e}^{-t}$$
$$y_2(t) = -\tfrac{2}{3}\,\mathrm{e}^{-4t} + \tfrac{2}{3}\,\mathrm{e}^{-t}.$$

Taking the first eigenvalue ($\lambda_1 = -4$), eqn (2.50) gives

$$\begin{bmatrix} -4+2 & -1 \\ -2 & -4+3 \end{bmatrix} \begin{bmatrix} m_{11} \\ m_{21} \end{bmatrix} = 0$$

whence $$-2m_{11} - m_{21} = 0$$

and $$-2m_{11} - m_{21} = 0.$$

Note that the same equation is obtained twice, so that only the ratio between m_{11} and m_{21} is available. This is evident from eqn (2.51) because if any $\mathbf{m}_1$ satisfies the equation, so does any scalar multiple of $\mathbf{m}_1$.

Thus $$\mathbf{m}_1 = m_{11} \begin{bmatrix} 1 \\ -2 \end{bmatrix}.$$

Taking the second eigenvalue, eqn (2.50) gives

$$\begin{bmatrix} -1+2 & -1 \\ -2 & -1+3 \end{bmatrix} \begin{bmatrix} m_{12} \\ m_{22} \end{bmatrix} = 0$$

whence $$m_{12} - m_{22} = 0$$

$$-2m_{12} + 2m_{22} = 0$$

so that $$\mathbf{m}_2 = m_{12} \begin{bmatrix} 1 \\ 1 \end{bmatrix}.$$

Thus for this example, the modal matrix $\mathbf{M}$ is given by

$$\mathbf{M} = \begin{bmatrix} m_{11} & m_{12} \\ -2m_{11} & m_{12} \end{bmatrix}.$$

2.4 Dynamics of compartmental systems

As noted previously in this chapter, the state variable description is particularly appropriate for compartmental systems since these consist of sets of first-order differential equations (one per compartment) and the states are the compartmental quantities or concentrations. For a linear, time-invariant compartmental system with p compartments, $\mathbf{A}$ is $p \times p$ and has elements

$$a_{ij} = k_{ij}, \qquad i \neq j \tag{2.52}$$

$$a_{jj} = -\sum_{\substack{i=1 \\ i \neq j}}^{p} k_{ij} - k_{0j}. \tag{2.53}$$

Thus the compartmental form constrains all elements on the main diagonal of **A** to be non-positive and all other elements to be non-negative. Further

$$|a_{jj}| \geqslant \sum_{\substack{i=1 \\ i \neq j}}^{p} k_{ij} \tag{2.54}$$

the equality applying only if $k_{0j} = 0$. So, for example, the system

$$\dot{x}_1 = -x_1 + x_2 + u_1 \tag{2.55a}$$

$$\dot{x}_2 = x_1 + x_2 + u_2 \tag{2.55b}$$

is non-compartmental because a_{22} is positive, while the system

$$\dot{x}_1 = -x_1 + 2x_2 + u_1 \tag{2.56a}$$

$$\dot{x}_2 = 2x_1 - 2x_2 + u_2 \tag{2.56b}$$

is non-compartmental because $|a_{11}|$ is less than a_{21}. On the other hand, the model considered in Examples 2.3 to 2.5 is compartmental, as it satisfies both of the above requirements. For this system,

$$a_{11} = -2, \quad a_{21} = k_{21} = 2 \quad \text{so} \quad k_{01} = -a_{21} - a_{11} = 0$$

and $\quad a_{22} = -3, \quad a_{12} = k_{12} = 1 \quad \text{so} \quad k_{02} = -a_{12} - a_{22} = 2.$

The form of **A** for a compartmental system imposes some restrictions on the form of the dynamic responses. Firstly, the transition matrix $e^{\mathbf{A}t}$ is non-negative for all t (Bellman, 1960, p. 172). The essence of the proof is to take the largest-magnitude element on the main diagonal, $\hat{a}_{jj}$, and to express **A** as

$$\mathbf{A} = \hat{a}_{jj}\mathbf{I} + \mathbf{R} \tag{2.57}$$

where **R** contains only non-negative terms, by definition. Then

$$\begin{aligned} e^{\mathbf{A}t} &= e^{(\hat{a}_{jj}\mathbf{I} + \mathbf{R})t} \\ &= e^{\hat{a}_{jj}\mathbf{I}t} \cdot e^{\mathbf{R}t} \end{aligned} \tag{2.58}$$

the latter manipulation being possible because **I** commutes with any other matrix so that **IR** = **RI**. Although $\hat{a}_{jj}$ is negative, $e^{\hat{a}_{jj}\mathbf{I}t}$ is non-negative, being a diagonal matrix with all its principal diagonal terms non-negative. Since **R** has only non-negative elements, $e^{\mathbf{R}t}$ is also non-negative because there are only non-negative terms in the infinite series for $e^{\mathbf{R}t}$ (see eqn (2.42)). Thus, from eqn (2.58), $e^{\mathbf{A}t}$ is non-negative. For a compartmental system, all elements of **B** are non-negative, so that for a positive input vector $\mathbf{u}(t)$ and non-negative initial state $\mathbf{x}(0^-)$, *all the states of a compartmental system are non-negative*, from eqn (2.41).

Secondly, the eigenvalues of a compartmental system are real if, for all cycles of r compartments within a system, a cycle being a series of interconnected compartments $\alpha \to \beta \to \gamma \to \ldots \to \rho \to \alpha$

$$k_{\alpha\beta}k_{\beta\gamma} \ldots k_{\rho\alpha} = k_{\beta\alpha}k_{\gamma\beta} \ldots k_{\alpha\rho} \tag{2.59}$$

(Goldberg, 1956). This is a sufficient, rather than a necessary, condition for real eigenvalues so that if it does not hold, the eigenvalues may be real or complex.

One of the main implications of this result is that *for a two-compartment system, the eigenvalues must be real*, since $k_{12}k_{21} = k_{21}k_{12}$. This is also evident from the definition of the eigenvalues $|\lambda\mathbf{I} - \mathbf{A}| = 0$ giving

$$(\lambda - a_{11})(\lambda - a_{22}) - a_{12}a_{21} = 0 \tag{2.60}$$

whence $$\lambda_1, \lambda_2 = \tfrac{1}{2}\{(a_{11} + a_{22}) \pm [(a_{11} - a_{22})^2 + 4a_{12}a_{21}]^{1/2}\} \tag{2.61}$$

Now, $(a_{11} - a_{22})^2 \geqslant 0$ and since $a_{12} = k_{12}$ and $a_{21} = k_{21}$, the product $a_{12}a_{21} \geqslant 0$ so the part under the square root sign in eqn (2.61) cannot be negative and the eigenvalues must be real.

The other main implication concerns two types of compartmental system of considerable practical interest—mammillary systems and catenary-connected systems (Sheppard and Householder, 1951). A mammillary system (Fig. 2.9) is one in which a central compartment (no. 1) is surrounded by $(p - 1)$ peripheral compartments which exchange material only with the central compartment and not with each other. A catenary-connected system (Fig. 2.10) is one in which the compartments are arranged in a chain, each

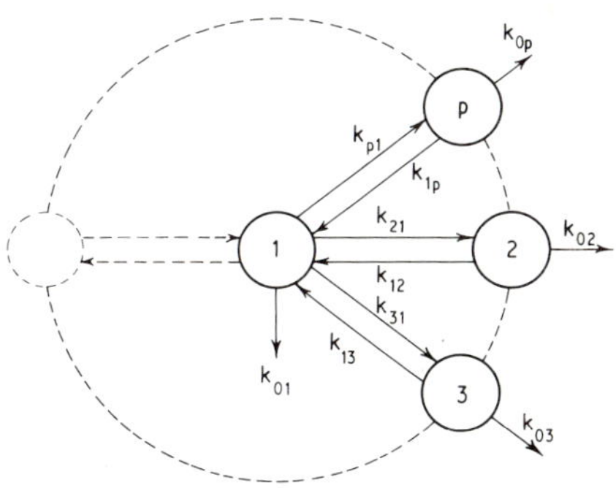

Figure 2.9 A mammillary compartmental system.

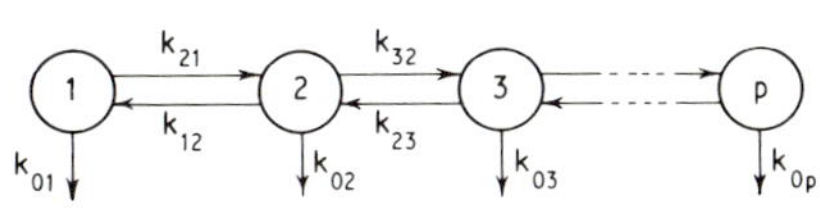

Figure 2.10 A catenary-connected compartmental system.

compartment exchanging material only with the immediately preceding and following compartments; in some applications, exchange is unidirectional. For both mammillary and catenary systems, the eigenvalues must be real because at least one pair of rate constants k_{ij} and k_{ji} must be zero, so that both sides of eqn (2.59) are zero. Hearon (1979) has shown that, for compartmental systems with real eigenvalues, in which $\mathbf{x}(0^-) = 0$ and an impulsive input is introduced to compartment r at $t = 0$, the subsequent decay of $x_r(t)$ is monotonic, i.e.

$$\frac{\mathrm{d}x_r}{\mathrm{d}t} \leqslant 0, \qquad t > 0. \tag{2.62}$$

This result does not necessarily hold for systems with complex eigenvalues.

For systems in which three or more compartments are connected in a cycle for which eqn (2.59) does not hold, complex eigenvalues are possible. Because $\mathbf{A}$ is real, all complex eigenvalues must occur as complex conjugate pairs:

$$\lambda_m, \lambda_n = -\alpha \pm j\beta$$

where $j = \sqrt{-1}$, α is real and positive and β is real. These give rise to terms of the form $E\,\mathrm{e}^{-\alpha t} \sin(\beta t + \phi)$ in the state responses $\mathbf{x}(t)$. (Here, E and ϕ are constants.) However, such oscillatory terms are in practice heavily damped by the $\mathrm{e}^{-\alpha t}$ term, and as we have seen above, no state response (compartmental quantity) may cross zero. A particular form of compartmental system giving rise to complex eigenvalues, the unilateral circulation system, will be considered in more detail in Chapter 4.

An experiment of practical interest in physiological applications is the washout test, in which a continuous infusion is made for a long time into one compartment until the system is in a steady state. The infusion is then withdrawn at $t = 0$, and the washout curves are the plots of $\mathbf{x}(t)$ for $t > 0$. From eqn (2.41), the washout curves are given by

$$\mathbf{x}(t) = \mathrm{e}^{\mathbf{A}t}\,\mathbf{x}(0^-) \tag{2.63}$$

where $\mathbf{x}(0^-)$ are the compartmental quantities just before the infusion is withdrawn. Hearon (1968) has shown that for any compartmental system, including those with complex eigenvalues, all the washout curves decay monotonically to zero, i.e.

$$\frac{\mathrm{d}\mathbf{x}(t)}{\mathrm{d}t} \leqslant 0, \qquad t > 0. \tag{2.64}$$

The most general form of **B**:

$$\mathbf{B} = \begin{bmatrix} b_{11} & b_{12} & \dots & b_{1m} \\ \vdots & & & \\ b_{p1} & b_{p2} & \dots & b_{pm} \end{bmatrix}$$

allows for proportions of the same input to be introduced to more than one compartment. In almost all compartmental applications, each input is introduced to one compartment only, so that the off-diagonal elements of **B** are zero. The vector **u** in the general system eqn (2.30) is, for compartmental systems, a vector of input rates to the p compartments:

$$\mathbf{u} = [u_1 \quad u_2 \dots u_p]^T.$$

For an impulsive input of magnitude D, for example, the input rate would be $D \cdot \delta(t)$, where $\delta(t)$ is the Dirac delta function, while for a step (constant continuous infusion) of magnitude k_i per unit time, the input rate would be k_i, $t > 0$.

For the diagonal form of **B**, which will be assumed in the remainder of this book, the product **Bu** is the vector

$$\mathbf{Bu} = [b_1 u_1 \; b_2 u_2 \dots b_p u_p]^T \tag{2.65}$$

where $u_1 \dots u_p$ are the input rates to compartments 1 to p and $b_1 \dots b_p$ are the input gains, i.e. those fractions of the inputs administered which appear in the compartmental system. For most experimental situations, $b_1 = b_2 = \dots = b_p = 1$, but an important exception occurs in metabolic systems and pharmacokinetics when a substance is administered via an extravascular route. Then, some of the dose administered does not at any stage appear in the compartmental system; the fraction which does is called the bioavailability fraction and is often not known beforehand.

Similarly, the most general form of **C**:

$$\mathbf{C} = \begin{bmatrix} c_{11} & c_{12} & \dots & c_{1p} \\ \vdots & & & \\ c_{n1} & c_{n2} & \dots & c_{np} \end{bmatrix}$$

allows for proportions of different states to be included in any one measurement. In most compartmental applications, each observation (measurement) is of one state only, so that **Cx** is a vector:

$$\mathbf{Cx} = [c_1 x_1 \; c_2 x_2 \dots c_p x_p]^T. \tag{2.66}$$

An exception to this may occur when external scanning is being used; then, the measurement of radioactivity may include a portion from more than one compartment. In the remainder of this book though **Cx** will be taken as the vector form of eqn (2.66).

2.5 Frequency-domain characterization of system dynamics

In this chapter, emphasis has deliberately been placed on the time-domain characterization of system dynamics, and the corresponding expressions in the Laplace operator s. There is also a considerable body of theory concerning frequency-domain characterization which gives no additional information, but which presents the same information in a different manner. This is sometimes more appropriate, particularly when sinusoidal input functions are feasible. Such functions are not feasible for most compartmental system applications, for which the inputs are more usually impulse and/or step functions. Details of frequency-domain characterizations will therefore not be given in this book, and the interested reader is instead referred to a standard text, for example Richards (1979).

3 Analysis of Systems with One and Two Compartments

In this chapter, the dynamic response of linear, time-invariant compartmental systems with one and two compartments will be considered.

3.1 One-compartment system

In some practical situations, the system being modelled can be approximated by a model with only one compartment. In physiological systems, for example, the compartment would contain the systemic blood and well perfused tissue, and the effect on the kinetics of less well perfused tissue would be negligible.

The one-compartment system is shown in Fig. 3.1. The differential equation for the quantity x_1 in the compartment is

$$\dot{x}_1(t) = -k_{01}x_1(t) + b_1u_1(t) \tag{3.1}$$

so that

$$\frac{X_1(s)}{U_1(s)} = \frac{b_1}{s + k_{01}}. \tag{3.2}$$

The observation is

$$y_1(t) = c_1x_1(t) \tag{3.3}$$

where c_1 is the observation gain. If the observation (i.e. measurement) is of concentration, for example, and $x_1(t)$ is a quantity,

$$c_1 = \frac{1}{V_1} \tag{3.4}$$

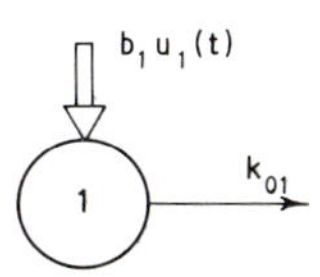

Figure 3.1 One-compartment model.

where V_1 is the *apparent volume of distribution* of the compartment. The overall transfer function $G(s)$ is thus

$$G(s) = \frac{Y_1(s)}{U_1(s)} = \frac{c_1 b_1}{s + k_{01}}. \tag{3.5}$$

3.1.1 *Impulsive input*

If a quantity D_1 is rapidly administered,

$$u_1(t) = D_1\,\delta(t) \tag{3.6}$$

where $\delta(t)$ is the unit delta function (see eqn (2.18)). Thus

$$U_1(s) = D_1 \tag{3.7}$$

and, by inverse Laplace transformation (assuming $x_1(0^-) = 0$),

$$y_1(t) = c_1 b_1 D_1\,\mathrm{e}^{-k_{01}t}, \qquad t \geqslant 0 \tag{3.8}$$

as shown in Fig. 3.2 (cf. Fig. 2.3). The zero-time intercept $y_1(0)$ is $c_1 b_1 D_1$, the time constant is $\dfrac{1}{k_{01}}$ and, from eqn (2.13), the half-life is $\dfrac{0.693}{k_{01}}$.

3.1.2 *Step input (constant continuous infusion)*

If, instead of being impulsive, the input is administered at a steady rate k_i per unit time, starting at $t = 0$, i.e.

$$u_1(t) = k_i, \qquad t \geqslant 0 \tag{3.9}$$

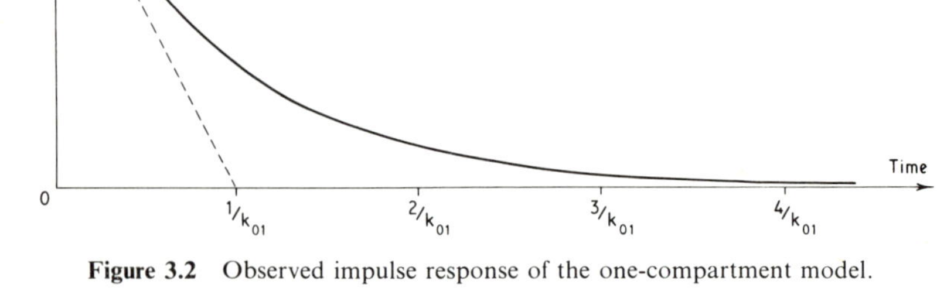

Figure 3.2 Observed impulse response of the one-compartment model.

then

$$U_1(s) = \frac{k_i}{s} \tag{3.10}$$

so that, from eqn (3.2) and eqn (3.3),

$$Y_1(s) = \frac{c_1 b_1 k_i}{s(s + k_{01})} = \frac{c_1 b_1 k_i}{k_{01}} \left(\frac{1}{s} - \frac{1}{s + k_{01}} \right). \tag{3.11}$$

Inverse Laplace transformation gives

$$y_1(t) = \frac{c_1 b_1 k_i}{k_{01}} (1 - e^{-k_{01} t}), \qquad t \geqslant 0 \tag{3.12}$$

as shown in Fig. 3.3 (cf. Fig. 2.4).

Note the general result here (for a linear system) that the unit step response (with $k_i = 1$) is the integral of the unit impulse response (with $D_1 = 1$); this is confirmed by integrating eqn (3.8) to give the unit step response

$$y_1(t) = \int_0^t c_1 b_1 \, e^{-k_{01} t} \, dt = \frac{c_1 b_1}{k_{01}} (1 - e^{-k_{01} t}), \qquad t \geqslant 0. \tag{3.13}$$

In a number of medical applications, it is preferable to administer a dose D_1 as a constant infusion over a time T_i, so that

$$u_1(t) = \frac{D_1}{T_i}, \qquad 0 \leqslant t < T_i \tag{3.14}$$

$$= 0, \qquad \text{otherwise.} \tag{3.15}$$

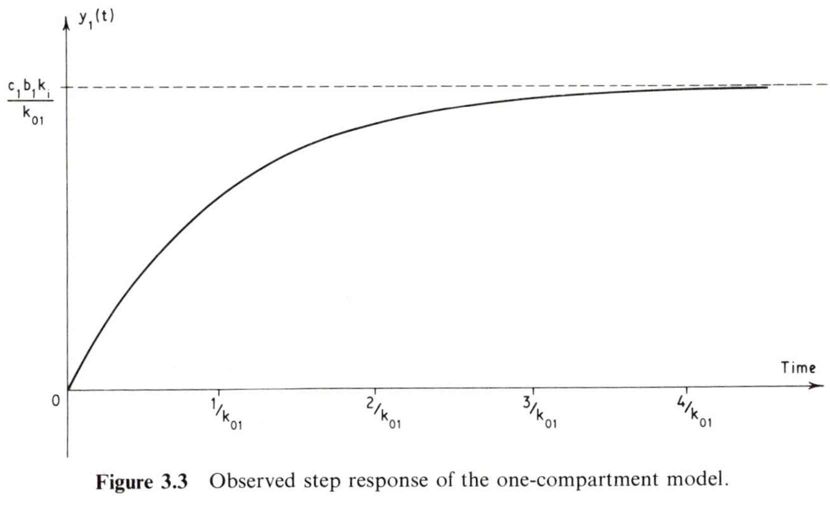

Figure 3.3 Observed step response of the one-compartment model.

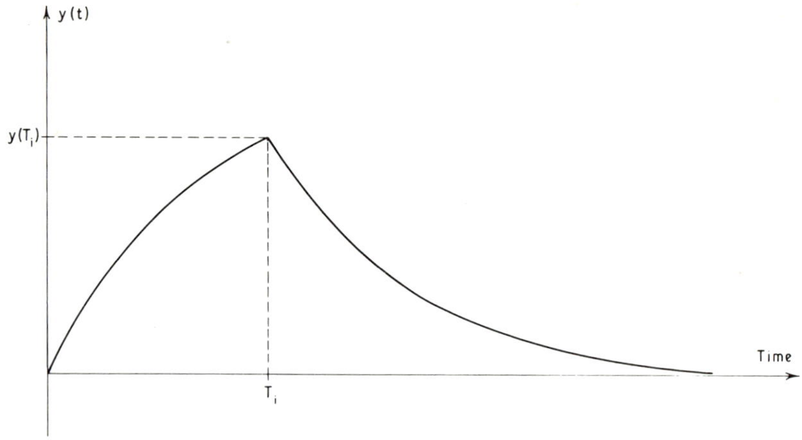

Figure 3.4 Observed response of the one-compartment model to a constant infusion over the time range $0 \leqslant t \leqslant T_i$.

From eqn (3.12), the measured quantity during the infusion is

$$y_1(t) = \frac{c_1 b_1}{k_{01}} \frac{D_1}{T_i} (1 - e^{-k_{01}t}), \qquad 0 \leqslant t < T_i \tag{3.16}$$

reaching a value at the end of the infusion of

$$y_1(T_i) = \frac{c_1 b_1}{k_{01}} \frac{D_1}{T_i} (1 - e^{-k_{01}T_i}). \tag{3.17}$$

After the infusion has stopped,

$$y_1(t) = y_1(T_i)\, e^{-k_{01}(t - T_i)}, \qquad t \geqslant T_i. \tag{3.18}$$

A plot of $y_1(t)$, as given by eqns (3.16) and (3.18), is shown in Fig. 3.4.

3.1.3 *The washout curve*

As noted in Section 2.4, a washout experiment is one in which a continuous infusion is made for a very long time until the system is in a steady state and is then stopped, the subsequent decay curves being the washout. If the input rate is k_i and $b_1 = 1$, then the steady-state value of x_1 is, from eqn (3.12),

$$x_{1ss} = \frac{k_i}{k_{01}} \tag{3.19}$$

so that the washout curve, with the input stopped at $t = 0$, is

$$x_1(t) = \frac{k_i}{k_{01}} e^{-k_{01}t}. \tag{3.20}$$

Comparing eqns (3.8) and (3.20), we see that the form of the impulse response is the same as that of the washout curve; this is not a general result and applies only to a one-compartment system.

3.1.4 *Repeated impulsive input*

In pharmacokinetics applications, doses are often repeated at regular intervals. Assume that the doses are impulsive, of magnitude D_1 and that the constant interval between doses is T_0; let the input fraction b_1 be unity. If the first dose is administered at $t = 0$ and $x_1(0^-) = 0$, then the quantity in the compartment between doses 1 and 2 is

$$x_1(t) = D_1 e^{-k_{01}t}, \qquad 0 \leqslant t < T_0. \tag{3.21}$$

The maximum value in this interval (at $t = 0^+$) is D_1 and the minimum value, just before the administration of the second dose, is $D_1 e^{-k_{01}T_0}$. After the second dose, but before the third dose,

$$x_1(t) = D_1 e^{-k_{01}t} + D_1 e^{-k_{01}(t-T_0)}, \qquad T_0 \leqslant t < 2T_0. \tag{3.22}$$

(The model is linear, so the principle of superposition applies.) The maximum value in this interval is $D_1(1 + e^{-k_{01}T_0})$ and the minimum value is $D_1 e^{-k_{01}T_0}(1 + e^{-k_{01}T_0})$.

By similar argument, in the interval between administration of doses n and $(n + 1)$, the maximum value of the compartmental quantity is

$$\begin{aligned} x_{1\max} &= D_1(1 + e^{-k_{01}T_0} + e^{-2k_{01}T_0} + \ldots + e^{-(n-1)k_{01}T_0}) \\ &= D_1 \frac{1 - e^{-nk_{01}T_0}}{1 - e^{-k_{01}T_0}}. \end{aligned} \tag{3.23}$$

In the same interval, the minimum quantity is

$$x_{1\min} = e^{-k_{01}T_0} \cdot x_{1\max} \tag{3.24}$$

As the number of doses becomes very large, the values of the maxima rise from D_1 towards $\dfrac{D_1}{1 - e^{-k_{01}T_0}}$ while the values of the minima rise from $D_1 e^{-k_{01}T_0}$ towards $\dfrac{D_1 e^{-k_{01}T_0}}{1 - e^{-k_{01}T_0}}$ (see Fig. 3.5).

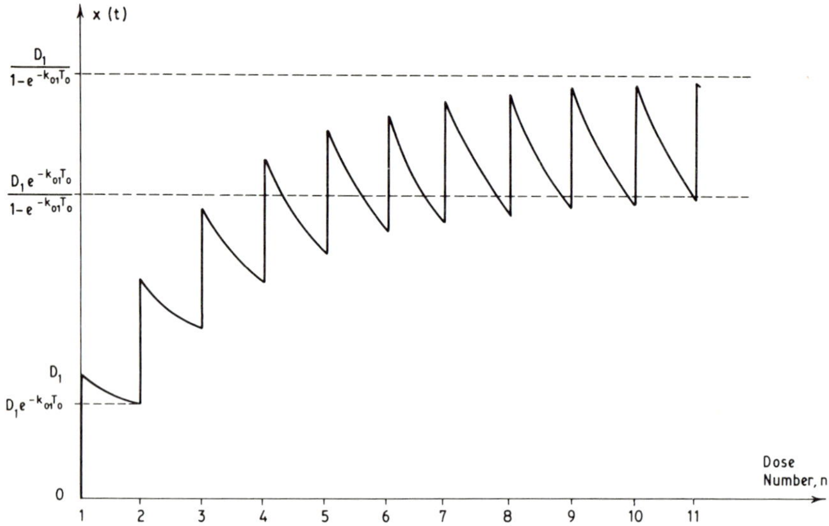

Figure 3.5 Compartmental quantity for a one-compartment model with impulsive inputs of magnitude D_1 repeated at regular intervals T_0.

3.2 Two-compartment systems

3.2.1 *The general two-compartment system*

The most general form of two-compartment system is shown in Fig. 3.6. For this system,

$$\mathbf{A} = \begin{bmatrix} a_{11} & a_{12} \\ a_{21} & a_{22} \end{bmatrix}$$

where
$$a_{11} = -(k_{01} + k_{21}), \qquad a_{12} = k_{12},$$
$$a_{21} = k_{21}, \qquad a_{22} = -(k_{02} + k_{12}).$$

The eigenvalues are given by eqn (2.61):

$$\lambda_1, \lambda_2 = \tfrac{1}{2}\{(a_{11} + a_{22}) \pm [(a_{11} - a_{22})^2 + 4a_{12}a_{21}]^{1/2}\}$$

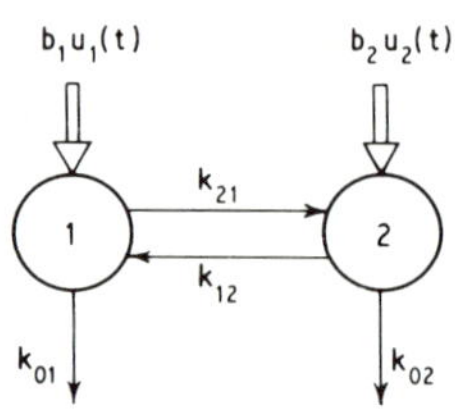

Figure 3.6 General two-compartment model.

and, from Section 2.4, are real. Further, unless

$$(a_{11} - a_{22})^2 + 4a_{12}a_{21} = 0,$$

the eigenvalues are distinct. Thus for repeated eigenvalues, a_{11} and a_{22} must be equal and $a_{12}a_{21}$ must be zero; the latter implies that either a_{12} or a_{21} must be zero (if they were both zero, the compartments would not be connected).

In this section, the most general case will be considered first and we will then proceed to special cases with one or more of the rate constants zero. As before, we are assuming that each input is applied to one compartment only so that $\mathbf{BU}(s) = [b_1 U_1(s) \quad b_2 U_2(s)]^T$, and that observations are of individual compartments only so that $y_1(t) = c_1 x_1(t)$ and $y_2(t) = c_2 x_2(t)$.

From the form of **A**,

$$(s\mathbf{I} - \mathbf{A}) = \begin{bmatrix} s - a_{11} & -a_{12} \\ -a_{21} & s - a_{22} \end{bmatrix}$$

$$\therefore \quad (s\mathbf{I} - \mathbf{A})^{-1} = \frac{1}{(s - \lambda_1)(s - \lambda_2)} \begin{bmatrix} s - a_{22} & a_{12} \\ a_{21} & s - a_{11} \end{bmatrix}. \tag{3.25}$$

The transfer function matrix is

$$\mathbf{G}(s) = \begin{bmatrix} G_{11}(s) & G_{12}(s) \\ G_{21}(s) & G_{22}(s) \end{bmatrix} = \begin{bmatrix} \dfrac{Y_1(s)}{U_1(s)} & \dfrac{Y_1(s)}{U_2(s)} \\[2ex] \dfrac{Y_2(s)}{U_1(s)} & \dfrac{Y_2(s)}{U_2(s)} \end{bmatrix}$$

$$= \begin{bmatrix} \dfrac{c_1 b_1 (s - a_{22})}{(s - \lambda_1)(s - \lambda_2)} & \dfrac{c_1 b_2 a_{12}}{(s - \lambda_1)(s - \lambda_2)} \\[2ex] \dfrac{c_2 b_1 a_{21}}{(s - \lambda_1)(s - \lambda_2)} & \dfrac{c_2 b_2 (s - a_{11})}{(s - \lambda_1)(s - \lambda_2)} \end{bmatrix}. \tag{3.26}$$

The Laplace transform of the observations is $\mathbf{Y}(s) = \mathbf{G}(s)\mathbf{U}(s)$ and the time response is given by $\mathbf{y}(t) = \mathscr{L}^{-1}[\mathbf{G}(s)\mathbf{U}(s)]$, assuming that $\mathbf{y}(0^-) = 0$.

Suppose, for example, that an impulsive input $D_1\,\delta(t)$ is introduced into compartment 1 and that there is no perturbation of compartment 2. Then $\mathbf{U}(s) = [D_1 \quad 0]^T$ and

$$\mathbf{Y}(s) = \begin{bmatrix} \dfrac{c_1 b_1 (s - a_{22}) \cdot D_1}{(s - \lambda_1)(s - \lambda_2)} \\[2ex] \dfrac{c_2 b_1 a_{21} D_1}{(s - \lambda_1)(s - \lambda_2)} \end{bmatrix}.$$

Taking partial fractions (see Appendix 1),

$$\mathbf{Y}(s)=\begin{bmatrix} c_1b_1D_1\left(\dfrac{\lambda_1-a_{22}}{\lambda_1-\lambda_2}\cdot\dfrac{1}{s-\lambda_1}+\dfrac{\lambda_2-a_{22}}{\lambda_2-\lambda_1}\cdot\dfrac{1}{s-\lambda_2}\right)\\ \dfrac{c_2b_1a_{21}D_1}{\lambda_1-\lambda_2}\left(\dfrac{1}{s-\lambda_1}-\dfrac{1}{s-\lambda_2}\right)\end{bmatrix}.$$

Inverse Laplace transformation (see Appendix 1) gives:

$$y_1(t)=c_1b_1D_1\left(\frac{\lambda_1-a_{22}}{\lambda_1-\lambda_2}\,\mathrm{e}^{\lambda_1 t}+\frac{\lambda_2-a_{22}}{\lambda_2-\lambda_1}\,\mathrm{e}^{\lambda_2 t}\right),\qquad t\geqslant 0 \quad (3.27\text{a})$$

$$y_2(t)=\frac{c_2b_1a_{21}D_1}{\lambda_1-\lambda_2}\,(\mathrm{e}^{\lambda_1 t}-\mathrm{e}^{\lambda_2 t}),\qquad t\geqslant 0. \quad (3.27\text{b})$$

If, instead, a constant continuous infusion (step function) of rate k_i per unit time is introduced into compartment 1 with no (external) input to compartment 2, so that

$$\mathbf{U}(s)=\begin{bmatrix}\dfrac{k_i}{s} & 0\end{bmatrix}^T,$$

$$\mathbf{Y}(s)=\begin{bmatrix}\dfrac{c_1b_1(s-a_{22})k_i}{s(s-\lambda_1)(s-\lambda_2)}\\ \dfrac{c_2b_1a_{21}k_i}{s(s-\lambda_1)(s-\lambda_2)}\end{bmatrix}$$

$$=\begin{bmatrix}c_1b_1k_i\left(-\dfrac{a_{22}}{\lambda_1\lambda_2}\dfrac{1}{s}+\dfrac{\lambda_1-a_{22}}{\lambda_1(\lambda_1-\lambda_2)}\dfrac{1}{s-\lambda_1}+\dfrac{\lambda_2-a_{22}}{\lambda_2(\lambda_2-\lambda_1)}\dfrac{1}{s-\lambda_2}\right)\\ c_2b_1a_{21}k_i\left(\dfrac{1}{\lambda_1\lambda_2}\dfrac{1}{s}+\dfrac{1}{\lambda_1(\lambda_1-\lambda_2)}\dfrac{1}{s-\lambda_1}+\dfrac{1}{\lambda_2(\lambda_2-\lambda_1)}\dfrac{1}{s-\lambda_2}\right)\end{bmatrix}.$$

Inverse Laplace transformation gives, for $t\geqslant 0$,

$$y_1(t)=c_1b_1k_i\left(\frac{-a_{22}}{\lambda_1\lambda_2}+\frac{\lambda_1-a_{22}}{\lambda_1(\lambda_1-\lambda_2)}\,\mathrm{e}^{\lambda_1 t}+\frac{\lambda_2-a_{22}}{\lambda_2(\lambda_2-\lambda_1)}\,\mathrm{e}^{\lambda_2 t}\right) \quad (3.28\text{a})$$

$$y_2(t)=c_2b_1a_{21}k_i\left(\frac{1}{\lambda_1\lambda_2}+\frac{1}{\lambda_1(\lambda_1-\lambda_2)}\,\mathrm{e}^{\lambda_1 t}+\frac{1}{\lambda_2(\lambda_2-\lambda_1)}\,\mathrm{e}^{\lambda_2 t}\right). \quad (3.28\text{b})$$

The reader is invited to confirm that the unit step responses ($k_i = 1$ in eqns (3.28)) are the same as the integrals of the unit impulse responses ($D_1 = 1$ in eqns (3.27)).

For a third experiment with an impulsive input $D_2\,\delta(t)$ introduced to compartment 2 and no (external) input to compartment 1, so that $\mathbf{U}(s) = [0 \quad D_2]^T$,

$$\mathbf{Y}(s) = \begin{bmatrix} \dfrac{c_1 b_2 a_{12} D_2}{(s-\lambda_1)(s-\lambda_2)} \\[2ex] \dfrac{c_2 b_2 D_2 (s-a_{11})}{(s-\lambda_1)(s-\lambda_2)} \end{bmatrix}$$

so that

$$y_1(t) = \frac{c_1 b_2 a_{12} D_2}{\lambda_1 - \lambda_2}(e^{\lambda_1 t} - e^{\lambda_2 t}), \qquad t \geqslant 0 \tag{3.29a}$$

$$y_2(t) = c_2 b_2 D_2 \left(\frac{\lambda_1 - a_{11}}{\lambda_1 - \lambda_2} e^{\lambda_1 t} + \frac{\lambda_2 - a_{11}}{\lambda_2 - \lambda_1} e^{\lambda_2 t} \right), \qquad t \geqslant 0 \tag{3.29b}$$

while for a step input to compartment 2 $\left(\mathbf{U}(s) = \begin{bmatrix} 0 & \dfrac{k_i}{s} \end{bmatrix}^T \right)$,

$$y_1(t) = c_1 b_2 a_{12} k_i \left(\frac{1}{\lambda_1 \lambda_2} + \frac{1}{\lambda_1(\lambda_1 - \lambda_2)} e^{\lambda_1 t} + \frac{1}{\lambda_2(\lambda_2 - \lambda_1)} e^{\lambda_2 t} \right), \qquad t \geqslant 0 \tag{3.30a}$$

$$y_2(t) = c_2 b_2 k_i \left(-\frac{a_{11}}{\lambda_1 \lambda_2} + \frac{\lambda_1 - a_{11}}{\lambda_1(\lambda_1 - \lambda_2)} e^{\lambda_1 t} + \frac{\lambda_2 - a_{11}}{\lambda_2(\lambda_2 - \lambda_1)} e^{\lambda_2 t} \right), \qquad t \geqslant 0. \tag{3.30b}$$

Example 3.1. *Consider a linear, time-invariant two-compartment system with* $k_{12} = k_{21} = k_{01} = k_{02} = 1$. *Determine the states* $x_1(t)$ *and* $x_2(t)$, $t \geqslant 0$, *for the following cases:*

(i). $u_1(t) = D_1\,\delta(t);\ u_2(t) = 0$
(ii). $u_1(t) = k_i,\ t \geqslant 0;\ u_2(t) = 0$
(iii). $u_1(t) = D_1\,\delta(t);\ u_2(t) = D_2\,\delta(t)$
(iv). $u_1(t) = k_i,\ t \geqslant 0;\ u_2(t) = D_2\,\delta(t)$
(v). $u_1(t) = D_1\,\delta(t);\ u_2(t) = k_i,\ t \geqslant 0.$

Both input fractions, b_1 *and* b_2, *may be taken as* 1.

Solution. The state equations are:

$$\dot{x}_1 = -2x_1 + x_2 + u_1(t)$$

$$\dot{x}_2 = x_1 - 2x_2 + u_2(t).$$

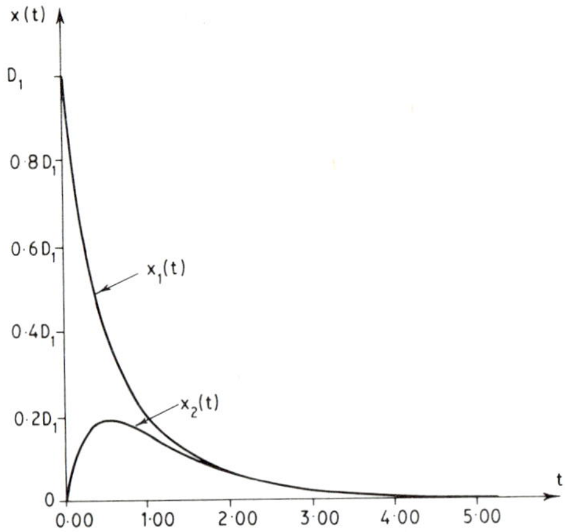

Figure 3.7 Responses of model of Example 3.1 to impulsive perturbation of compartment 1.

The eigenvalues are, from eqn (2.61),

$$\lambda_1, \lambda_2 = \tfrac{1}{2}(-4 \pm (4)^{1/2}) = -3, -1$$

so that
$$(s\mathbf{I} - \mathbf{A})^{-1} = \begin{bmatrix} \dfrac{s+2}{(s+3)(s+1)} & \dfrac{1}{(s+3)(s+1)} \\ \dfrac{1}{(s+3)(s+1)} & \dfrac{s+2}{(s+3)(s+1)} \end{bmatrix}.$$

Case (*i*). $\mathbf{U}(s) = [D_1 \quad 0]^T$.

Recalling that $x_1(t) = \dfrac{y_1(t)}{c_1}$ and $x_2(t) = \dfrac{y_2(t)}{c_2}$, we have, from eqns (3.27),

$$x_1(t) = \tfrac{1}{2}D_1(\mathrm{e}^{-3t} + \mathrm{e}^{-t})$$
$$x_2(t) = \tfrac{1}{2}D_1(-\mathrm{e}^{-3t} + \mathrm{e}^{-t})$$

which are shown in Fig. 3.7.

Case (*ii*). $\mathbf{U}(s) = \begin{bmatrix} \dfrac{k_i}{s} & 0 \end{bmatrix}^T$ so that from eqns (3.28),

$$x_1(t) = k_i(\tfrac{2}{3} - \tfrac{1}{6}\,\mathrm{e}^{-3t} - \tfrac{1}{2}\,\mathrm{e}^{-t})$$
$$x_2(t) = k_i(\tfrac{1}{3} + \tfrac{1}{6}\,\mathrm{e}^{-3t} - \tfrac{1}{2}\,\mathrm{e}^{-t})$$

which are shown in Fig. 3.8.

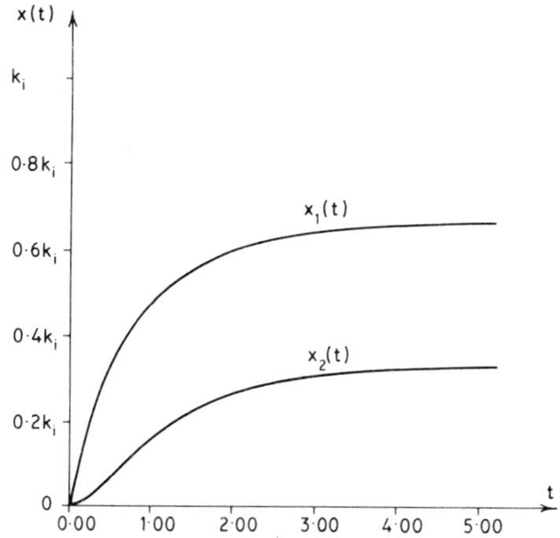

Figure 3.8 Responses of model of Example 3.1 to step perturbation of compartment 1.

Case (iii). $\mathbf{U}(s) = [D_1 \quad D_2]^T$ so that

$$\mathbf{X}(s) = \begin{bmatrix} \dfrac{D_1(s+2) + D_2}{(s+3)(s+1)} \\ \dfrac{D_1 + D_2(s+2)}{(s+3)(s+1)} \end{bmatrix}.$$

Inverse Laplace transformation gives

$$x_1(t) = \tfrac{1}{2}[(D_1 - D_2)\,\mathrm{e}^{-3t} + (D_1 + D_2)\,\mathrm{e}^{-t}]$$
$$x_2(t) = \tfrac{1}{2}[(D_2 - D_1)\,\mathrm{e}^{-3t} + (D_1 + D_2)\,\mathrm{e}^{-t}].$$

Note that if $D_1 = D_2$, only one of the exponentials appears in the states; this is not a general result and arises in this case from the symmetry of the model and inputs. With other values for the rate constants, particular sizes of perturbations can usually be found which result in only one exponential appearing in the states.

Case (iv). $\mathbf{U}(s) = \begin{bmatrix} \dfrac{k_i}{s} & D_2 \end{bmatrix}^T$ so that

$$\mathbf{X}(s) = \begin{bmatrix} \dfrac{k_i(s+2)}{s(s+3)(s+1)} + \dfrac{D_2}{(s+3)(s+1)} \\ \dfrac{k_i}{s(s+3)(s+1)} + \dfrac{D_2(s+2)}{(s+3)(s+1)} \end{bmatrix}$$

whence $$x_1(t) = \tfrac{2}{3}k_i - (\tfrac{1}{6}k_i + \tfrac{1}{2}D_2)\,\mathrm{e}^{-3t} - (\tfrac{1}{2}k_i - \tfrac{1}{2}D_2)\,\mathrm{e}^{-t}$$
$$x_2(t) = \tfrac{1}{3}k_i + (\tfrac{1}{6}k_i + \tfrac{1}{2}D_2)\,\mathrm{e}^{-3t} - (\tfrac{1}{2}k_i - \tfrac{1}{2}D_2)\,\mathrm{e}^{-t}.$$

Note that if k_i is numerically equal to D_2, the e^{-t} will not appear in the states. As for case (iii) this is not a general result but may occur with other values of rate constants for particular sizes of perturbations.

Case (v). $\mathbf{U}(s) = \begin{bmatrix} D_1 & \dfrac{k_i}{s} \end{bmatrix}^T$ so that

$$\mathbf{X}(s) = \begin{bmatrix} \dfrac{D_1(s+2)}{(s+3)(s+1)} + \dfrac{k_i}{s(s+3)(s+1)} \\[2ex] \dfrac{D_1}{(s+3)(s+1)} + \dfrac{k_i(s+2)}{s(s+3)(s+1)} \end{bmatrix}$$

whence $$x_1(t) = \tfrac{1}{3}k_i + (\tfrac{1}{2}D_1 + \tfrac{1}{6}k_i)\,\mathrm{e}^{-3t} + (\tfrac{1}{2}D_1 - \tfrac{1}{2}k_i)\,\mathrm{e}^{-t}$$
$$x_2(t) = \tfrac{2}{3}k_i - (\tfrac{1}{2}D_1 + \tfrac{1}{6}k_i)\,\mathrm{e}^{-3t} + (\tfrac{1}{2}D_1 - \tfrac{1}{2}k_i)\,\mathrm{e}^{-t}.$$

Expressions for $x_1(t)$ in a two-compartment system with impulsive inputs repeated at regular intervals to one of the compartments will be found in Gibaldi and Perrier (1975), p. 119.

3.2.2 *Repeated eigenvalues*

As we have seen above, a_{11} must equal a_{22} and $a_{12}a_{21}$ must be zero for repeated eigenvalues in a two-compartment system. Consider a system in which $a_{12} = 0$ and $a_{11} = a_{22}$, for which $a_{11} = -(k_{01} + k_{21}) = a_{22} = -k_{02}$.

From eqn (2.61), the eigenvalues are

$$\lambda_1, \lambda_2 = \tfrac{1}{2}(a_{11} + a_{22}) = -k_{02}. \tag{3.31}$$

For this system, $$\mathbf{A} = \begin{bmatrix} a_{11} & a_{12} \\ a_{21} & a_{22} \end{bmatrix} = \begin{bmatrix} -k_{02} & 0 \\ k_{21} & -k_{02} \end{bmatrix}$$

and

$$(s\mathbf{I} - \mathbf{A})^{-1} = \frac{1}{(s+k_{02})^2}\begin{bmatrix} s+k_{02} & 0 \\ k_{21} & s+k_{02} \end{bmatrix} = \begin{bmatrix} \dfrac{1}{s+k_{02}} & 0 \\[2ex] \dfrac{k_{21}}{(s+k_{02})^2} & \dfrac{1}{s+k_{02}} \end{bmatrix}. \tag{3.32}$$

If an impulsive input $D_1\,\delta(t)$ is applied to compartment 1 (with no external input to compartment 2), then

$$y_1(t) = c_1 b_1 D_1\, e^{-k_{02}t}, \qquad t \geqslant 0 \tag{3.33a}$$

$$y_2(t) = c_2 b_1 D_1 k_{21} t\, e^{-k_{02}t}, \qquad t \geqslant 0 \tag{3.33b}$$

while the responses to an impulsive input $D_2\,\delta(t)$ applied to compartment 2 are:

$$y_1(t) = 0 \tag{3.34a}$$

$$y_2(t) = c_2 b_2 D_2\, e^{-k_{02}t}, \qquad t \geqslant 0 \tag{3.34b}$$

($y_1(t)$ is obviously zero in this case because there is no link from compartment 2 to compartment 1).

If, instead, a constant continuous infusion, of rate k_i per unit time, is introduced into compartment 1, with no external input to compartment 2, $\mathbf{U}(s) = \begin{bmatrix} \dfrac{k_i}{s} & 0 \end{bmatrix}^T$, and

$$Y_1(s) = \frac{c_1 b_1 k_i}{s(s + k_{02})}$$

whence

$$y_1(t) = \frac{c_1 b_1 k_i}{k_{02}}(1 - e^{-k_{02}t}), \qquad t \geqslant 0 \tag{3.35a}$$

while

$$Y_2(s) = \frac{c_2 b_1 k_i k_{21}}{s(s + k_{02})^2}$$

whence

$$y_2(t) = \frac{c_2 b_1 k_i k_{21}}{k_{02}^2}(1 - e^{-k_{02}t}) - \frac{c_2 b_1 k_i k_{21}}{k_{02}}\, t\, e^{-k_{02}t}, \qquad t \geqslant 0 \tag{3.35b}$$

(see Appendix 1).

Example 3.2. *For the two-compartment system with* $k_{12} = 0$, $k_{21} = k_{01} = 1$ *and* $k_{02} = 2$, *find* $x_1(t)$ *and* $x_2(t)$ *for* $t \geqslant 0$ *for the following cases:*

(i). $u_1(t) = D_1\,\delta(t);\ u_2(t) = 0$
(ii). $u_1(t) = k_i,\ t \geqslant 0;\ u_2(t) = 0.$

The input fraction $b_1 = 1$ *in both cases.*

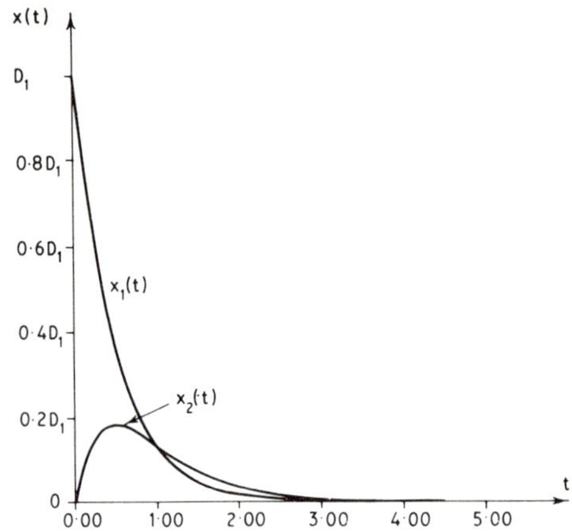

Figure 3.9 Responses of model of Example 3.2 to impulsive perturbation of compartment 1.

Solution. Case (*i*). From eqns (3.33),

$$x_1(t) = D_1 \,\mathrm{e}^{-2t}$$
$$x_2(t) = D_1 t \,\mathrm{e}^{-2t}$$

(see Fig. 3.9).

Case (*ii*). From eqns (3.35),

$$x_1(t) = \tfrac{1}{2}k_i(1 - \mathrm{e}^{-2t})$$
$$x_2(t) = k_i[\tfrac{1}{4}(1 - \mathrm{e}^{-2t}) - \tfrac{1}{2}t \,\mathrm{e}^{-2t}]$$

(see Fig. 3.10).

3.2.3 *Source compartments*

A source compartment is one which does not receive material from any other compartment; it may or may not excrete ma\ˌ:rial to the environment. All source compartments respond as a one-compartment system, no matter how many compartments there are in the rest of the system, so that only one of the modes (single-exponential decays) appears in the response of a source compartment to a perturbation.

In a two-compartment system, compartment 1 is a source compartment if $k_{12} = 0$. Although the presence of a source compartment is the only way in which a two-compartment system can have repeated eigenvalues (as in Example 3.2) repeated eigenvalues can only occur if, in addition, $a_{11} = a_{22}$. Let us consider an example with distinct eigenvalues.

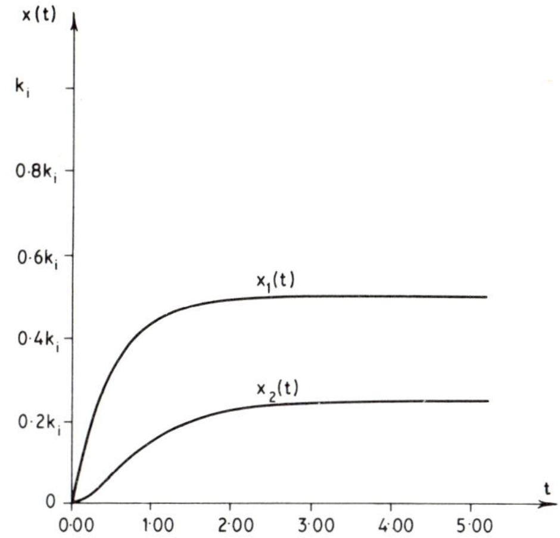

Figure 3.10 Responses of model of Example 3.2 to step perturbation of compartment 1.

Example 3.3. *Let* $k_{12} = 0, k_{21} = k_{01} = k_{02} = 1$. *Determine* $x_1(t)$ *and* $x_2(t)$ *for* $t \geqslant 0$ *if* $u_1(t) = D_1\,\delta(t)$, $u_2(t) = 0$ *and* $b_1 = 1$.

Solution. For this system,

$$\mathbf{A} = \begin{bmatrix} -2 & 0 \\ 1 & -1 \end{bmatrix}$$

which has eigenvalues $\lambda_1, \lambda_2 = \frac{1}{2}(-3 \pm 1) = -2, -1$. Substituting in eqns (3.27),

$$x_1(t) = D_1\,\mathrm{e}^{-2t}$$
$$x_2(t) = D_1(\mathrm{e}^{-t} - \mathrm{e}^{-2t})$$

(see Fig. 3.11).

In many applications, the source compartment does not excrete material to the environment. We will now evaluate a further example with $k_{01} = 0$.

Example 3.4. *For the two-compartment system with* $k_{12} = k_{01} = 0$, $k_{21} = 1$ *and* $k_{02} = 2$, *determine* $x_1(t)$ *and* $x_2(t)$ *for* $t \geqslant 0$ *if* $u_1(t) = D_1\,\delta(t)$, $u_2(t) = 0$ *and* $b_1 = 1$.

Solution. For this system,

$$\mathbf{A} = \begin{bmatrix} -1 & 0 \\ 1 & -2 \end{bmatrix}$$

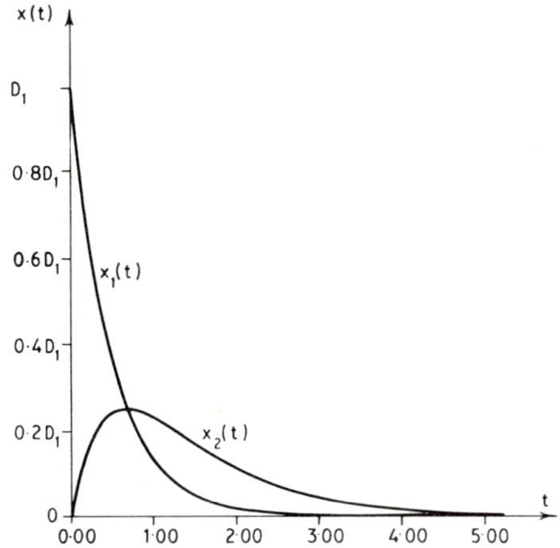

Figure 3.11 Responses of model of Example 3.3 to impulsive perturbation of compartment 1.

with eigenvalues $\lambda_1, \lambda_2 = -2, -1$. Substituting in eqns (3.27),

$$x_1(t) = D_1 \, e^{-t}$$

$$x_2(t) = D_1(e^{-t} - e^{-2t})$$

(see Fig. 3.12).

If, instead, k_{02} had been 1, there would have been repeated eigenvalues with $\lambda_1 = \lambda_2 = -1$. For the same input, $x_1(t)$ would still have been the same, but from eqn (3.33b), $x_2(t)$ would have been $D_1 t \, e^{-t}$.

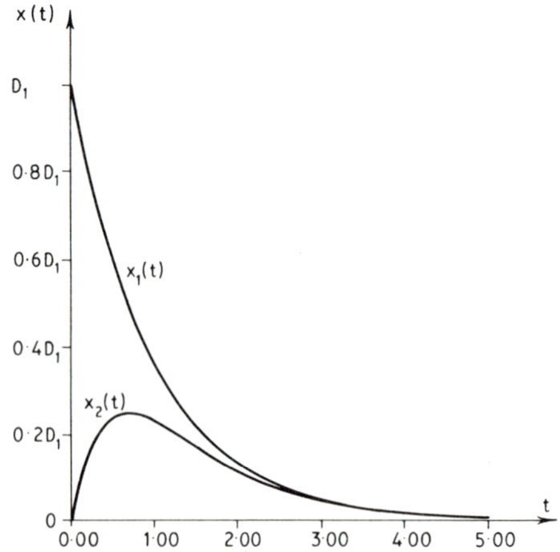

Figure 3.12 Responses of model of Example 3.4 to impulsive perturbation of compartment 1.

3.2.4 *Sink compartments*

A sink (or trap) compartment is one which does not excrete material either to the environment or to any other compartment. It results in a zero on the principal diagonal of **A** and a zero eigenvalue. Physically, a sink compartment acts as a pure integrator of the flows into it.

Consider a two-compartment system in which compartment 2 is a sink, i.e. $k_{12} = k_{02} = 0$ so that $a_{22} = 0$. For this system,

$$\mathbf{A} = \begin{bmatrix} a_{11} & 0 \\ a_{12} & 0 \end{bmatrix}$$

is singular. From $|\lambda \mathbf{I} - \mathbf{A}| = 0$, the eigenvalues are

$$\lambda_1, \lambda_2 = a_{11}, 0$$

and

$$(s\mathbf{I} - \mathbf{A})^{-1} = \frac{1}{s(s - a_{11})} \begin{bmatrix} s & 0 \\ a_{21} & s - a_{11} \end{bmatrix}. \tag{3.36}$$

If

$$\mathbf{U}(s) = [D_1 \quad 0]^T,$$

$$Y_1(s) = \frac{c_1 b_1 D_1}{s - a_{11}}$$

whence

$$y_1(t) = c_1 b_1 D_1 \, e^{-(k_{01} + k_{21})t}, \qquad t \geqslant 0 \tag{3.37a}$$

and

$$Y_2(s) = \frac{c_2 b_1 D_1 a_{21}}{s(s - a_{11})}$$

whence

$$y_2(t) = \frac{c_2 b_1 D_1 k_{21}}{k_{01} + k_{21}} (1 - e^{-(k_{01} + k_{21})t}), \qquad t \geqslant 0. \tag{3.37b}$$

If, on the other hand, $\mathbf{U}(s) = [0 \quad D_2]^T$, then

$$Y_2(s) = \frac{c_2 b_2 D_2}{s}$$

whence

$$y_2(t) = c_2 b_2 D_2, \qquad t \geqslant 0$$

as expected since the dose remains in compartment 2.

Example 3.5. *For the two-compartment system with* $k_{12} = k_{02} = 0$, $k_{21} = k_{01} = 1$, *determine* $x_1(t)$ *and* $x_2(t)$ *for* $t \geqslant 0$ *if* $u_1(t) = D_1\,\delta(t)$, $u_2(t) = 0$ *and* $b_1 = 1$.

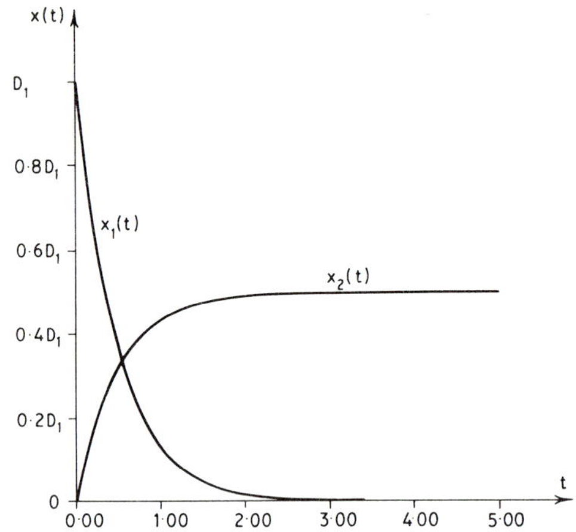

Figure 3.13 Responses of model of Example 3.5 to impulsive perturbation of compartment 1.

Solution. For this system,

$$\mathbf{A} = \begin{bmatrix} -2 & 0 \\ 1 & 0 \end{bmatrix}$$

and $\lambda_1, \lambda_2 = -2, 0$.

Thus, from eqns (3.37),

$$x_1(t) = D_1\,\mathrm{e}^{-2t}$$

$$x_2(t) = \tfrac{1}{2}D_1(1 - \mathrm{e}^{-2t})$$

(which are shown in Fig. 3.13).

3.2.5 *Closed systems*

A closed system is one in which no compartment excretes material to the environment; it may or may not contain sink compartments. All closed systems have a zero eigenvalue.

A two-compartment closed system has $k_{01} = k_{02} = 0$, so that $a_{11} = -k_{21}$ and $a_{22} = -k_{12}$. For this system,

$$\mathbf{A} = \begin{bmatrix} -k_{21} & k_{12} \\ k_{21} & -k_{12} \end{bmatrix}.$$

From $|\lambda\mathbf{I} - \mathbf{A}| = 0$, the eigenvalues are

$$\lambda_1, \lambda_2 = -(k_{12} + k_{21}), 0$$

and
$$(s\mathbf{I} - \mathbf{A})^{-1} = \frac{1}{s(s + k_{12} + k_{21})}\begin{bmatrix} s + k_{12} & k_{12} \\ k_{21} & s + k_{21} \end{bmatrix}. \tag{3.38}$$

If
$$u_1(t) = D_1\,\delta(t) \quad \text{and} \quad u_2(t) = 0,$$

$$Y_1(s) = \frac{c_1 b_1 D_1 (s + k_{12})}{s(s + k_{12} + k_{21})}$$

whence
$$y_1(t) = \frac{c_1 b_1 D_1}{k_{12} + k_{21}}(k_{12} + k_{21}\,\mathrm{e}^{-(k_{12}+k_{21})t}), \qquad t \geqslant 0 \tag{3.39a}$$

and
$$Y_2(s) = \frac{c_2 b_1 D_1 k_{21}}{s(s + k_{12} + k_{21})}$$

whence
$$y_2(t) = \frac{c_2 b_1 D_1 k_{21}}{k_{12} + k_{21}}(1 - \mathrm{e}^{-(k_{12}+k_{21})t}), \qquad t \geqslant 0. \tag{3.39b}$$

Note that in the steady state, the quantity administered to the system is shared between the compartments in the ratio $\dfrac{k_{12}}{k_{21}}$.

If, in addition, compartment 2 is a sink compartment, with $k_{12} = 0$, then the responses to the same form of perturbation are

$$y_1(t) = c_1 b_1 D_1\,\mathrm{e}^{-k_{21}t}, \qquad t \geqslant 0 \tag{3.40a}$$

$$y_2(t) = c_2 b_1 D_1 (1 - \mathrm{e}^{-k_{21}t}), \qquad t \geqslant 0. \tag{3.40b}$$

Example 3.6. *For the two-compartment system with* $k_{01} = k_{02} = 0$, $k_{12} = k_{21} = 1$, *determine* $x_1(t)$ *and* $x_2(t)$ *for* $t \geqslant 0$ *if* $u_1(t) = D_1\,\delta(t)$, $u_2(t) = 0$, $b_1 = 1$.

Solution. From eqns (3.39),

$$x_1(t) = \tfrac{1}{2}D_1(1 + \mathrm{e}^{-2t})$$

$$x_2(t) = \tfrac{1}{2}D_1(1 - \mathrm{e}^{-2t})$$

which are shown in Fig. 3.14.

3.2.6 *The washout curves*

Consider the general two-compartment system with distinct eigenvalues and suppose that a constant continuous infusion of rate k_i per unit time has been made into compartment 1 (no external input to compartment 2), until the

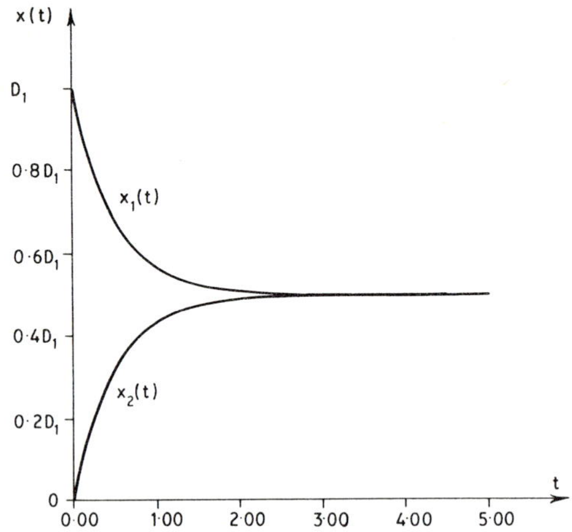

Figure 3.14 Responses of model of Example 3.6 to impulsive perturbation of compartment 1.

system is in a steady state. From eqns (3.28), the steady-state compartmental quantities are

$$x_{1ss} = -\frac{b_1 k_i a_{22}}{\lambda_1 \lambda_2} \tag{3.41a}$$

$$x_{2ss} = \frac{b_1 k_i a_{21}}{\lambda_1 \lambda_2}. \tag{3.41b}$$

The washout curves are the compartmental quantities after the infusion has stopped (at $t = 0$), and from eqn (2.63) are given by

$$\mathbf{x}(t) = \mathrm{e}^{\mathbf{A}t}\mathbf{x}(0^-)$$

where

$$\mathrm{e}^{\mathbf{A}t} = \mathscr{L}^{-1}(s\mathbf{I} - \mathbf{A})^{-1}$$

and $\mathbf{x}(0^-)$ is the vector consisting of x_{1ss} and x_{2ss}, given by eqns (3.41).

From eqn (3.25),

$$(s\mathbf{I} - \mathbf{A})^{-1} = \frac{1}{(s-\lambda_1)(s-\lambda_2)}\begin{bmatrix} s - a_{22} & a_{12} \\ a_{21} & s - a_{11} \end{bmatrix}$$

so that

$$x_1(t) = \frac{b_1 k_i}{\lambda_1 \lambda_2}\mathscr{L}^{-1}\left[\frac{-a_{22}(s - a_{22}) + a_{12}a_{21}}{(s-\lambda_1)(s-\lambda_2)}\right], \qquad t \geqslant 0 \tag{3.42a}$$

and

$$x_2(t) = \frac{b_1 k_i}{\lambda_1 \lambda_2} \mathscr{L}^{-1} \left[\frac{-a_{21}a_{22} + (s - a_{11})a_{21}}{(s - \lambda_1)(s - \lambda_2)} \right], \qquad t \geqslant 0. \qquad (3.42b)$$

Taking partial fractions and inverse Laplace transforming,

$$x_1(t) = \frac{b_1 k_i}{\lambda_1 \lambda_2 (\lambda_1 - \lambda_2)} [(-a_{22}(\lambda_1 - a_{22}) + a_{12}a_{21})\,\mathrm{e}^{\lambda_1 t}$$

$$- (-a_{22}(\lambda_2 - a_{22}) + a_{12}a_{21})\,\mathrm{e}^{\lambda_2 t}], \qquad t \geqslant 0 \qquad (3.43a)$$

$$x_2(t) = \frac{b_1 k_i}{\lambda_1 \lambda_2 (\lambda_1 - \lambda_2)} [(-a_{21}a_{22} + (\lambda_1 - a_{11})a_{21})\,\mathrm{e}^{\lambda_1 t}$$

$$- (-a_{21}a_{22} + (\lambda_2 - a_{11})a_{21})\,\mathrm{e}^{\lambda_2 t}], \qquad t \geqslant 0. \qquad (3.43b)$$

Example 3.7. *Find the washout curves for the compartmental system of Example 3.1 (with $k_{12} = k_{21} = k_{01} = k_{02} = 1$), assuming that, prior to $t = 0$, a constant continuous infusion of* 1 *unit per unit time had been made to compartment* 1, *with $b_1 = 1$.*

Solution. For this system, $a_{11} = -2$, $a_{22} = -2$, $\lambda_1 = -3$ and $\lambda_2 = -1$. The initial conditions are, from eqns (3.41), $x_1(0^-) = \frac{2}{3}$, $x_2(0^-) = \frac{1}{3}$. Substituting in eqns (3.43),

$$x_1(t) = \frac{1}{(3)(-2)} [(2(-1) + 1)\,\mathrm{e}^{-3t} - (2(1) + 1)\,\mathrm{e}^{-t}]$$

$$= \tfrac{1}{6}\,\mathrm{e}^{-3t} + \tfrac{1}{2}\,\mathrm{e}^{-t}, \qquad t \geqslant 0$$

$$x_2(t) = \frac{1}{(3)(-2)} [(2 + (-1)(1))\,\mathrm{e}^{-3t} - (2 + (1)(1))\,\mathrm{e}^{-t}]$$

$$= -\tfrac{1}{6}\,\mathrm{e}^{-3t} + \tfrac{1}{2}\,\mathrm{e}^{-t}, \qquad t \geqslant 0$$

which are shown in Fig. 3.15.

3.2.7 *Summary of relationships for two-compartment systems*

For a two-compartment system with

$$\mathbf{A} = \begin{bmatrix} a_{11} & a_{12} \\ a_{21} & a_{22} \end{bmatrix}$$

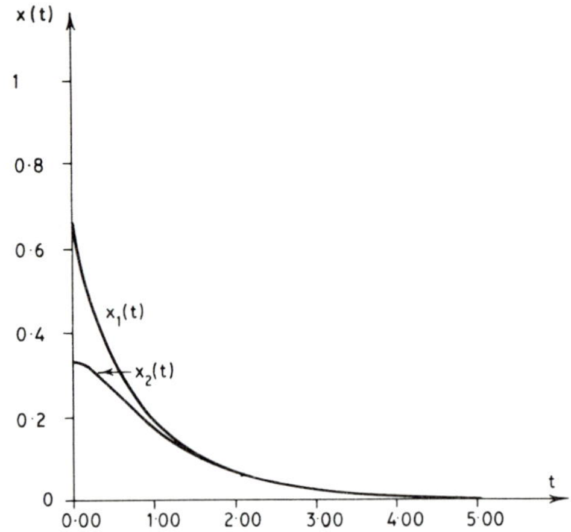

Figure 3.15 Washout curves for the two-compartment model of Example 3.7.

where $a_{11} = -(k_{01} + k_{21})$, $a_{12} = k_{12}$, $a_{21} = k_{21}$, $a_{22} = -(k_{02} + k_{12})$, the eigenvalues must be real. They are also distinct unless $a_{11} = a_{22}$ and $a_{12}a_{21} = 0$. For the system,

$$(s\mathbf{I} - \mathbf{A})^{-1} = \frac{1}{(s - \lambda_1)(s - \lambda_2)} \begin{bmatrix} s - a_{22} & a_{12} \\ a_{21} & s - a_{11} \end{bmatrix}$$

where the eigenvalues are given by

$$\lambda_1, \lambda_2 = \tfrac{1}{2}\{(a_{11} + a_{22}) \pm [(a_{11} - a_{22})^2 + 4a_{12}a_{21}]^{1/2}\}.$$

The responses of the states to $u_1(t) = \delta(t)$, $u_2(t) = 0$, $b_1 = 1$ are

$$x_1(t) = \frac{\lambda_1 - a_{22}}{\lambda_1 - \lambda_2}\, e^{\lambda_1 t} + \frac{\lambda_2 - a_{22}}{\lambda_2 - \lambda_1}\, e^{\lambda_2 t}, \qquad t \geqslant 0$$

$$x_2(t) = \frac{a_{21}}{\lambda_1 - \lambda_2}\,(e^{\lambda_1 t} - e^{\lambda_2 t}), \qquad t \geqslant 0.$$

If a_{12} $(=k_{12}) = 0$ and $(k_{01} + k_{21}) = k_{02}$, the repeated eigenvalue is $\lambda = -k_{02}$ and the corresponding state impulse responses are

$$x_1(t) = e^{\lambda t} \qquad t \geqslant 0$$

$$x_2(t) = a_{21} t\, e^{\lambda t}, \qquad t \geqslant 0.$$

Responses to a unit step input applied to compartment 1 may be found by integration of the unit impulse responses, so that for distinct eigenvalues,

$$x_1(t) = -\frac{a_{22}}{\lambda_1\lambda_2} + \frac{\lambda_1 - a_{22}}{\lambda_1(\lambda_1 - \lambda_2)}\,e^{\lambda_1 t} + \frac{\lambda_2 - a_{22}}{\lambda_2(\lambda_2 - \lambda_1)}\,e^{\lambda_2 t}, \qquad t \geqslant 0$$

$$x_2(t) = a_{21}\left(\frac{1}{\lambda_1\lambda_2} + \frac{1}{\lambda_1(\lambda_1 - \lambda_2)}\,e^{\lambda_1 t} + \frac{1}{\lambda_2(\lambda_2 - \lambda_1)}\,e^{\lambda_2 t}\right), \qquad t \geqslant 0.$$

while for repeated eigenvalues,

$$x_1(t) = \frac{1}{\lambda}(1 - e^{\lambda t}), \qquad t \geqslant 0$$

$$x_2(t) = a_{21}\left(\frac{1}{\lambda^2} + \frac{1}{\lambda}\,t\,e^{\lambda t} - \frac{1}{\lambda^2}\,e^{\lambda t}\right), \qquad t \geqslant 0.$$

If compartment 1 is a source compartment, i.e. $k_{12} = 0$, then $\lambda_1, \lambda_2 = a_{11}, a_{22}$ and the unit impulse responses are

$$x_1(t) = e^{a_{11}t}, \qquad t \geqslant 0$$

$$x_2(t) = \frac{a_{21}}{a_{11} - a_{22}}(e^{a_{11}t} - e^{a_{22}t}), \qquad t \geqslant 0.$$

If compartment 2 is a sink compartment, i.e. $k_{02} = k_{12} = 0$, then one of the eigenvalues is zero and the unit impulse responses are

$$x_1(t) = e^{a_{11}t}, \qquad t \geqslant 0$$

$$x_2(t) = \frac{a_{21}}{a_{11}}(e^{a_{11}t} - 1), \qquad t \geqslant 0.$$

For a closed system (k_{01} and k_{02} both zero), one of the eigenvalues is zero and the other is $-(k_{12} + k_{21})$. Since $a_{22} = -k_{12}$, the unit impulse responses are

$$x_1(t) = \frac{k_{12}}{k_{12} + k_{21}} + \frac{k_{21}}{k_{12} + k_{21}}\,e^{-(k_{12}+k_{21})t}, \qquad t \geqslant 0$$

$$x_2(t) = \frac{k_{21}}{k_{12} + k_{21}} - \frac{k_{21}}{k_{12} + k_{21}}\,e^{-(k_{12}+k_{21})t}, \qquad t \geqslant 0.$$

Results which carry over to compartmental systems with any number of compartments are:

(i). a source compartment behaves as a first-order system;
(ii). a sink compartment results in a zero eigenvalue;
(iii). a closed system has a zero eigenvalue.

4 Analysis of Systems with Three or More Compartments

4.1 Three-compartment systems

The third-order, time-invariant linear system has $\mathbf{A}$ of the form:

$$\mathbf{A} = \begin{bmatrix} a_{11} & a_{12} & a_{13} \\ a_{21} & a_{22} & a_{23} \\ a_{31} & a_{32} & a_{33} \end{bmatrix} \tag{4.1}$$

so that

$$(s\mathbf{I} - \mathbf{A}) = \begin{bmatrix} (s - a_{11}) & -a_{12} & -a_{13} \\ -a_{21} & (s - a_{22}) & -a_{23} \\ -a_{31} & -a_{32} & (s - a_{33}) \end{bmatrix}$$

and

$$(s\mathbf{I} - \mathbf{A})^{-1} =$$

$$\frac{1}{\Delta(s)} \begin{bmatrix} (s-a_{22})(s-a_{33})-a_{23}a_{32} & a_{12}(s-a_{33})+a_{13}a_{32} & a_{13}(s-a_{22})+a_{12}a_{23} \\ a_{21}(s-a_{33})+a_{23}a_{31} & (s-a_{11})(s-a_{33})-a_{13}a_{31} & a_{23}(s-a_{11})+a_{13}a_{21} \\ a_{31}(s-a_{22})+a_{21}a_{32} & a_{32}(s-a_{11})+a_{12}a_{31} & (s-a_{11})(s-a_{22})-a_{12}a_{21} \end{bmatrix} \tag{4.2}$$

where

$$\Delta(s) = (s - \lambda_1)(s - \lambda_2)(s - \lambda_3), \tag{4.3}$$

the eigenvalues λ_1, λ_2 and λ_3 being the roots of the equation:

$$\begin{aligned} &\lambda^3 - (a_{11} + a_{22} + a_{33})\lambda^2 \\ &\quad + (a_{11}a_{22} + a_{22}a_{33} + a_{33}a_{11} - a_{12}a_{21} - a_{13}a_{31} - a_{23}a_{32})\lambda \\ &\quad - a_{11}a_{22}a_{33} + a_{11}a_{23}a_{32} + a_{22}a_{13}a_{31} + a_{33}a_{12}a_{21} \\ &\quad - a_{12}a_{23}a_{31} - a_{13}a_{32}a_{21} = 0. \end{aligned} \tag{4.4}$$

From a compartmental system,

$$a_{11} = -(k_{01} + k_{21} + k_{31}), \qquad a_{12} = k_{12}, \qquad \text{etc.} \tag{4.5}$$

The general three-compartment model is shown in Fig. 4.1.

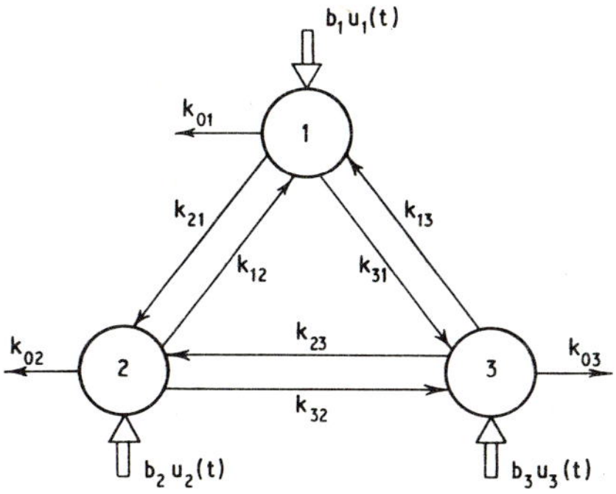

Figure 4.1 General linear, time invariant three-compartment model.

The transfer function matrix $\mathbf{G}(s) = \mathbf{C}(s\mathbf{I} - \mathbf{A})^{-1}\mathbf{B}$ gives Laplace transforms of responses to unit impulse inputs. Thus, for a unit impulsive input to compartment 1 only, the states of the compartments for $t \geqslant 0$ are given by the inverse Laplace transforms of

$$X_1(s) = \frac{(s - a_{22})(s - a_{33}) - a_{23}a_{32}}{(s - \lambda_1)(s - \lambda_2)(s - \lambda_3)} \tag{4.6a}$$

$$X_2(s) = \frac{a_{21}(s - a_{33}) + a_{23}a_{31}}{(s - \lambda_1)(s - \lambda_2)(s - \lambda_3)} \tag{4.6b}$$

$$X_3(s) = \frac{a_{31}(s - a_{22}) + a_{21}a_{32}}{(s - \lambda_1)(s - \lambda_2)(s - \lambda_3)}. \tag{4.6c}$$

Example 4.1. *For the general three-compartment model with all rate constants equal to* 1, *determine* $x_1(t)$, $x_2(t)$ *and* $x_3(t)$ $(t \geqslant 0)$ *for a unit impulse input to compartment* 1 $(b_1 = 1)$ *and no inputs to compartments* 2 *and* 3, *i.e.* $\mathbf{BU}(s) = [1 \quad 0 \quad 0]^T$.

Solution. For this system, $a_{11} = a_{22} = a_{33} = -3$ and all other elements of $\mathbf{A} = 1$. From eqn (4.4), the eigenvalues are the solutions of:

$$\lambda^3 + 9\lambda^2 + 24\lambda + 16 = 0$$

i.e.
$$(\lambda + 1)(\lambda + 4)^2 = 0$$

giving
$$\lambda_1 = \lambda_2 = -4, \lambda_3 = -1.$$

Note that there is a repeated eigenvalue; there are many more ways in which repeated eigenvalues can occur for a three-compartment system than there were for a two-compartment system.

From eqns (4.6),

$$X_1(s) = \frac{(s+3)^2 - 1}{(s+4)^2(s+1)} = \frac{s^2 + 6s + 8}{(s+4)^2(s+1)} = \frac{s+2}{(s+4)(s+1)}$$

$$= \frac{2/3}{s+4} + \frac{1/3}{s+1}$$

whence $$x_1(t) = \tfrac{2}{3}\,e^{-4t} + \tfrac{1}{3}\,e^{-t}$$

and $$X_2(s) = \frac{(s+3)+1}{(s+4)^2(s+1)} = \frac{1}{(s+4)(s+1)}$$

$$= \frac{-1/3}{s+4} + \frac{1/3}{s+1}$$

whence $$x_2(t) = -\tfrac{1}{3}\,e^{-4t} + \tfrac{1}{3}\,e^{-t}.$$

Clearly with these values of the rate constants, $x_3(t) = x_2(t)$ by comparison of eqns (4.6b) and (4.6c). The responses $x_1(t)$ and $x_2(t)$ are shown in Fig. 4.2.

Although this is a three-compartment system, there are only two exponentials in the state responses; the $t\,e^{-4t}$ term which might have been expected, as a result of the repeated eigenvalue of -4 does not appear. This is due to the symmetry of the model with these particular values of the rate constants and is not a general result.

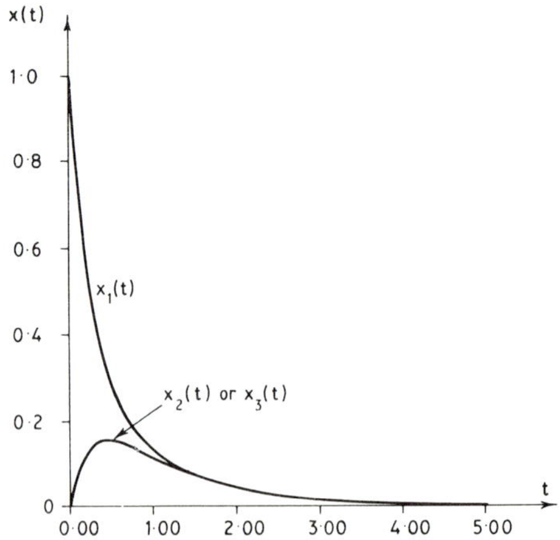

Figure 4.2 Responses of model of Example 4.1 to unit impulsive perturbation of compartment 1.

The general three-compartment model (with all links present) is only of limited practical interest. In the remainder of Section 4.1, therefore, we will instead consider some special cases.

4.1.1 *The mammillary system* (Fig. 4.3)

For the mammillary three-compartment system, $a_{23} = a_{32} = 0$ and, from eqn (2.59), the eigenvalues must be real. The system is described by eqns (4.2) to (4.4) with $a_{23} = a_{32} = 0$.

In many applications using a mammillary model, there is prior knowledge that excretion to the environment is from one compartment only. For example, if excretion is from the central compartment only ($k_{02} = k_{03} = 0$), then

$$a_{22} = -k_{12} \quad \text{and} \quad a_{33} = -k_{13}. \tag{4.7}$$

Thus

$$(s\mathbf{I} - \mathbf{A})^{-1} = \frac{1}{\Delta(s)} \begin{bmatrix} (s + k_{12})(s + k_{13}) & k_{12}(s + k_{13}) & k_{13}(s + k_{12}) \\ k_{21}(s + k_{13}) & (s - a_{11})(s + k_{13}) - k_{13}k_{31} & k_{21}k_{13} \\ k_{31}(s + k_{12}) & k_{12}k_{31} & (s - a_{11})(s + k_{12}) - k_{12}k_{21} \end{bmatrix} \tag{4.8a}$$

and now λ_1, λ_2 and λ_3 are the roots of

$$\begin{aligned} &\lambda^3 + (k_{01} + k_{21} + k_{31} + k_{12} + k_{13})\lambda^2 \\ &\quad + (k_{01}k_{12} + k_{31}k_{12} + k_{13}k_{01} + k_{13}k_{21} + k_{13}k_{12})\lambda \\ &\quad + k_{01}k_{12}k_{13} = 0. \end{aligned} \tag{4.8b}$$

Example 4.2. *For a particular mammillary three-compartment system with excretion from the central compartment only,* $k_{21} = 1.709$, $k_{12} = 2.065$, $k_{31} = 1.255$, $k_{13} = 0.325$ *and* $k_{01} = 0.746$. *Determine* $x_1(t)$, $x_2(t)$ *and* $x_3(t)$ $(t \geqslant 0)$ *if* $u_1(t) = \delta(t)$, $u_2(t) = u_3(t) = 0$ *and* $b_1 = 1$.

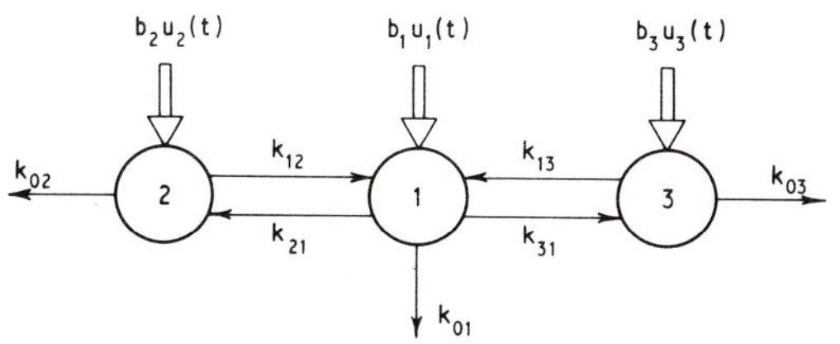

Figure 4.3 Mammillary model with three compartments.

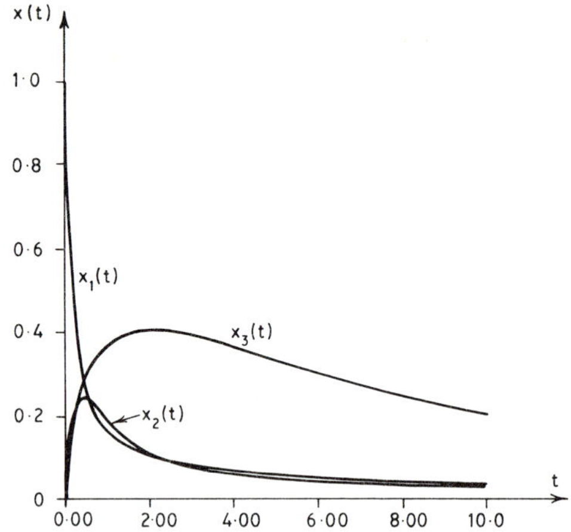

Figure 4.4 Responses of model of Example 4.2 to unit impulsive perturbation of compartment 1.

Solution. From eqn (4.8b), the eigenvalues are the solutions of:

$$\lambda^3 + 6.1\lambda^2 + 5.6\lambda + 0.5 = 0$$

i.e.

$$(\lambda + 5)(\lambda + 1)(\lambda + 0.1) = 0$$

$$\lambda_1 = -5, \lambda_2 = -1, \lambda_3 = -0.1.$$

Hence $\mathbf{X}(s)$, given by the first column of $(s\mathbf{I} - \mathbf{A})^{-1}$, *is*

$$\frac{1}{(s+5)(s+1)(s+0.1)}\begin{bmatrix}(s+2.065)(s+0.325)\\ 1.709(s+0.325)\\ 1.255(s+2.065)\end{bmatrix}.$$

Taking partial fractions (see Appendix 1),

$$X_1(s) = \frac{0.7}{s+5} + \frac{0.2}{s+1} + \frac{0.1}{s+0.1}$$

whence $x_1(t) = 0.7\,e^{-5t} + 0.2\,e^{-t} + 0.1\,e^{-0.1t}$

and similarly,

$$x_2(t) = -0.408\,e^{-5t} + 0.320\,e^{-t} + 0.088\,e^{-0.1t}$$

and $x_3(t) = -0.188\,e^{-5t} - 0.371\,e^{-t} + 0.559\,e^{-0.1t}$.

The responses are shown in Fig. 4.4. Note that both $x_2(t)$ and $x_3(t)$ are zero when $t = 0^+$ (0^+ denoting time just after the impulsive input), as they should be since material cannot instantaneously be transported from the input compartment to either of the other two.

4.1.2 *The catenary system* (Fig. 4.5)

By inspection of Figs 4.3 and 4.5, it may be seen that the structure of the three-compartment catenary model is the same as that of the three-compartment mammillary model, the only difference being that the numbering of compartments 1 and 2 is interchanged. (This is not so for models with four or more compartments.) The form of $(s\mathbf{I} - \mathbf{A})^{-1}$ and the equation for the eigenvalues may be obtained by putting $a_{13} = a_{31} = 0$ in eqns (4.2) and (4.4).

In many practical applications of catenary models, input is to one of the end compartments only (no. 1, say) and there is prior knowledge that excretion to the environment is from the other end compartment (no. 3). For such a model,

$$a_{11} = -k_{21}$$

$$a_{22} = -(k_{12} + k_{32})$$

and

$$a_{33} = -(k_{23} + k_{03}).$$

In that case, eqn (4.2) simplifies to

$$(s\mathbf{I} - \mathbf{A})^{-1} =$$

$$\frac{1}{\Delta(s)}\begin{bmatrix} (s+k_{12}+k_{32})(s+k_{23}+k_{03})-k_{23}k_{32} & k_{12}(s+k_{23}+k_{03}) & k_{12}k_{23} \\ k_{21}(s+k_{23}+k_{03}) & (s+k_{21})(s+k_{23}+k_{03}) & k_{23}(s+k_{21}) \\ k_{21}k_{32} & k_{32}(s+k_{21}) & (s+k_{21})(s+k_{12}+k_{32})-k_{12}k_{21} \end{bmatrix} \tag{4.9a}$$

where $\Delta(s) = (s - \lambda_1)(s - \lambda_2)(s - \lambda_3)$, the eigenvalues being the roots of

$$\lambda^3 + (k_{21} + k_{12} + k_{32} + k_{03} + k_{23})\lambda^2$$
$$+ (k_{21}k_{32} + k_{12}k_{23} + k_{12}k_{03} + k_{32}k_{03} + k_{23}k_{21} + k_{03}k_{21})\lambda$$
$$+ k_{21}k_{32}k_{03} = 0. \tag{4.9b}$$

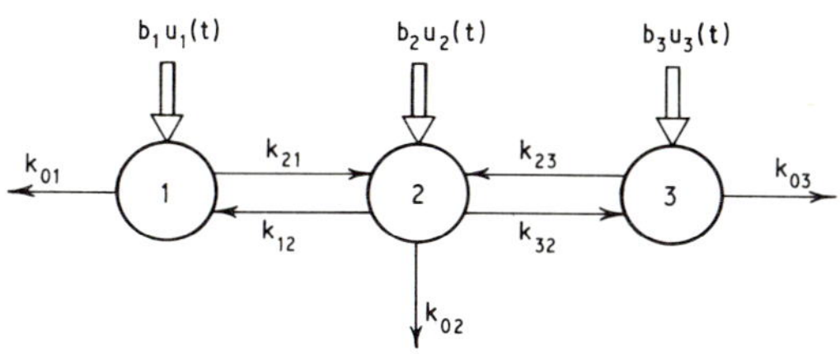

Figure 4.5 Catenary model with three compartments.

Example 4.3. *For a particular catenary three-compartment system with excretion from compartment 3 only,* $k_{21} = 3.71$, $k_{12} = 1.061$, $k_{23} = 0.340$, $k_{32} = 0.825$ *and* $k_{03} = 0.164$. *Determine* $x_1(t)$, $x_2(t)$ *and* $x_3(t)$ $(t \geqslant 0)$ *if* $u_1(t) = \delta(t)$, $u_2(t) = u_3(t) = 0$ *and* $b_1 = 1$.

Solution. The eigenvalues are the roots of

$$\lambda^3 + 6.1\lambda^2 + 5.6\lambda + 0.5 = 0$$

i.e.

$$\lambda_1 = -5, \lambda_2 = -1, \lambda_3 = -0.1.$$

Hence

$$\mathbf{X}(s) = \frac{1}{(s+5)(s+1)(s+0.1)} \begin{bmatrix} (s+1.886)(s+0.504) - 0.281 \\ 3.71(s+0.504) \\ 3.06 \end{bmatrix}$$

giving

$$x_1(t) = 0.7\,\mathrm{e}^{-5t} + 0.2\,\mathrm{e}^{-t} + 0.1\,\mathrm{e}^{-0.1t}$$

$$x_2(t) = -0.851\,\mathrm{e}^{-5t} + 0.511\,\mathrm{e}^{-t} + 0.340\,\mathrm{e}^{-0.1t}$$

$$x_3(t) = 0.156\,e^{-5t} - 0.850\,e^{-t} + 0.694\,e^{-0.1t}$$

which are shown in Fig. 4.6. Note that $x_2(0^+) = x_3(0^+) = 0$, as in Example 4.2, but in the present example, $\mathrm{d}x_3/\mathrm{d}t(0^+) = 0$ because there is no direct link from the input compartment to compartment 3. The transfer function $X_3(s)/U_1(s)$ has numerator degree three less than the denominator degree.

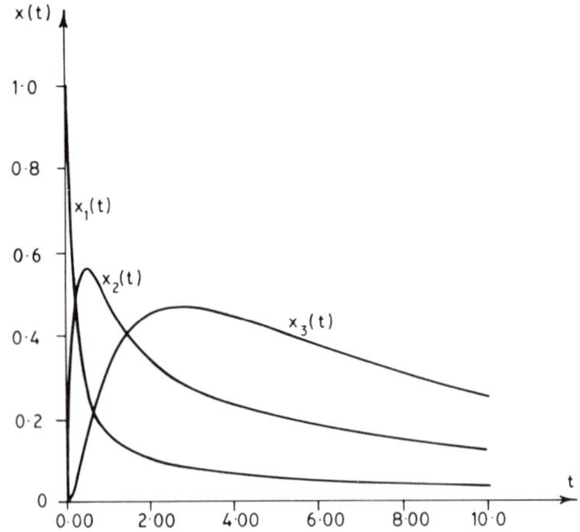

Figure 4.6 Responses of model of Example 4.3 to unit impulsive perturbation of compartment 1.

For any such transfer function, the impulse response will have both initial value and intial slope zero, a point which is discussed further in Appendix 1.

4.1.3 *Closed systems*

All closed systems have a zero eigenvalue. Material which is introduced into the system remains in the system, being shared between the compartments in proportions dependent on the inter-compartment rate constants. For a three-compartment closed system, $k_{01} = k_{02} = k_{03} = 0$, so that $a_{11} = -(k_{21} + k_{31})$, $a_{22} = -(k_{12} + k_{32})$ and $a_{33} = -(k_{13} + k_{23})$. The constant term in eqn (4.4) is zero so that $\lambda = 0$ is one of the roots of eqn (4.4).

As an example of a closed system, we will now work an example of a closed catenary system; further examples appear in Section 4.1.6.

Example 4.4. *A closed catenary system has* $k_{12} = k_{21} = 2$ *and* $k_{23} = k_{32} = 1$. *Find the states* $(t \geqslant 0)$ *if* $u_1(t) = \delta(t)$, $u_2(t) = u_3(t) = 0$ *and* $b_1 = 1$.

Solution. The eigenvalues are, from eqn (4.9b), the roots of

$$\lambda(\lambda^2 + 6\lambda + 6) = 0$$

i.e.

$$\lambda_1 = 0, \quad \lambda_2 = -1.268, \quad \lambda_3 = -4.732.$$

From eqn (4.9a),

$$\mathbf{X}(s) = \frac{1}{s(s+1.268)(s+4.732)} \begin{bmatrix} (s+3)(s+1)-1 \\ 2(s+1) \\ 2 \end{bmatrix}$$

giving

$$x_1(t) = \tfrac{1}{3}(1 + e^{-1.268t} + e^{-4.732t})$$

$$x_2(t) = \tfrac{1}{3} + 0.122\, e^{-1.268t} - 0.455\, e^{-4.732t}$$

$$x_3(t) = \tfrac{1}{3} - 0.455\, e^{-1.268t} + 0.122\, e^{-4.732t}$$

which are shown in Fig. 4.7. As in the previous example.

$$x_2(0^+) = x_3(0^+) = \frac{dx_3}{dt}(0^+) = 0.$$

4.1.4 *Source compartments*

As in the two-compartment system considered in Section 3.2.3, only one of the modes (exponential decays) appears in the response of a source compartment. By way of example, consider a catenary three-compartment

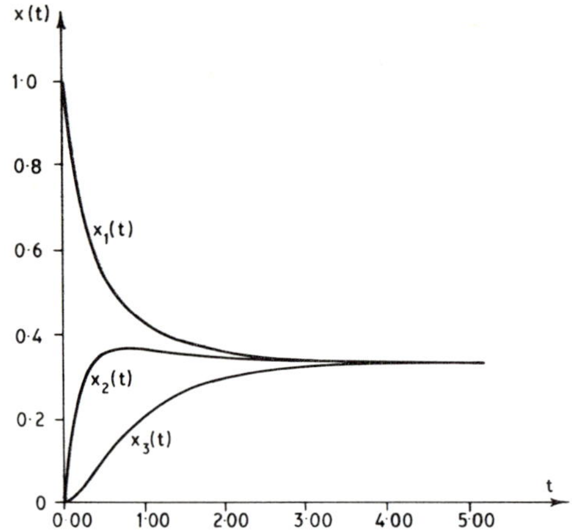

Figure 4.7 Responses of model of Example 4.4 to unit impulsive perturbation of compartment 1.

system with unidirectional exchange from compartment 1 to compartment 2 and from compartment 2 to compartment 3, as shown in Fig. 4.8. For this system, eqns (4.9) apply, with the additional simplification that $k_{12} = k_{23} = 0$, so that

$$(s\mathbf{I} - \mathbf{A})^{-1} =$$

$$\frac{1}{\Delta(s)}\begin{bmatrix} (s + k_{32})(s + k_{03}) & 0 & 0 \\ k_{21}(s + k_{03}) & (s + k_{21})(s + k_{03}) & 0 \\ k_{21}k_{32} & k_{32}(s + k_{21}) & (s + k_{21})(s + k_{32}) \end{bmatrix} \tag{4.10a}$$

where $\Delta(s) = (s - \lambda_1)(s - \lambda_2)(s - \lambda_3)$, the eigenvalues being the roots of $|\lambda\mathbf{I} - \mathbf{A}| = 0$, i.e.

$$(\lambda + k_{21})(\lambda + k_{32})(\lambda + k_{03}) = 0. \tag{4.10b}$$

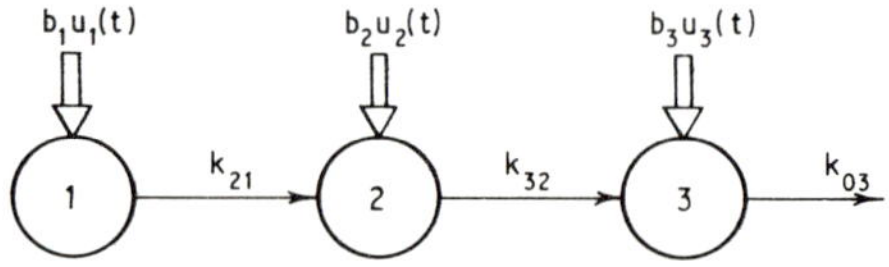

Figure 4.8 Catenary three-compartment model with unidirectional exchange.

If $u_1(t) = \delta(t)$, $u_2(t) = u_3(t) = 0$ and $b_1 = 1$,

$$\mathbf{X}(s) = \frac{1}{(s + k_{21})(s + k_{32})(s + k_{03})} \begin{bmatrix} (s + k_{32})(s + k_{03}) \\ k_{21}(s + k_{03}) \\ k_{21}k_{32} \end{bmatrix}$$

so that

$$x_1(t) = \mathrm{e}^{-k_{21}t} \tag{4.11a}$$

$$x_2(t) = \frac{k_{21}}{k_{32} - k_{21}} (\mathrm{e}^{-k_{21}t} - \mathrm{e}^{-k_{32}t}) \tag{4.11b}$$

$$x_3(t) = \frac{k_{21}k_{32}}{(k_{32} - k_{21})(k_{03} - k_{21})} \mathrm{e}^{-k_{21}t} + \frac{k_{21}k_{32}}{(k_{21} - k_{32})(k_{03} - k_{32})} \mathrm{e}^{-k_{32}t} + \frac{k_{21}k_{32}}{(k_{21} - k_{03})(k_{32} - k_{03})} \mathrm{e}^{-k_{03}t}. \tag{4.11c}$$

Note that $x_2(0^+) = x_3(0^+) = \dfrac{\mathrm{d}x_3}{\mathrm{d}t}(0^+) = 0$, as usual for a catenary three-compartment system with input to compartment 1 only. Unit impulse responses of this system for the specific values $k_{21} = 3$, $k_{32} = 2$ and $k_{03} = 1$ are shown in Fig. 4.9.

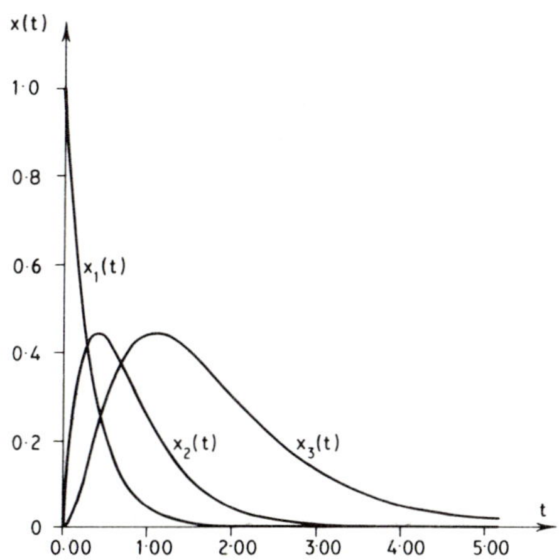

Figure 4.9 Responses of model of Fig. 4.8 with $k_{21} = 3$, $k_{32} = 2$ and $k_{03} = 1$ to unit impulsive perturbation of compartment 1.

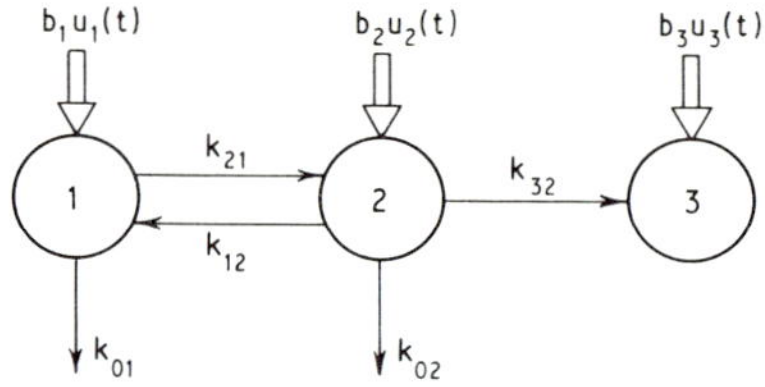

Figure 4.10 Three-compartment model with a sink compartment.

4.1.5 *Sink compartments*

The presence of a sink compartment anywhere in a three-compartment system results in one of the eigenvalues being zero. Consider by way of example the system shown in Fig. 4.10 in which compartment 3 is a sink compartment. As far as compartments 1 and 2 are concerned, the system is exactly the same as the general *two*-compartment system considered in Section 3.2.1, with $(k_{32} + k_{02})$ replacing k_{02}. If $u_3(t) = 0$, the state of compartment 3 is given by

$$x_3(t) = k_{32} \int_0^t x_2(t)\,\mathrm{d}t. \tag{4.12}$$

For this system,

$$(s\mathbf{I} - \mathbf{A})^{-1} = \begin{bmatrix} \dfrac{s - a_{22}}{(s - \lambda_2)(s - \lambda_3)} & \dfrac{a_{12}}{(s - \lambda_2)(s - \lambda_3)} & 0 \\ \dfrac{a_{21}}{(s - \lambda_2)(s - \lambda_3)} & \dfrac{s - a_{11}}{(s - \lambda_2)(s - \lambda_3)} & 0 \\ \dfrac{a_{21}a_{32}}{s(s - \lambda_2)(s - \lambda_3)} & \dfrac{a_{32}(s - a_{11})}{s(s - \lambda_2)(s - \lambda_3)} & \dfrac{1}{s} \end{bmatrix} \tag{4.13}$$

the two non-zero eigenvalues λ_2 and λ_3 being the roots of $\lambda^2 - (a_{11} + a_{22})\lambda + (a_{11}a_{22} - a_{12}a_{21}) = 0$. By inspection, the responses $x_1(t)$ and $x_2(t)$ to an input to compartment 1 and/or compartment 2 are the same as for a two-compartment system (eqn (3.25)) while the response $x_3(t)$ is $k_{32} \int_0^t x_2(t)\,\mathrm{d}t$, as expected.

More generally, Foster and Jacquez (1975) have shown that if m of the eigenvalues of an open p-compartment system are zero, then there are m sink compartments.

4.1.6 *The unilateral circulation model*

The examples considered so far in this chapter have all had real eigenvalues because their rate constants have satisfied Goldberg's condition, eqn (2.59).

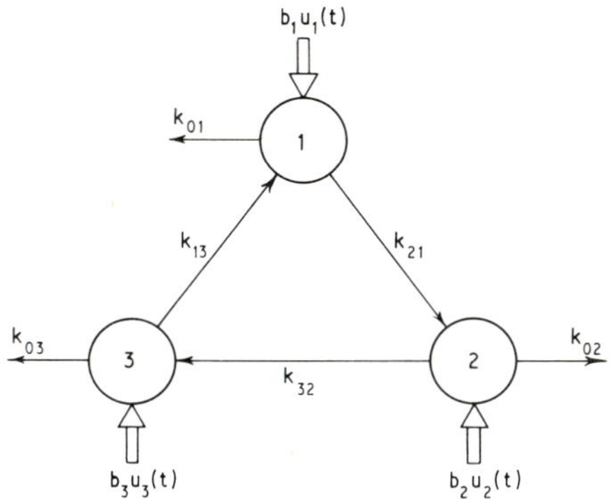

Figure 4.11 Unilateral circulation model with three compartments.

An interesting system for which this condition does not hold is the unilateral circulation model (Schoenfeld (1963)) shown in Fig. 4.11, in which material circulates round the loop in one direction only. If the system is closed ($k_{01} = k_{02} = k_{03} = 0$) then, from eqn (4.4), the eigenvalues are the roots of

$$\lambda^3 + (k_{21} + k_{32} + k_{13})\lambda^2 + (k_{21}k_{32} + k_{32}k_{13} + k_{13}k_{21})\lambda = 0 \quad (4.14)$$

so $\lambda = 0$ is a solution and the other two eigenvalues are

$$\lambda_2, \lambda_3 = \tfrac{1}{2}\{-(k_{21} + k_{32} + k_{13}) \pm [(k_{21} + k_{32} + k_{13})^2 - 4(k_{21}k_{32} + k_{32}k_{13} + k_{13}k_{21})]^{1/2}\}. \quad (4.15)$$

For a compartmental system, any complex eigenvalues must occur as complex conjugate pairs ($-\alpha \pm j\beta$, where α is real and positive and β is real) because all the terms in $|\lambda\mathbf{I} - \mathbf{A}|$ are real, $\mathbf{A}$ being real.

There is thus a complex conjugate pair if

$$4(k_{21}k_{32} + k_{32}k_{13} + k_{13}k_{21}) > (k_{21} + k_{32} + k_{13})^2 \quad (4.16)$$

and these will give rise to terms of the form $E\,\mathrm{e}^{-\alpha t}\sin(\beta t + \phi)$ in the impulse responses, where E and ϕ are constants and

$$\alpha = \tfrac{1}{2}(k_{21} + k_{32} + k_{13}) \quad (4.17a)$$

$$\beta^2 = (k_{21}k_{32} + k_{32}k_{13} + k_{13}k_{21}) - \tfrac{1}{4}(k_{21} + k_{32} + k_{13})^2. \quad (4.17b)$$

As we have seen in Section 2.1.2, a second-order system is at its most oscillatory when the damping ratio ζ is a minimum. In standard notation, the denominator of a second-order system, in terms of the eigenvalues, is, as in eqn (2.22),

$$\lambda^2 + 2\zeta\omega_0\lambda + \omega_0^2$$

where ω_0 is the undamped natural frequency (rad s^{-1}) of the oscillations obtained when ζ is zero. Thus for the complex pair of eigenvalues in the unilateral circulation system,

$$2\zeta\omega_0 = (k_{21} + k_{32} + k_{13})$$

and

$$\omega_0^2 = (k_{21}k_{32} + k_{32}k_{13} + k_{13}k_{21})$$

whence $$\zeta = \tfrac{1}{2}(k_{21} + k_{32} + k_{13})(k_{21}k_{32} + k_{32}k_{13} + k_{13}k_{21})^{-1/2}. \qquad (4.18)$$

(It is readily shown that ζ is also equal to $\dfrac{\alpha}{(\alpha^2 + \beta^2)^{1/2}}$ where α and β^2 are as given in eqns (4.17).)

To find the minimum possible ζ for this system, we will differentiate with respect to k_{21}, k_{32} and k_{13} in turn.

$$\begin{aligned}\frac{\partial\zeta}{\partial k_{21}} &= \tfrac{1}{2}(k_{21}k_{32} + k_{32}k_{13} + k_{13}k_{21})^{-1/2} \\ &\quad - \tfrac{1}{4}(k_{21} + k_{32} + k_{13})(k_{21}k_{32} + k_{32}k_{13} + k_{13}k_{21})^{-3/2}(k_{32} + k_{13}) \\ &= \frac{2(k_{21}k_{32} + k_{32}k_{13} + k_{13}k_{21}) - (k_{21} + k_{32} + k_{13})(k_{32} + k_{13})}{4(k_{21}k_{32} + k_{32}k_{13} + k_{13}k_{21})^{3/2}}.\end{aligned}$$

For minimum ζ, the numerator must be zero, so that

$$\begin{aligned}2(k_{21}k_{32} + k_{32}k_{13} + k_{13}k_{21}) &= k_{21}k_{32} + k_{32}^2 + k_{13}k_{32} \\ &\quad + k_{21}k_{13} + k_{32}k_{13} + k_{13}^2\end{aligned}$$

i.e.

$$k_{21}(k_{32} + k_{13}) = k_{32}^2 + k_{13}^2$$

whence $$k_{21} = \frac{k_{32}^2 + k_{13}^2}{k_{32} + k_{13}}. \qquad (4.19a)$$

Similarly, $\dfrac{\partial\zeta}{\partial k_{32}} = 0$ gives

$$k_{32} = \frac{k_{21}^2 + k_{13}^2}{k_{21} + k_{13}} \qquad (4.19b)$$

and $\dfrac{\partial\zeta}{\partial k_{13}} = 0$ gives

$$k_{13} = \frac{k_{21}^2 + k_{32}^2}{k_{21} + k_{32}} \qquad (4.19c)$$

Equations (4.19) hold simultaneously if $k_{21} = k_{32} = k_{13}$ when, from eqn (4.18), $\zeta = \dfrac{\sqrt{3}}{2}$.

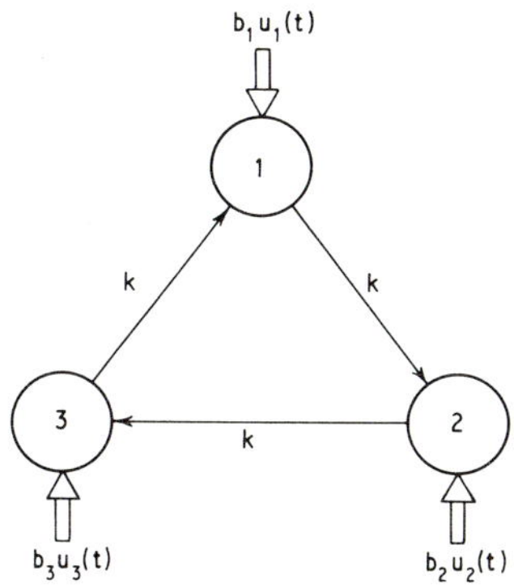

Figure 4.12 Closed three-compartment unilateral circulation model with three compartments and with equal rate constants.

If $k_{21} = k_{32} = k_{13} = k$, so that the model is as in Fig. 4.12,

$$(s\mathbf{I} - \mathbf{A})^{-1} = \frac{1}{\Delta(s)} \begin{bmatrix} (s+k)^2 & k^2 & k(s+k) \\ k(s+k) & (s+k)^2 & k^2 \\ k^2 & k(s+k) & (s+k)^2 \end{bmatrix} \tag{4.20}$$

where $\Delta(s) = s(s - \lambda_2)(s - \lambda_3)$ and the non-zero eigenvalues are the roots of

$$\lambda^2 + 3k\lambda + 3k^2 = 0 \tag{4.21a}$$

i.e.

$$\lambda_2, \lambda_3 = \tfrac{1}{2}[-3k \pm jk\sqrt{3}]. \tag{4.21b}$$

The three eigenvalues thus lie equally spaced on a circle in the complex plane, with centre $(-k, 0)$ and radius k, as shown in Fig. 4.13.

For this model, if $u_1(t) = \delta(t)$, $u_2(t) = u_3(t) = 0$ and $b_1 = 1$, the Laplace transforms of the states are, from eqns (4.20) and (4.21a),

$$\mathbf{X}(s) = \frac{1}{s(s^2 + 3ks + 3k^2)} \begin{bmatrix} (s+k)^2 \\ k(s+k) \\ k^2 \end{bmatrix} \tag{4.22a}$$

so that

$$X_1(s) = \frac{1}{3}\left[\frac{1}{s} + \frac{2s + 3k}{s^2 + 3ks + 3k^2}\right]$$

giving

$$x_1(t) = \frac{1}{3}\left(1 + 2\,\mathrm{e}^{-3kt/2} \cos \frac{k\sqrt{3}}{2}\, t\right) \tag{4.22b}$$

(see Appendix 1).

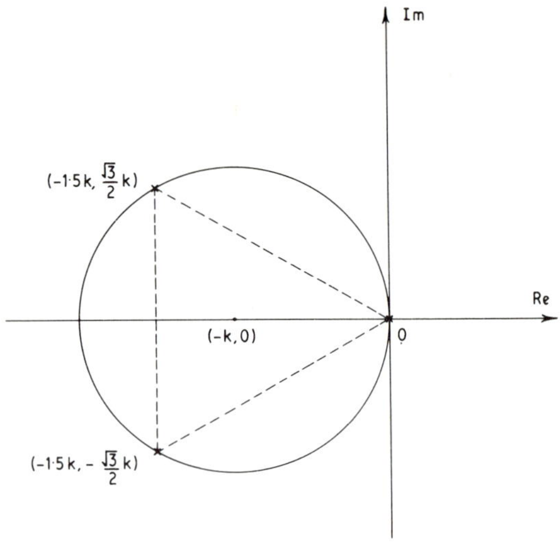

Figure 4.13 Positions, on the complex plane, of the three eigenvalues of the model of Fig. 4.12.

Similarly,
$$x_2(t) = \frac{1}{3}\left[1 - e^{-3kt/2}\left(\cos\frac{k\sqrt{3}}{2}t - \sqrt{3}\sin\frac{k\sqrt{3}}{2}t\right)\right]. \qquad (4.22c)$$

and
$$x_3(t) = \frac{1}{3}\left[1 - e^{-3kt/2}\left(\cos\frac{k\sqrt{3}}{2}t + \sqrt{3}\sin\frac{k\sqrt{3}}{2}t\right)\right]. \qquad (4.22d)$$

The responses are shown in Fig. 4.14. Note that $x_1(t)$ does not decrease monotonically from its peak value of 1 to its final value of $\frac{1}{3}$ and that there is a small undershoot. By differentiation, it is readily shown that the first minimum of $x_1(t)$ is at $t = \dfrac{4\pi}{3k\sqrt{3}}$ at which time $x_1 = \frac{1}{3}(1 - 0.027)$. The oscillatory nature of this type of system becomes more noticeable as the number of compartments increases, as we will see in Section 4.2.5.

Two modifications to the above model are also of interest; firstly when the model is open and secondly when there is also flow in the opposite direction in the closed system. Consider an open system, with excretion to the environment from one compartment only, no. 1 say, with rate constant k_{01}. Then

$$\mathbf{A} = \begin{bmatrix} -k - k_{01} & 0 & k \\ k & -k & 0 \\ 0 & k & -k \end{bmatrix} \qquad (4.23a)$$

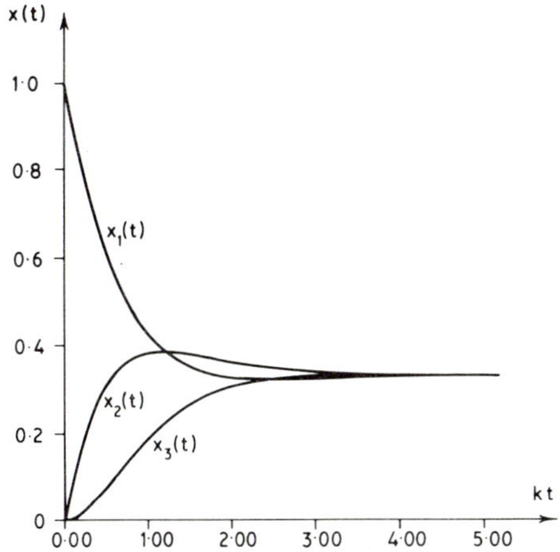

Figure 4.14 Responses of model of Fig. 4.12 to unit impulsive perturbation of compartment 1.

and the eigenvalues are the roots of

$$(\lambda + k)^2(\lambda + k + k_{01}) = k^3$$

i.e.

$$(\lambda + k)^3 + k_{01}(\lambda + k)^2 = k^3. \tag{4.23b}$$

The following proof that ζ is greater for this system than that for $k_{01} = 0$ is due to J. P. Norton of the University of Birmingham. It is a trivial matter to generalize it to a system with any number of compartments. Differentiating eqn (4.23b) with respect to k_{01},

$$3(\lambda + k)^2 \frac{\partial \lambda}{\partial k_{01}} + 2k_{01}(\lambda + k)\frac{\partial \lambda}{\partial k_{01}} + (\lambda + k)^2 = 0$$

so

$$\frac{\partial \lambda}{\partial k_{01}} = -\frac{\lambda + k}{3(\lambda + k) + 2k_{01}}. \tag{4.24}$$

Now, for λ in the upper half of the complex plane and k_{01} real and positive,

$$0 < \angle(3(\lambda + k) + 2k_{01}) < \angle(\lambda + k)$$

so that

$$\pi < \angle\left(\frac{\partial \lambda}{\partial k_{01}}\right) < \pi + \angle(\lambda + k).$$

This is illustrated in Fig. 4.15. A sufficiently small increase in k_{01} from zero moves the eigenvalue from its position P in a direction such that the new

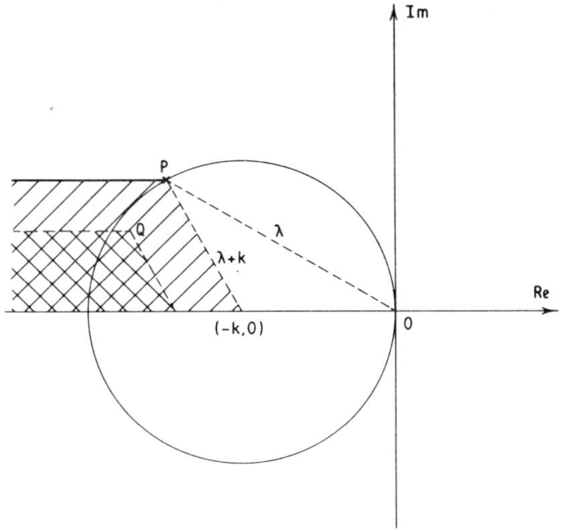

Figure 4.15 Change in position on the complex plane of complex eigenvalue of the model of Fig. 4.13 as excretion to the environment is introduced.

position is within the lightly-shaded region. Similarly, a small increase in k_{01} from a non-zero value moves the eigenvalue from a position Q in a direction such that the new position is within the heavily-shaded region. The behaviour of the other pole of the complex conjugate pair is, of course, the mirror image of this in the real axis. Note that, should the eigenvalue approach the real axis, $\angle\left(\frac{\partial\lambda}{\partial k_{01}}\right)$ would approach $\pm\pi$ and the real axis would not be crossed. Movement from P into the shaded region corresponds to increasing ζ since from Fig. 4.15,

$$\zeta = \frac{\text{Re}}{(\text{Re}^2 + \text{Im}^2)^{1/2}} = \cos^{-1}(\pi - \angle\lambda).$$

As this is true whatever the (positive real) value of k_{01}, the argument may be applied repetitively for successive increments in k_{01} from zero, to show that the total effect of non-zero k_{01} is to increase the damping ratio from that for k_{01} zero.

Example 4.5. *For an open unilateral circulation system with $k_{01} = k_{02} = k_{03} = k_{21} = k_{32} = k_{13} = k$, find the eigenvalues and confirm that ζ is greater than for the corresponding closed system. Find $x_1(t)$ $(t \geqslant 0)$ if $u_1(t) = \delta(t)$, $u_2(t) = u_3(t) = 0$ and $b_1 = 1$.*

Solution. From eqn (4.4), the eigenvalues are the roots of

$$\lambda^3 + 6k\lambda^2 + 12k^2\lambda + 7k^3 = 0$$

i.e.
$$(\lambda + k)(\lambda^2 + 5k\lambda + 7k^2) = 0$$

so that $\lambda_1 = -k$ and λ_2, λ_3 are the roots of

$$\lambda^2 + 5k\lambda + 7k^2 = 0$$

i.e.
$$\lambda_2, \lambda_3 = -2.5k \pm jk\frac{\sqrt{3}}{2}.$$

Comparing these values with those for the closed system, we see that the imaginary part has remained the same $\left(k\frac{\sqrt{3}}{2}\right)$ while the magnitude of the real part has increased from $1.5k$ to $2.5k$. Thus the damping factor has increased and for the open system is given by

$$\zeta = \frac{2.5k}{k[(2.5)^2 + 0.75]^{1/2}} = \frac{2.5}{(7)^{1/2}} = 0.945$$

so that the open system is indeed less oscillatory than the closed system.

From eqn (4.2), for the given perturbation,

$$X_1(s) = \frac{(s + 2k)^2}{(s + k)(s^2 + 5ks + 7k^2)}$$

whence
$$x_1(t) = \tfrac{1}{3}\,\mathrm{e}^{-kt} + \tfrac{2}{3}\,\mathrm{e}^{-2.5kt}\cos\frac{k\sqrt{3}}{2}t$$

which is shown in Fig. 4.16 together with the corresponding $x_2(t)$ and $x_3(t)$.

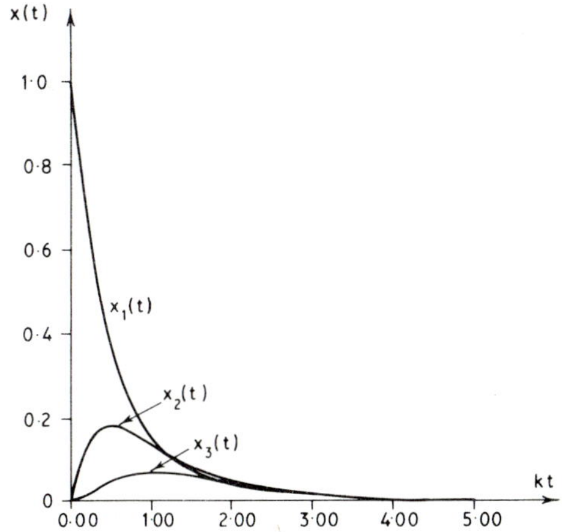

Figure 4.16 Responses of model of Example 4.5 to unit impulsive perturbation of compartment 1.

The damping ($e^{-2.5kt}$) of the oscillatory term is considerably greater than for the closed system and no oscillation can be detected on the decay of $x_1(t)$.

The second modification of interest is when there is counter-flow. A proof along similar lines to that above gives a local result that if there is a small amount of counter-flow between two of the compartments, then ζ is increased (and so the system becomes less oscillatory). However a global result would not be possible, for if there were equal counter-flow in the three links with rate constant $\gg k$, the system would be equally oscillatory. Apart from this, it seems likely that the closed system with equal flows produces the most oscillatory responses. If equal counter-flow with rate constant k were present, the eigenvalues would be real (from eqn (2.59)). The eigenvalues are also real if there is counter-flow with rate constant k in only one of the inter-compartment links; the eigenvalues are then 0, $-2k$ and $-2k$.

4.2 Systems with more than three compartments

4.2.1 *The general p-compartment system*

For the general linear time-invariant model with p compartments,

$$\mathbf{A} = \begin{bmatrix} a_{11} & a_{12} & \cdots & a_{1p} \\ a_{21} & a_{22} & \cdots & a_{2p} \\ \vdots & & & \\ a_{p1} & a_{p2} & \cdots & a_{pp} \end{bmatrix}$$

where $\quad a_{11} = -(k_{21} + k_{31} + \ldots + k_{p1} + k_{01}), \quad a_{12} = k_{12} \quad \text{etc.} \qquad (4.25)$

Shipley and Clark (1972, pp. 218–220) give formulae enabling gains and exponents of an impulse response to be calculated from known rate constants for a general four-compartment system. In general, though, analysis of compartmental systems with more than three compartments proves lengthy and tedious and it is usually desirable to resort to a computer simulation to give the shape of the responses to a perturbation.

Rather than continue with the general system, some specific examples will be considered in the remainder of this chapter and for each one, many of the elements of **A** are zero.

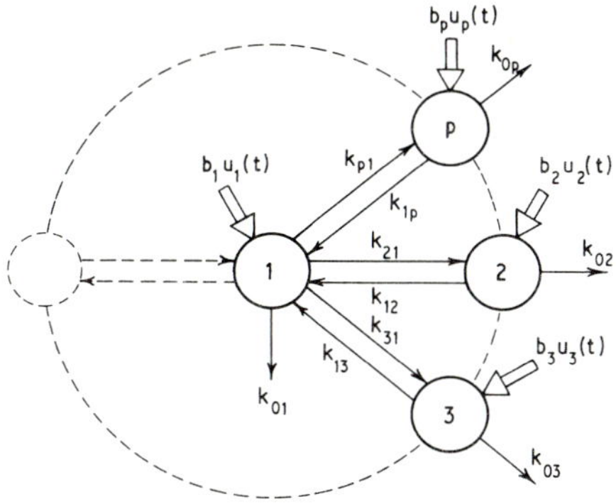

Figure 4.17 Mammillary model with p compartments.

4.2.2 *The mammillary system* (Fig. 4.17)

All elements of **A** except those in the first row, the first column and along the main diagonal are zero and the main diagonal elements are given by

$$a_{11} = -(k_{21} + k_{31} + \ldots + k_{p1} + k_{01}) \tag{4.26a}$$

$$a_{ii} = -(k_{1i} + k_{0i}), \quad 2 \leqslant i \leqslant p. \tag{4.26b}$$

The eigenvalues are all real (from eqn (2.59)).

4.2.3 *The catenary system* (Fig. 4.18)

For $p > 3$, the catenary system is different in structure from the mammillary system. All elements of **A** except those along the main diagonal and the diagonals immediately above and below are zero. The main diagonal elements are given by

$$a_{11} = -(k_{21} + k_{01}) \tag{4.27a}$$

$$a_{ii} = -(k_{(i-1)i} + k_{(i+1)i} + k_{0i}), \qquad 2 \leqslant i \leqslant (p-1) \tag{4.27b}$$

$$a_{pp} = -(k_{(p-1)p} + k_{0p}). \tag{4.27c}$$

Again, from eqn (2.59), the eigenvalues are all real.

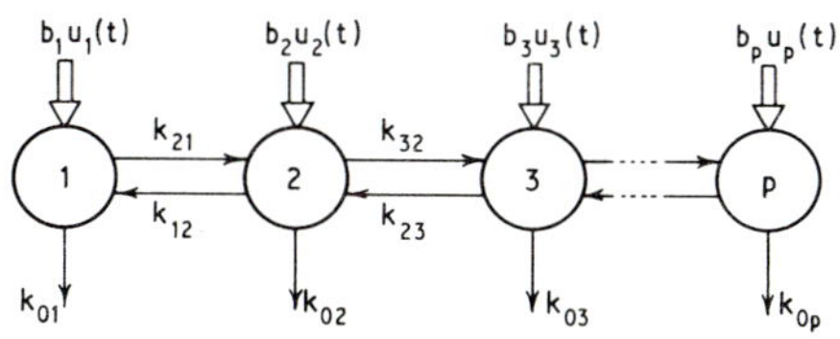

Figure 4.18 Catenary model with p compartments.

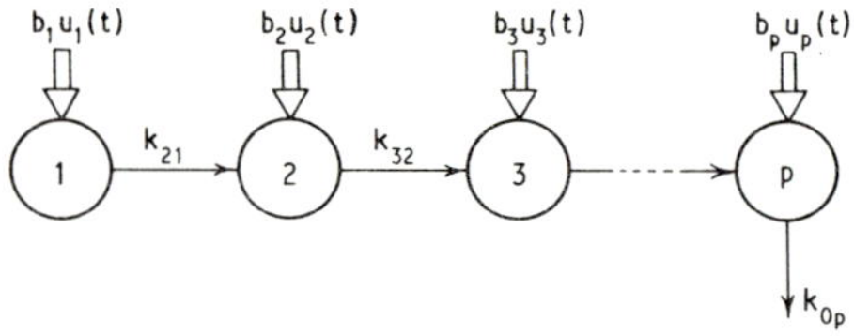

Figure 4.19 Catenary model with unidirectional exchange and excretion from the final compartment only.

Consider now the special case of the unidirectional catenary system with excretion from compartment p only, as shown in Fig. 4.19. For such a system, $k_{12} = k_{23} = \ldots = k_{(p-1)p} = 0$ and $k_{01} = k_{02} = \ldots = k_{0(p-1)} = 0$. If the input is to compartment 1 only (with $b_1 = 1$), the Laplace transforms of the states can be written in terms of transfer functions:

$$\frac{X_1(s)}{U_1(s)} = \frac{1}{s + k_{21}} \tag{4.28a}$$

$$\frac{X_2(s)}{U_1(s)} = \frac{X_2(s)}{X_1(s)} \cdot \frac{X_1(s)}{U_1(s)} = \frac{k_{21}}{(s + k_{32})(s + k_{21})} \tag{4.28b}$$

$$\frac{X_3(s)}{U_1(s)} = \frac{X_3(s)}{X_2(s)} \cdot \frac{X_2(s)}{X_1(s)} \cdot \frac{X_1(s)}{U_1(s)} = \frac{k_{32} \cdot k_{21}}{(s + k_{43})(s + k_{32})(s + k_{21})} \tag{4.28c}$$

. . . .

$$\frac{X_p(s)}{U_1(s)} = \frac{X_p(s)}{X_{p-1}(s)} \cdot \frac{X_{p-1}(s)}{X_{p-2}(s)} \cdots \frac{X_1(s)}{U_1(s)}$$

$$= \frac{k_{p(p-1)} k_{(p-1)(p-2)} \ldots k_{32} k_{21}}{(s + k_{0p})(s + k_{p(p-1)}) \ldots (s + k_{21})}. \tag{4.28d}$$

If $u_1(t)$ were an impulsive input, for example, $x_1(t)$ would consist of one exponential and, assuming distinct rate constants, $x_2(t)$ would consist of two exponentials and so on, with $x_p(t)$ consisting of p exponentials.

4.2.4 *Interconnected subsystems*

Several compartmental systems in the literature consist of two or more subsystems with at most two unidirectional connections between them. For example, the model of Fig. 4.20 is a typical two-compartment drug, two-compartment metabolite pharmacokinetic model (Gibaldi and Perrier, 1975)

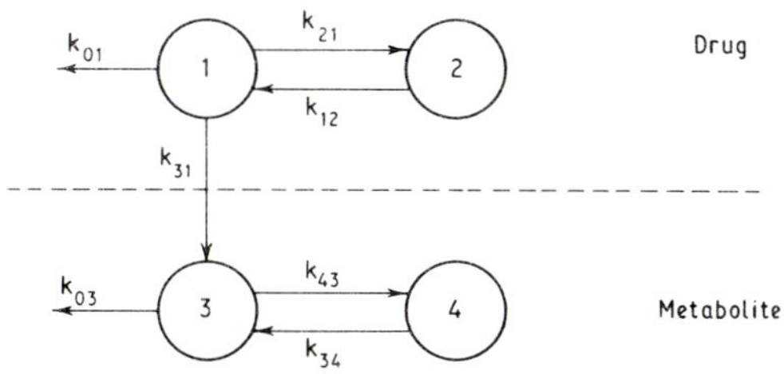

Figure 4.20 Typical two-compartment drug, two-compartment metabolite model.

with compartments 1 and 2 modelling the drug kinetics, compartments 3 and 4 modelling the metabolite kinetics and the link from 1 to 3 representing biotransformation from drug to metabolite. This type of model will be considered further in Application Example 4 in Chapter 11. The system of Fig. 4.21 has been used to model the kinetics of bromosulfophthalein (BSP) (Milanese and Molino, 1975). Compartments 1, 2 and 3 model the kinetics of free BSP, while compartments 4, 5 and 6 model the kinetics of conjugated BSP, the link from compartment 2 to compartment 4 representing conjugation in the liver. Compartments 1 and 5 contain the blood, 2 and 4 contain the liver cells and compartments 3 and 6 represent the intrahepatic biliary system.

The technique of dealing with interconnected subsystems will be illustrated with a third example—of a model of thyroid hormone metabolism (DiStefano *et al.*, 1975; DiStefano and Mori, 1977). In the model, shown in Fig. 4.22, compartments 1, 2 and 3 are a three-compartment mammillary model for triiodothyronine (T_3) and compartments 4, 5 and 6 are a three-compartment mammillary model for thyroxine (T_4). Links from compartment 5 to compartment 2 and from compartment 6 to compartment 3

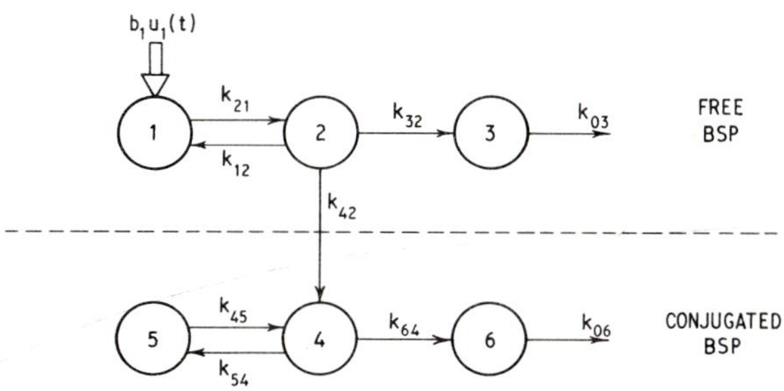

Figure 4.21 Model for the kinetics of bromosulphophthalein (after Milanese and Molino, 1975).

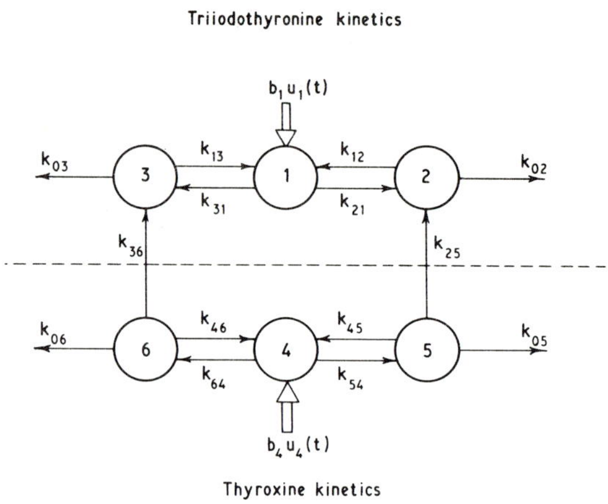

Figure 4.22 Model describing thyroid hormone metabolism (after DiStefano *et al.*, 1975; DiStefano and Mori, 1977).

represent biotransformation from T_4 to T_3; there is no biotransformation from T_3 to T_4. Thus, for the overall system,

$$\mathbf{A} = \begin{bmatrix} a_{11} & a_{12} & a_{13} & 0 & 0 & 0 \\ a_{21} & a_{22} & 0 & 0 & a_{25} & 0 \\ a_{31} & 0 & a_{33} & 0 & 0 & a_{36} \\ 0 & 0 & 0 & a_{44} & a_{45} & a_{46} \\ 0 & 0 & 0 & a_{54} & a_{55} & 0 \\ 0 & 0 & 0 & a_{64} & 0 & a_{66} \end{bmatrix} \tag{4.29}$$

where

$$a_{11} = -(k_{21} + k_{31}) \tag{4.30a}$$

$$a_{22} = -(k_{12} + k_{02}) \tag{4.30b}$$

$$a_{33} = -(k_{13} + k_{03}) \tag{4.30c}$$

$$a_{44} = -(k_{54} + k_{64}) \tag{4.30d}$$

$$a_{55} = -(k_{45} + k_{25} + k_{05}) \tag{4.30e}$$

$$a_{66} = -(k_{46} + k_{36} + k_{06}) \tag{4.30f}$$

$$a_{12} = k_{12}, a_{13} = k_{13} \text{ etc.} \tag{4.30g}$$

Consider experiments in which the central compartments (1 and 4) only are perturbed. For a unit impulsive input to compartment 1 (with $b_1 = 1$), the compartmental quantity in compartment 1 is, from eqns (4.6a) and (4.4)

$$X_1(s) = \frac{(s - a_{22})(s - a_{33})}{s^3 + \gamma_1 s^2 + \delta_1 s + \varepsilon_1} \tag{4.31}$$

where
$$\gamma_1 = -(a_{11} + a_{22} + a_{33}) \tag{4.32a}$$
$$\delta_1 = a_{11}a_{22} + a_{22}a_{33} + a_{33}a_{11} - a_{12}a_{21} - a_{13}a_{31} \tag{4.32b}$$
$$\varepsilon_1 = -a_{11}a_{22}a_{33} + a_{12}a_{21}a_{33} + a_{13}a_{31}a_{22}. \tag{4.32c}$$

Similarly, for a unit impulsive perturbation of compartment 4 (with $b_4 = 1$),

$$X_4(s) = \frac{(s - a_{55})(s - a_{66})}{s^3 + \gamma_2 s^2 + \delta_2 s + \varepsilon_s} \tag{4.33}$$

where
$$\gamma_2 = -(a_{44} + a_{55} + a_{66}) \tag{4.34a}$$
$$\delta_2 = a_{44}a_{55} + a_{55}a_{66} + a_{66}a_{44} - a_{45}a_{54} - a_{46}a_{64} \tag{4.34b}$$
$$\varepsilon_2 = -a_{44}a_{55}a_{66} + a_{45}a_{54}a_{66} + a_{46}a_{64}a_{55}. \tag{4.34c}$$

To determine $X_1(s)$ in this case, we note that there are two inputs to the T_3 part of the model—an input at rate $k_{25}x_5(t)$ to compartment 2 and an input at rate $k_{36}x_6(t)$ to compartment 3. From eqns (4.6b) and (4.6c), the Laplace transforms of these input rates are:

$$U_2(s) = a_{25}X_5(s) = a_{25}\frac{a_{54}(s - a_{66})}{s^3 + \gamma_2 s^2 + \delta_2 s + \varepsilon_2} \tag{4.35a}$$

and

$$U_3(s) = a_{36}X_6(s) = a_{36}\frac{a_{64}(s - a_{55})}{s^3 + \gamma_2 s^2 + \delta_2 s + \varepsilon_2}. \tag{4.35b}$$

From eqn (4.7a), the response in compartment 1 to these inputs is

$$X_1(s) = \frac{a_{12}(s - a_{33})}{s^3 + \gamma_1 s^2 + \delta_1 s + \varepsilon_1} U_2(s) + \frac{a_{13}(s - a_{22})}{s^3 + \gamma_1 s^2 + \delta_1 s + \varepsilon_1} U_3(s)$$
$$= \frac{a_{25}a_{54}a_{12}(s - a_{33})(s - a_{66}) + a_{36}a_{64}a_{13}(s - a_{22})(s - a_{55})}{(s^3 + \gamma_1 s^2 + \delta_1 s + \varepsilon_1)(s^3 + \gamma_2 s^2 + \delta_2 s + \varepsilon_2)}. \tag{4.36}$$

This approach, analysing the T_3 and T_4 subsystems separately and then connecting them via the two links, is a good deal simpler than trying to analyse the model as a sixth-order system. Responses to simultaneous perturbations of compartments 1 and 4 can also be found readily by superposition.

4.2.5 *The unilateral circulation model*

In Section 4.1.6, we saw that an oscillatory solution could occur for a three-compartment unilateral circulation model and that for a closed system, the damping ratio ζ was least (i.e. the solution most oscillatory) when all the rate constants were equal. Let us now pursue that line of inquiry and investigate the behaviour of a closed p-compartment unilateral circulation model with all rate constants equal to k, as shown in Fig. 4.23. For this system,

$$\mathbf{A} = \begin{bmatrix} -k & 0 & 0 & \ldots & 0 & k \\ k & -k & 0 & \ldots & 0 & 0 \\ 0 & k & -k & \ldots & 0 & 0 \\ \cdot & \cdot & \cdot & \cdot & \cdot & \cdot \\ 0 & 0 & 0 & \ldots & k & -k \end{bmatrix}. \tag{4.37}$$

The eigenvalues are the solutions of

$$\begin{vmatrix} \lambda + k & 0 & 0 & \ldots & 0 & -k \\ -k & \lambda + k & 0 & \ldots & 0 & 0 \\ 0 & -k & \lambda + k & \ldots & 0 & 0 \\ \cdot & \cdot & \cdot & & \cdot & \cdot \\ 0 & 0 & 0 & \ldots & -k & \lambda + k \end{vmatrix} = 0 \tag{4.38}$$

i.e.

$$(\lambda + k)^p - k^p = 0$$

$$\lambda + k = (k^p)^{1/p} = k(1)^{1/p}$$

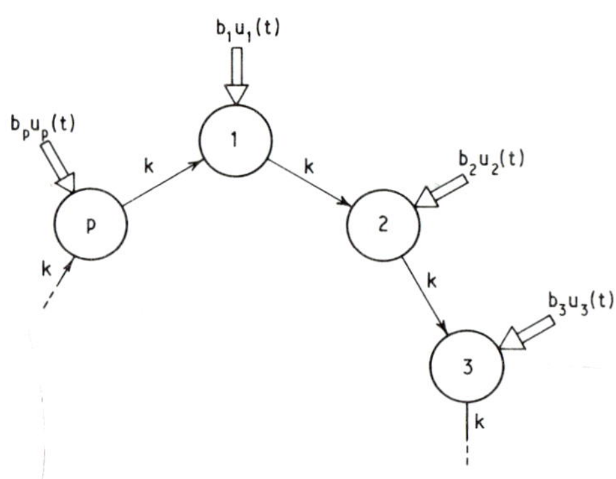

Figure 4.23 Closed unilateral circulation model with p compartments and with all rate constants equal.

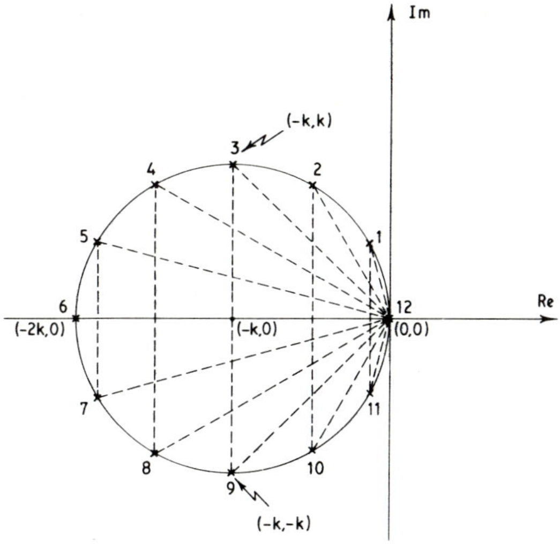

Figure 4.24 Positions, on the complex plane, of the eigenvalues of the model of Fig. 4.23 (shown for $p = 12$; values of m (eqn 4.39) are shown round circle).

whence $$\lambda = -k + k\left(\cos\frac{2\pi m}{p} + j\sin\frac{2\pi m}{p}\right), \qquad m = 1, 2 \ldots p. \tag{4.39}$$

The eigenvalues thus lie on a unit circle in the complex plane, with centre $-k$ and radius k, as shown in Fig. 4.24. One eigenvalue is at the origin and for p even, one eigenvalue is equal to $-2k$; the remaining eigenvalues form complex conjugate pairs.

The complex pair with $m = 1$ in eqn (4.39) gives rise to an oscillatory term with damping ratio ζ given by

$$\zeta = \frac{\text{Re}}{(\text{Re}^2 + \text{Im}^2)^{1/2}} = \frac{1 - \cos\dfrac{2\pi}{p}}{\left[\left(1 - \cos\dfrac{2\pi}{p}\right)^2 + \sin^2\dfrac{2\pi}{p}\right]^{1/2}}$$

(see Fig. 4.25) so that

$$\zeta = \frac{1 - \cos\dfrac{2\pi}{p}}{\left(2 - 2\cos\dfrac{2\pi}{p}\right)^{1/2}} = \sin\frac{\pi}{p}. \tag{4.40}$$

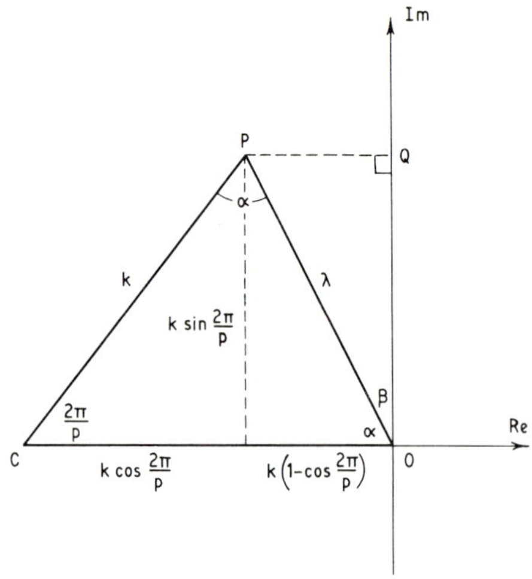

Figure 4.25 Position, on the complex plane, of one of the eigenvalues (with $m = 1$) of the model of Fig. 4.23.

$\Big($That $\zeta = \sin\dfrac{\pi}{p}$ is evident from the geometry. Since OPC is an isosceles triangle, $\angle\text{COP} = \angle\text{CPO} = \dfrac{\pi}{2} - \dfrac{\pi}{p}$. Thus $\beta = \dfrac{\pi}{p}$ and $\zeta = \dfrac{\text{QP}}{\text{OP}} = \sin\beta = \sin\dfrac{\pi}{p}$.$\Big)$

As $p \to \infty$, therefore, the value of ζ associated with this pair of eigenvalues ($m = 1$ and $m = (p - 1)$) approaches zero so that they give rise to an oscillatory term which, in the limit, is undamped. However, as we have already seen in Chapter 2, no compartmental response to a positive input and/or positive initial conditions can go negative (Bellman, 1960), so that the amplitudes of any terms with very small values of ζ must become vanishingly small as $p \to \infty$ (Thron, 1972).

We will now consider the case $p = 4$ in more detail. From eqn (4.39), the eigenvalues are:

$$\lambda_1 = -k + k(\cos 0 + j \sin 0) = 0 \tag{4.41a}$$

$$\lambda_2 = -k + k\left(\cos\frac{\pi}{2} + j \sin\frac{\pi}{2}\right) = -k + jk \tag{4.41b}$$

$$\lambda_3 = -k + k(\cos \pi + j \sin \pi) = -2k \tag{4.41c}$$

$$\lambda_4 = -k + k\left(\cos\frac{3\pi}{2} + j\sin\frac{3\pi}{2}\right) = -k - jk. \tag{4.41d}$$

For the complex pair $\lambda_2, \lambda_4 = -k \pm jk$, $\zeta = \dfrac{\sqrt{2}}{2}$ $\left(\text{cf. } \dfrac{\sqrt{3}}{2} \text{ for the three-compartment system}\right)$.

$$\mathbf{sI} - \mathbf{A} = \begin{bmatrix} s+k & 0 & 0 & -k \\ -k & s+k & 0 & 0 \\ 0 & -k & s+k & 0 \\ 0 & 0 & -k & s+k \end{bmatrix} \tag{4.42}$$

so that

$$(\mathbf{sI} - \mathbf{A})^{-1} = \frac{1}{\Delta(s)}\begin{bmatrix} (s+k)^3 & k^3 & k^2(s+k) & k(s+k)^2 \\ k(s+k)^2 & (s+k)^3 & k^3 & k^2(s+k) \\ k^2(s+k) & k(s+k)^2 & (s+k)^3 & k^3 \\ k^3 & k^2(s+k) & k(s+k)^2 & (s+k)^3 \end{bmatrix} \tag{4.43}$$

where

$$\begin{aligned}\Delta(s) &= (s+k)^4 - k^4 \\ &= s(s+2k)(s+k-jk)(s+k+jk) \\ &= s(s+2k)(s^2+2ks+2k^2).\end{aligned} \tag{4.44}$$

For an impulsive input to compartment 1, $u_1(t) = \delta(t)$ (with $b_1 = 1$), and no input to compartments 2, 3 and 4,

$$\begin{aligned}X_1(s) &= \frac{(s+k)^3}{s(s+2k)(s^2+2ks+2k^2)} \\ &= \frac{1}{4}\left[\frac{1}{s} + \frac{1}{s+2k} + \frac{2s+2k}{s^2+2ks+2k^2}\right] \\ &= \frac{1}{4}\left[\frac{1}{s} + \frac{1}{s+2k} + \frac{2(s+k)}{(s+k)^2+k^2}\right]\end{aligned}$$

so that

$$x_1(t) = \tfrac{1}{4} + \tfrac{1}{4}e^{-2kt} + \tfrac{1}{2}e^{-kt}\cos kt, \tag{4.45}$$

which is shown in Fig. 4.26(a). The oscillation about the final value of 0.25 is evident. The responses $x_2(t)$, $x_3(t)$ and $x_4(t)$ are shown in Figs 4.26(b), (c) and (d).

(a)

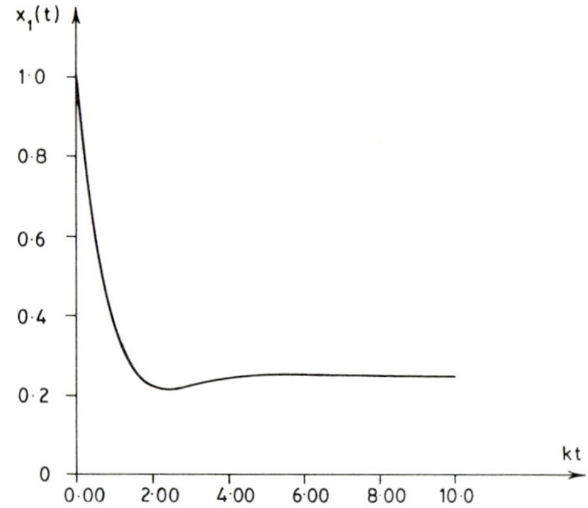

(c)

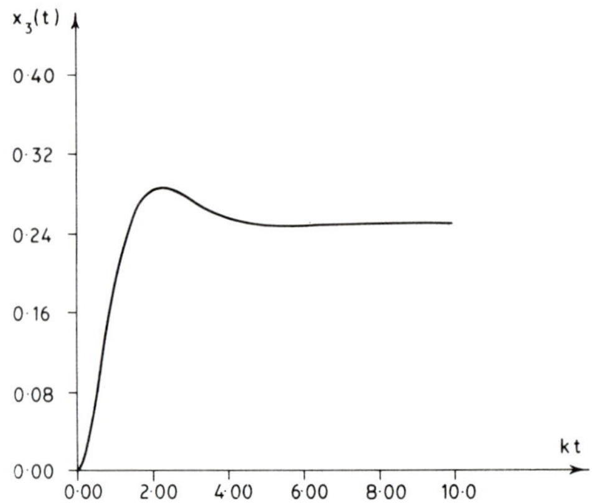

Figure 4.26 Responses of model of Fig. 4.23 with $p = 4$ to unit impulsive perturbation of compartment 1 (a) $x_1(t)$; (b) $x_2(t)$; (c) $x_3(t)$; (d) $x_4(t)$.

By considering the form of the modal matrix **M** (the matrix having the eigenvectors of **A** as its columns), Thron (1972) has shown that the response of the input compartment to a unit impulsive input is

$$x_1(t) = \frac{1}{p}(e^{\lambda_1 t} + e^{\lambda_2 t} + \ldots + e^{\lambda_p t}). \tag{4.46}$$

(Thron also obtains expressions for the other compartmental quantities; he numbers them in reverse order to that of Fig. 4.23.) Thus, for the three-compartment system, with

(b)

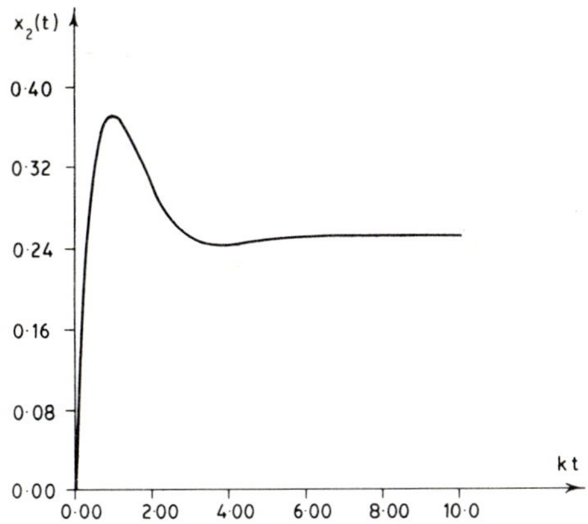

(d)

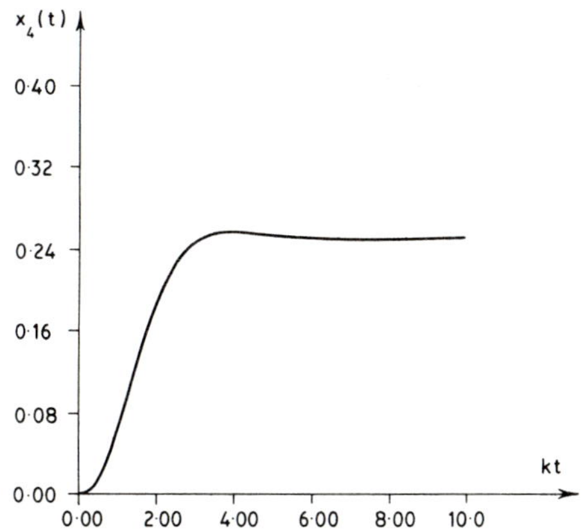

$$\lambda_1 = -1.5k + j\frac{\sqrt{3}}{2}k, \quad \lambda_2 = 1.5k - j\frac{\sqrt{3}}{2}k, \quad \lambda_3 = 0,$$

$$x_1(t) = \frac{1}{3}\left(\exp\left(-1.5k + j\frac{\sqrt{3}}{2}k\right)t + \exp\left(-1.5k - j\frac{\sqrt{3}}{2}k\right)t + 1\right)$$

$$= \frac{1}{3}\left[1 + e^{-1.5kt}\left(\exp\left(j\frac{\sqrt{3}}{2}kt\right) + \exp\left(-j\frac{\sqrt{3}}{2}kt\right)\right)\right]$$

$$= \frac{1}{3} + \frac{2}{3}e^{-1.5kt}\cos\frac{k\sqrt{3}}{2}t,$$

as found in eqn (4.22b).

The solution of eqn (4.45) is obtained for $p = 4$, while for $p = 5$,

$$x_1(t) = \tfrac{1}{5} + \tfrac{2}{5}\,e^{-0.691kt}\cos(0.951kt) + \tfrac{2}{5}\,e^{-1.809kt}\cos(0.588kt) \tag{4.47}$$

and for $p = 6$,

$$x_1(t) = \frac{1}{6} + \frac{1}{6}e^{-2kt} + \frac{1}{3}e^{-0.5kt}\cos\frac{k\sqrt{3}}{2}t + \frac{1}{3}e^{-1.5kt}\cos\frac{k\sqrt{3}}{2}t. \tag{4.48}$$

Computed responses for $p = 10$, 25 and 50 are shown in Figs 4.27(a), (b) and (c).

The question of which system is most oscillatory as p is increased is difficult to pose. With two or more pairs of complex eigenvalues for $p \geqslant 5$, it is possible that one system may produce large oscillations early in the response, which die away fairly quickly, while another may produce an oscillation which, while smaller initially, dies away only slowly. The most oscillatory eigenvalue pair will give a term in the impulse response:

$$\frac{2}{p}\exp\left(-k + k\cos\frac{2\pi}{p}\right)t\,\cos\left(k\sin\frac{2\pi}{p}\right)t$$
$$= \frac{2}{p}\exp\left(-2k\sin^2\frac{\pi}{p}\right)t\,\cos\left(2k\sin\frac{\pi}{p}\cos\frac{\pi}{p}\right)t \tag{4.49}$$

so that, as p becomes large and $\sin\frac{\pi}{p} \to \frac{\pi}{p}$ and $\cos\frac{\pi}{p} \to 1$, the most oscillatory term approximates to

$$\frac{2}{p}\exp\left(-2k\frac{\pi^2}{p^2}\right)t\,\cos\left(2k\frac{\pi}{p}\right)t$$

with period $\dfrac{p}{2k\pi}\cdot 2\pi = \dfrac{p}{k}\cdot$

Landahl (1972) wrote the impulse response of the perturbed compartment as an infinite series:

$$x_1(t) = e^{-kt}\left(1 + \frac{(kt)^p}{p!} + \frac{(kt)^{2p}}{(2p)!} + \ldots\right)\cdot \tag{4.50}$$

Figure 4.27 (*opposite*) Responses $x_1(t)$ of the model of Fig. 4.23 to unit impulsive perturbation of compartment 1, (a) $p = 10$; (b) $p = 25$; (c) $p = 50$.

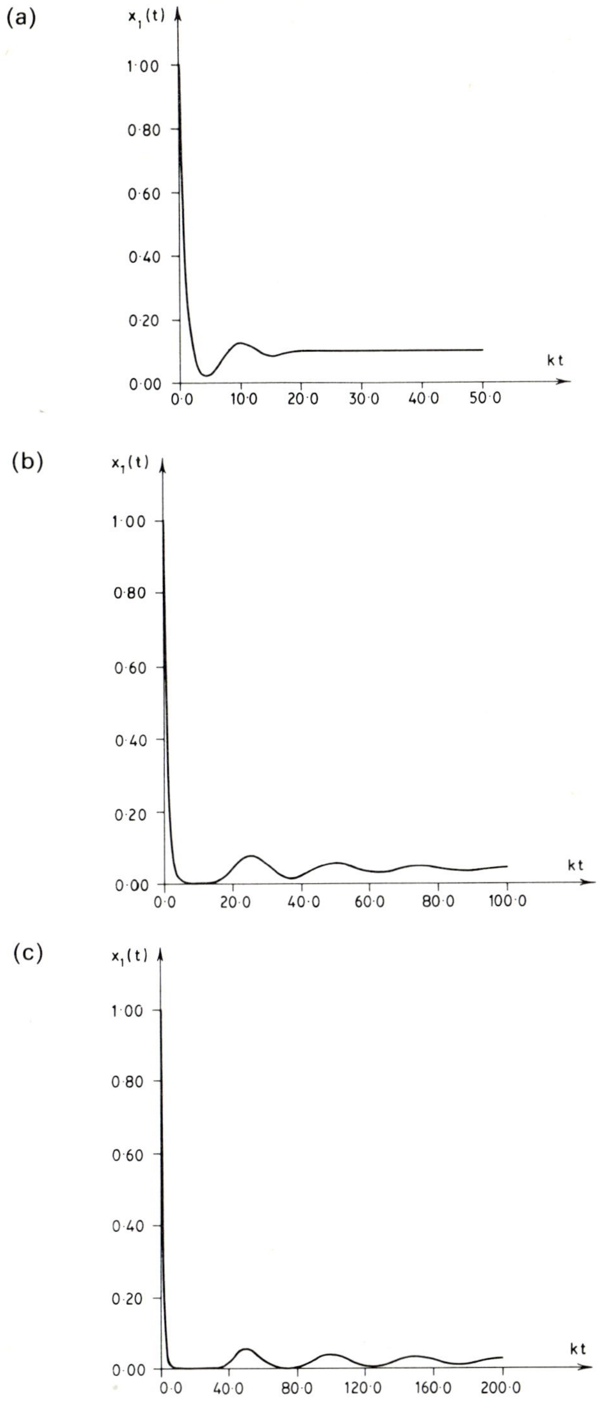
(a)
x1(t)
1·00
0·80
0·60
0·40
0·20
0·00
kt
0·0
10·0
20·0
30·0
40·0
50·0
(b)
x1(t)
1·00
0·80
0·60
0·40
0·20
0·00
kt
0·0
20·0
40·0
60·0
80·0
100·0
(c)
x1(t)
1·00
0·80
0·60
0·40
0·20
0·00
kt
0·0
40·0
80·0
120·0
160·0
200·0

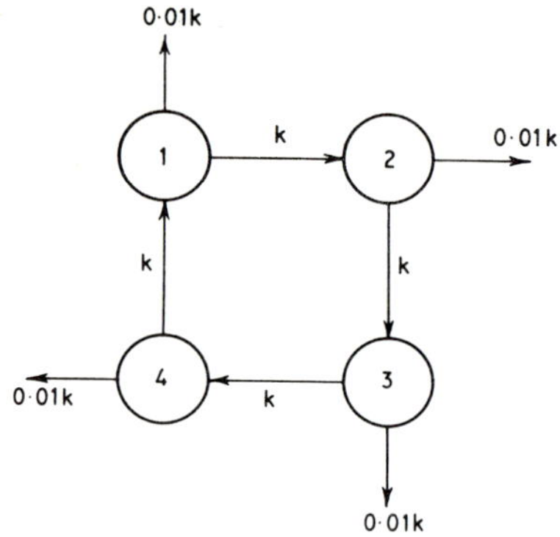

Figure 4.28 "Slightly open" unilateral circulation model with $p = 4$.

This form is helpful in seeing certain trends as p becomes large. Landahl found by simulation that the greatest difference between the first minimum of the impulse response and the first maximum occurs for $p = 11$.

The proof given in Section 4.1.6 that making the system open results in less oscillatory eigenvalues also applies for $p > 3$, so that the damping ratio ζ for each pair of complex eigenvalues increases as the excretion to the environment is increased. The impulse response of the perturbed compartment for the "slightly open" four-compartment system of Fig. 4.28 is shown in Fig. 4.29. Although the impulse response is oscillatory, we know from eqn (2.64) that the washout curve decays monotonically to zero (Hearon, 1968), and

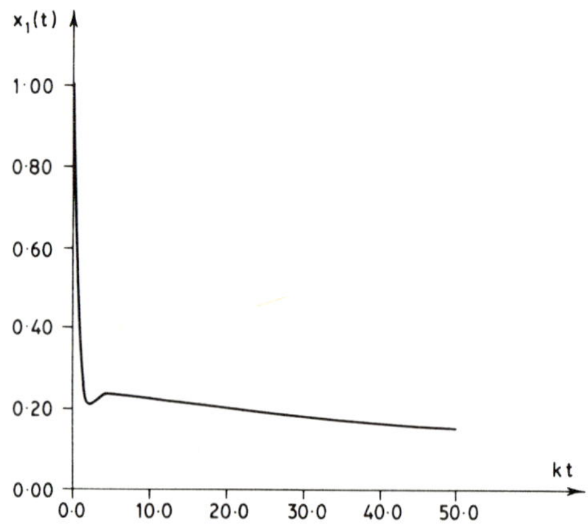

Figure 4.29 Response $x_1(t)$ to unit impulsive perturbation of compartment 1 of the model of Fig. 4.28.

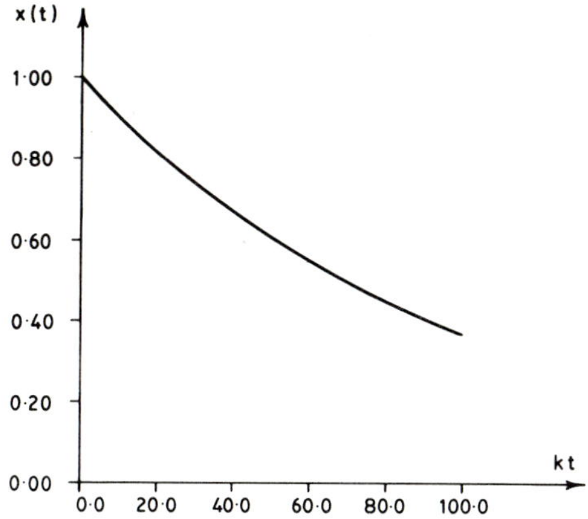

Figure 4.30 Washout curve for any of the compartmental quantities of the model of Fig. 4.28.

Fig. 4.30 shows the washout curve for any of the states of the model of Fig. 4.28 from the initial state $\mathbf{x}(0) = [1\ \ 1\ \ 1\ \ 1]^T$; the corresponding impulse response of Fig. 4.29 was from an initial state $\mathbf{x}(0) = [1\ \ 0\ \ 0\ \ 0]^T$.

Finally, returning to the closed system, as for the model with $p = 3$, it is only possible to obtain a local result that counterflow reduces ζ for each pair of complex eigenvalues. It seems unlikely, though, that a more oscillatory linear, time-invariant compartmental model can be found than the closed unilateral circulation model with equal rate constants.

5 The Concept of Identifiability

5.1 The identifiability problem

In Chapters 3 and 4, we have been concerned with finding the responses of the states (the compartmental quantities) to an external perturbation of the system, given the structure of the system (i.e. the pattern of non-zero inter-compartmental links) and the values of the rate constants. This is often referred to as the direct problem of compartmental analysis (see, for example, Rubinow and Winzer, 1971) but it is not the problem usually faced by the experimenter, who has available measurements of one or more of the states and, sometimes, prior knowledge of some of the parameters, and who then has to determine the values of the unknown parameters. This latter problem of identifiability, often referred to as the inverse problem (Rubinow and Winzer, 1971), is bound up with the form of the assumed model, prior knowledge of the model parameters, the possible perturbations and observations and the quality of the experimental observations.

An important initial part of any identifiability analysis is concerned with solutions for the model parameters when perfect (noise-free) observations are available. If, from this initial analysis, the parameters are found to be unidentifiable, the experimenter has to determine whether a solution could be obtained by perturbing more or different inputs, observing more or different states or using further prior knowledge of some of the parameters; if none of these is feasible, simplifications to the structure of the model which are reasonable in the context of the system being investigated must be considered.

Identifiability with perfect observations was originally formalized by Bellman and Åström (1970) under the name of structural identifiability, although individual systems had been considered a good deal earlier than this (Shipley *et al.*, 1959). In this context, a structural property is one that applies for almost all values of the parameters. By "almost all", particular (isolated) combinations of parameter values giving rise to unusual dynamic behaviour are excluded. For example, we found in Example 4.1 that impulse responses of the general three-compartment model with all rate constants equal contained only two exponentials, but this particular case would not invalidate the structural identifiability analysis which would assume a tri-exponential response for this model.

A number of definitions of terms associated with identifiability have appeared over the years, but these have led to some criticism. Recently, Cobelli and DiStefano have proposed some rather formal definitions (Cobelli and DiStefano, 1980; DiStefano and Cobelli, 1980) but several aspects of their definitions have been criticized (LeCourtier and Walter, 1981). There seems to the author of this book a need to keep terms and definitions as simple as possible, and in this book, we shall mainly use those given by Bellman and Åström (1970).

There are three possible outcomes to the first stage of an identifiability analysis (assuming perfect data) (Bellman and Åström, 1970; Glover and Willems, 1974):

(i). The parameters of an assumed model can be estimated uniquely and the model is *globally* (*uniquely*) *identifiable* from the experiment.
(ii). Any of a finite number of alternative estimates for the model parameters fits the data and the model is *locally identifiable*.
(iii). Any of an infinite number of estimates fits the data and the model is *unidentifiable* from the experiment.

Although we shall use the above definitions of Bellman and Åström, the adjective "structural" which they used for this part of an identifiability analysis is not completely appropriate in the context of compartmental models, because, as we will see in Chapter 6, the identifiability may depend on whether the input fractions and observation gains are known, or whether there is prior knowledge of non-zero values of any rate constants and, when more than one input is applied simultaneously, on the shape of the perturbations. In this book, therefore, we will use the term *deterministic identifiability*.

For a linear, time-invariant compartmental model, the parameters to be identified are the rate constants. Some authors (e.g. Cobelli and DiStefano, 1980) include compartmental volumes as parameters to be identified, but it is possible to determine all such volumes in only very few cases, so that they will not be included here as parameters to be identified; if any are available from a particular experiment, then that is a bonus. For nonlinear and time-varying systems, parameters characterizing the nonlinearity or the time variation also have to be identified. For a pathway with Michaelis–Menten kinetics, for example, with flow rate from i to j given by $\dfrac{V_m x_i(t)}{K_m + x_i(t)}$ the parameters V_m and K_m would be required, while for a pathway with a sinusoidally-varying rate coefficient, so that the flow rate from i to j is given by $k_{ji}(1 + \gamma \sin \omega t) \cdot x_i$, the constants k_{ji} and γ would be required; the radian frequency may be known (for example if it is known that the time variation is circadian) or it may have to be identified as well.

It is important to stress again that deterministic identifiability is concerned only with ideal observations. For example, if a compartmental response is of the form $A\,e^{-\alpha t} + B\,e^{-\beta t}$, then A, α, B and β are assumed to be known exactly. Imperfect data resulting from noisy and sampled observations impose many further restrictions on identifiability. These affect the accuracy with which parameters can be estimated, and this second stage of the identifiability analysis will be called *numerical identifiability*.

In Chapter 8, we will see, for the linear, time-invariant case, that numerical identifiability places a practical limitation on the number of compartments which can be fitted, in the absence of prior knowledge of some of the parameters. Numerical identifiability also poses severe problems for nonlinear systems (Chapter 9) and time-varying systems (Chapter 10) for which any measured response must be significantly different from that of the corresponding linear time-invariant model for the parameters characterizing the nonlinearity or the time variation to be estimated with reasonable accuracy.

5.2 Methods of testing for deterministic identifiability

As we saw in Chapter 2, a linear time-invariant compartmental system may be described by the equations:

$$\dot{\mathbf{x}}(t) = \mathbf{A}\mathbf{x}(t) + \mathbf{B}\mathbf{u}(t) \tag{5.1a}$$

$$\mathbf{y}(t) = \mathbf{C}\mathbf{x}(t) \tag{5.1b}$$

relating p state variables (compartmental quantities) $\mathbf{x}(t)$ to the forcing $\mathbf{Bu}(t)$ and the observations $\mathbf{y}(t)$ to the state $\mathbf{x}(t)$. There are several methods available for testing for numerical identifiability and some of these will now be reviewed briefly.

The longest-established method, the Laplace transform method, compares the Laplace transforms of the observed variables, given by

$$\mathbf{Y}(s) = \mathbf{C}(s\mathbf{I} - \mathbf{A})^{-1}\mathbf{B}\mathbf{U}(s) \tag{5.2}$$

with the Laplace transforms of the corresponding experimental responses, in terms of the amplitudes and decay constants of the exponentials, the latter being assumed to be known exactly. The expression $\mathbf{C}(s\mathbf{I} - \mathbf{A})^{-1}\mathbf{B}$ is in terms of rate constants, observation gains and input fractions, and the deterministic identifiability problem, as defined here, concerns whether the rate constants can be identified from the proposed configuration of perturbations and observations. The resulting set of polynomial equations is checked for unique solutions (global identifiability), non-unique solutions (local identifiability) or no solutions. An algebraic trial solution has to be attempted, for almost all values of the coefficients (rather than specific values).

The method is often referred to as the transfer function approach (e.g. Cobelli and DiStefano, 1980) and for a single input, the identifiability problem does indeed boil down to determining whether rate constants can be determined from the transfer function matrix $\mathbf{C}(s\mathbf{I} - \mathbf{A})^{-1}\mathbf{B}$. The only requirement is that the input perturbation should have a Laplace transform (i.e. that $\mathbf{U}(s)$ should exist). However, when more than one input is applied simultaneously, identifiability may depend on the shape of the two inputs, and it is then essential to examine the form of the observations $\mathbf{Y}(s)$ rather than individual transfer functions.

The Laplace transform method is conceptually very simple and it is by far the most widely used test for deterministic identifiability of compartmental systems. It will be applied to a number of linear, time-invariant compartmental systems in Chapter 6. The method suffers from the following drawbacks:

(i). The equations in the rate constants are nonlinear (they are of degree up to p) so that for $p \geqslant 3$, it is sometimes difficult to see whether there are redundant equations or multiple solutions.
(ii). If the parameters are not globally identifiable from a particular experiment, it is often not clear how to modify the combination of model structure, perturbation and observation to achieve a unique solution; a trial solution has to be evaluated for each modification in turn.
(iii). It is difficult to generalize conclusions from specific cases.

These drawbacks apply also to two other methods. In the first of these (Pohjanpalo, 1978), the observed response waveforms are expanded about $t = 0^+$ (immediately after application of the perturbation) as Taylor series, successive terms of the series being expressed as functions of the model unknowns. For linear systems, the method seems to offer no advantage over the Laplace transform approach. It does prove useful, though, for nonlinear or time-varying systems (where the Laplace transform approach is no longer applicable) and it will be considered further in Chapters 9 and 10. In the second method (Rubinow and Winzer, 1971) the rate constants are related directly to the amplitudes and exponents of normal modes appearing in the observations (for a discussion of modal analysis, see Power and Simpson, 1978, Chapter 2). The family of system matrices compatible with the observations is found and is then reduced by applying constraints such as non-negativity of all rate constants and known zero rate constants.

A further method based on normal modes has been designed to overcome the drawbacks listed above (Norton, 1980a; Norton, Brown and Godfrey, 1980). The method assembles a set of equations, initially bilinear, in the rows of the modal matrix **M** (which has the eigenvectors as its columns—see Section 2.3.3) and the columns of its inverse **N**. Such equations are provided by $\mathbf{MN} = \mathbf{I}$ and further such equations are given by any known rate constants. The relations between perturbed inputs and observed quantities determine some of the unknown rows and columns so that some of the bilinear equations are reduced to linear equations. If the complete set can be solved, then the system is identifiable. In contrast to the Laplace transform method, the equations are always of low degree, but there are many more equations to be examined. The more systematic nature of the normal-mode method means that it is to be preferred to the Laplace transform approach when examining several different cases with the same number of compartments. In individual cases, though, it is very difficult to forecast which of the two methods will involve less work.

Grewal and Glover (1976) developed a method of deterministic identifiability analysis which tests whether the Markov parameter matrix

$$[(\mathbf{CB})^{\mathrm{T}}(\mathbf{CAB})^{\mathrm{T}} \ldots (\mathbf{CA}^{2p-1}\mathbf{B})^{\mathrm{T}}]^{\mathrm{T}}$$

determines the model parameters, their definition of identifiability being based on distinguishability of the outputs corresponding to distinct parameter vectors, for some admissible inputs and initial conditions. The method is algebraically complicated, and even with a modification (Walter, Le Cardinal and Bertrand, 1976) which gives in a simpler form the identities needed for checking output distinguishability, it is still not easy to apply for $p \geqslant 3$. Also, the approach only checks for local identifiability.

Other methods for analysing deterministic identifiability continue to appear (e.g. Delforge (1980, 1981); Travis and Haddock (1981)). When applied to compartmental systems, they may prove easier in particular cases than the Laplace transform approach, but they do not appear to offer any great advantage in general.

5.3 The magnitude of the deterministic identifiability problem

Although most methods of deterministic identifiability analysis allow for a given input to be introduced into more than one compartment and for a given observation to be a function of more than one state, this is unusual in practice. As stated in Chapter 2, we will assume that each input is applied to one compartment only so that

$$\mathbf{Bu}(t) = [b_1 u_1(t) \quad b_2 u_2(t) \ldots]^{\mathrm{T}}$$

and that each observation is of one state only so that

$$\mathbf{y}(t) = [c_1 x_1(t) \quad c_2 x_2(t) \ldots]^{\mathrm{T}}.$$

Even with these restrictions, the problem of deterministic identifiability is a large one. For example, Norton (1982) has shown that there are 826 essentially distinct non-degenerate three-compartment cases, with **B** and **C** known and with direct input to one compartment only and direct observation of one or more compartments. Identifiability analysis of all 826 cases is reported by Norton (1982), who notes that the variety of causes for non-uniqueness is "wide enough to make it very difficult to design an experiment for unique identifiability by excluding features which might lead to non-uniqueness".

A general and comprehensive deterministic identifiability analysis certainly proves a good deal more difficult than one might expect.

5.4 Controllability, observability and connectability

Readers well versed in control engineering concepts might think that there would be a close link between deterministic identifiability and the familiar properties of controllability (essentially the ability to determine states from controls, given **B** and **A**) and observability (essentially the ability to infer states from outputs, given **C** and **A**). However, the ability to do either or both does not imply ability to infer parameters of **A**, **B** and **C** from outputs, and controllability and observability are neither necessary nor sufficient for unique identifiability (DiStefano, 1977; Cobelli, Lepschy and Romanin-Jacur, 1978b).

There is a closer relationship between deterministic identifiability and connectability (Davison, 1977) and this has led to an inquiry into what features of model structure are associated with non-unique identifiability, with the objective of developing topological criteria for uniqueness (Cobelli, Lepschy and Romanin-Jacur, 1979a). Briefly, a compartment is input and/or output connectable if there exists a path from a compartment directly entered by an input to the one being considered and/or a path from it to an observed compartment, and the authors use a sketch of the connections to detect isolation of the state variables from the inputs or observations, or the loss of independent equations in the Laplace transform approach due to poles at the origin, pole-zero cancellation or shared zeros. The objective is to provide a quick test for unidentifiability (Cobelli and DiStefano, 1980) but as noted by Norton (1982), the method does not account for all features which may prevent a solution. Redundancies may arise from less visible sources and even when enough independent equations remain, they may have non-unique solutions.

6 Laplace Transform Identifiability Analysis

There are two main objectives of this chapter, firstly to familiarize the reader with the application of the Laplace transform approach and secondly to illustrate a number of factors affecting identifiability. The approach is introduced as gently as possible in Section 6.1, dealing with the one-compartment system, and is then applied to two-compartment systems in Section 6.2. In Section 6.2.1, we see that the general two-compartment system is unidentifiable from perturbation of, and observation of a single compartment. Three methods by which this can be overcome are then discussed: incorporation of prior knowledge of one of the rate constants (Section 6.2.2); observation of both compartments (Section 6.2.3) and perturbation of both compartments (Section 6.2.4). The way in which local identifiability arises when one of the compartments is a source compartment is described in Section 6.2.5. Although we are concerned in this book with identifying rate constants, we conclude Section 6.2 with a discussion (in Section 6.2.6) of identification of the volumes of distribution of compartments, which are important in some biomedical applications.

The enormity of the deterministic identifiability problem has been discussed in Section 5.3, and for a catalogue of identifiability results for all 826 essentially distinct three-compartment systems with single compartment perturbation, the reader is referred to Norton (1982). In this book, we will confine attention to systems and experiments which illustrate further important points concerning identifiability or identifiability analysis. To set the scene, the general three-compartment system is considered in Section 6.3.1. An important source of local identifiability is symmetry of model and experiment (which cannot arise for a two-compartment system) and this is described for a three-compartment mammillary system, with perturbation of, and observation of, the central compartment only, in Section 6.3.2. It is shown in Section 6.3.3 how, for the same system (with the same prior knowledge) perturbation of, and observation of, one of the peripheral compartments (only) yields global identifiability, while in Section 6.3.4, we see how well camouflaged a redundant equation can be when we consider the same system with the central compartment (only) perturbed and with observation of the central compartment and of one of the peripheral compartments. In Section 6.3.5, we consider a system with three compartments, one a source compartment, and see how, although the algebra indicates three solutions, substitution of actual values makes one of the solutions inadmissible. The effect of input and observation gains on identifiability is clearly illustrated by the three-compartment catenary system with unidirectional exchange considered in Section 6.3.6.

Finally, some identifiability results for mammillary and catenary systems with more than three compartments are described in Section 6.4.

6.1 The one-compartment system

For the one-compartment system shown in Fig. 6.1, the parameter to be identified is k_{01} and the Laplace transform of the observation is, from Section 3.1,

$$Y_1(s) = \frac{c_1 b_1}{s + k_{01}} U_1(s). \tag{6.1}$$

With only one compartment to be perturbed, we can consider the transfer function

$$G_{11}(s) = \frac{Y_1(s)}{U_1(s)} = \frac{c_1 b_1}{s + k_{01}}. \tag{6.2}$$

The corresponding experimental transfer function is

$$G_{11}(s) = \frac{A}{s + \alpha} \tag{6.3}$$

where A and α are measurable constants. Comparing the denominators of eqns (6.2) and (6.3),

$$k_{01} = \alpha \tag{6.4}$$

so that the system is globally identifiable from this experiment.

It may be possible to extract some useful information from the numerator of the observed response. Suppose, for example, that the input is impulsive,

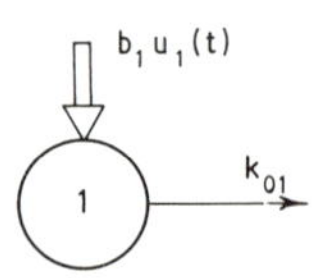

Figure 6.1 One-compartment model.

with $u_1(t) = D_1\,\delta(t)$, where D_1 is known. Then, comparing numerators,

$$D_1 c_1 b_1 = A. \tag{6.5}$$

In most cases, the input fraction b_1 is known (and equal to 1). If, in addition, the measurement is of concentration, so that

$$c_1 = \frac{1}{V_1} \tag{6.6}$$

where V_1 is the apparent volume of distribution of compartment 1, then V_1 is also globally identifiable from

$$V_1 = \frac{D_1}{A}. \tag{6.7}$$

6.2 Two-compartment systems (Fig. 6.2)

For any two-compartment system, the transfer function matrix

$$\mathbf{G}(s) = \begin{bmatrix} G_{11}(s) & G_{12}(s) \\ G_{21}(s) & G_{22}(s) \end{bmatrix} \triangleq \begin{bmatrix} \dfrac{Y_1(s)}{U_1(s)} & \dfrac{Y_1(s)}{U_2(s)} \\ \dfrac{Y_2(s)}{U_1(s)} & \dfrac{Y_2(s)}{U_2(s)} \end{bmatrix}$$

is, from Section 3.2.1,

$$\mathbf{G}(s) = \frac{1}{\Delta(s)} \begin{bmatrix} c_1 b_1 (s - a_{22}) & c_1 b_2 a_{12} \\ c_2 b_1 a_{21} & c_2 b_2 (s - a_{11}) \end{bmatrix} \tag{6.8}$$

where

$$\Delta(s) = (s - \lambda_1)(s - \lambda_2) = s^2 - (a_{11} + a_{22})s + a_{11}a_{22} - a_{12}a_{21}. \tag{6.9}$$

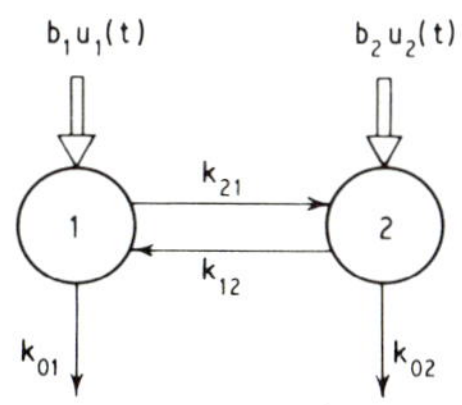

Figure 6.2 General two-compartment model.

The corresponding experimental transfer function is

$$\mathbf{G}(s) = \frac{1}{s^2 + \gamma s + \delta} \begin{bmatrix} K_1(s + \alpha) & K_2 \\ K_3 & K_4(s + \beta) \end{bmatrix} \tag{6.10}$$

where $K_1, K_2, K_3, K_4, \alpha, \beta, \gamma$ and δ are measurable constants. Thus, if either state response to a perturbation is observed, the quantities $(a_{11} + a_{22})$ and $(a_{11}a_{22} - a_{12}a_{21})$ can be identified from the denominator since, from eqns (6.9) and (6.10),

$$-(a_{11} + a_{22}) = \gamma \tag{6.11a}$$

and

$$a_{11}a_{22} - a_{12}a_{21} = \delta. \tag{6.11b}$$

What further can be identified from the numerator relationships depends on a number of factors which will become clear from Sections 6.2.1 to 6.2.6.

6.2.1 *Single-input, single-output (SISO) experiments*

In biomedical experiments at least, by far the most common experiment is to perturb and observe the same compartment (no. 1, say). Comparing eqns (6.8) and (6.10), this gives the following numerator equations:

$$c_1 b_1 = K_1 \tag{6.12a}$$

$$-a_{22} = \alpha \tag{6.12b}$$

in addition to the denominator eqns (6.11). Identifiable quantities involving rate constants are

$$a_{22} = -\alpha \tag{6.13a}$$

$$a_{11} = \alpha - \gamma \tag{6.13b}$$

$$a_{12}a_{21} = -\delta - \alpha(\alpha - \gamma) \tag{6.13c}$$

so that no individual rate constant is identifiable from this experiment. By analogy with Section 6.1, the apparent volume of distribution V_1 is globally identifiable from this experiment if the measurement is of concentration.

If, instead, compartment 2 (only) were observed (with compartment 1 (only) perturbed), eqns (6.12) would be replaced with the single equation

$$c_2 b_1 a_{21} = K_3 \tag{6.14}$$

so that a_{21} $(=k_{21})$ can be identified provided $c_2 b_1$ is known, but the remaining three rate constants (k_{12}, k_{02} and k_{01}) remain unidentifiable. Whether $c_2 b_1$ is known depends on the experiment. In biomedical applications, for example, the input fraction b_1 is often known (and usually equal to 1 for intravenous injection if compartment 1 contains the systemic blood) but observations are usually of concentration only in those compartments containing the plasma. If compartment 2 is the peripheral compartment, it is unlikely that the measurement will be of concentration. For instance if the input were a radioactive tracer administered to a compartment containing the systemic blood (no. 1 in this example) measurement of compartment 2 could be accomplished by externally counting the total radioactivity levels of the organs comprising compartment 2, using a Gamma Camera. Whether the observation gain c_2 would then be known with sufficient accuracy for meaningful identification of a_{21} is a matter of conjecture. (In Application Example 2 in Chapter 11, it is assumed that the observation gain of such a measurement *is* known with sufficient accuracy.) Note that none of the available equations for this experiment (eqns (6.11a), (6.11b) and (6.14)) involves the observation gain c_1, so V_1 is not identifiable.

It is clear that perturbation of one compartment and observation of one compartment is, on its own, insufficient to allow the four rate constants of the general model of Fig. 6.2 to be identified. In Sections 6.2.2 to 6.2.4, we will consider (in turn) three possible modifications to the experiment to see whether rate constants can be identified individually.

6.2.2 *SISO experiment with prior knowledge incorporated*

Suppose that only compartment 1 is perturbed and observed, and let us see how identifiability is affected when there is prior knowledge of one of the rate constants. Often the prior knowledge will be that a rate constant is zero, but for the moment, $k_{12} = 0$ or $k_{21} = 0$ will be excluded, as these constitute a special case to be considered in Section 6.2.5.

The available equations are (6.11) and (6.12); rewriting (6.11a), (6.11b) and (6.12b) in terms of the k's:

$$k_{02} + k_{12} = \alpha \tag{6.15}$$

$$(k_{01} + k_{21}) + (k_{02} + k_{12}) = \gamma \tag{6.16}$$

$$(k_{01} + k_{21})(k_{02} + k_{12}) - k_{12}k_{21} = \delta. \tag{6.17}$$

Consider prior knowledge of the four rate constants in turn.

(i) $\underline{k_{02} = \rho}$ (a known constant). From eqn (6.15),

$$k_{12} = \alpha - \rho. \tag{6.18a}$$

Then from eqns (6.16) and (6.17),

$$k_{01} + k_{21} = \gamma - \alpha$$

and

$$(\gamma - \alpha)\alpha - k_{12}k_{21} = \delta$$

whence $$k_{21} = \frac{(\gamma - \alpha)\alpha - \delta}{\alpha - \rho} \tag{6.18b}$$

(for $\rho \neq \alpha$, i.e. $k_{12} \neq 0$)

and $$k_{01} = \gamma - \alpha - k_{21}. \tag{6.18c}$$

(ii) $\underline{k_{12} = \rho}$ ($\neq 0$). From eqn (6.15),

$$k_{02} = \alpha - \rho. \tag{6.19a}$$

Then, from eqns (6.16) and (6.17),

$$k_{01} + k_{21} = \gamma - \alpha$$

and $$(\gamma - \alpha)\alpha - \rho k_{21} = \delta$$

whence $$k_{21} = \frac{(\gamma - \alpha)\alpha - \delta}{\rho} \tag{6.19b}$$

and $$k_{01} = \gamma - \alpha - k_{21}. \tag{6.19c}$$

(iii) $\underline{k_{01} = \rho}$. From eqns (6.15) and (6.16),

$$k_{21} = \gamma - \alpha - \rho. \tag{6.20a}$$

Then from eqn (6.17), $$k_{12} = \frac{(\gamma - \alpha)\alpha - \delta}{\gamma - \alpha - \rho} \tag{6.20b}$$

(for $\rho \neq \gamma - \alpha$, i.e. $k_{21} \neq 0$) and from eqn (6.15),

$$k_{02} = \alpha - k_{12}. \tag{6.20c}$$

(iv) $\underline{k_{21} = \rho\ (\neq 0\ )}$. From eqns (6.15) and (6.16),

$$k_{01} = \gamma - \alpha - \rho. \tag{6.21a}$$

From eqn (6.17), $$k_{12} = \frac{(\gamma - \alpha)\alpha - \delta}{\rho} \tag{6.21b}$$

and from eqn (6.15), $$k_{02} = \alpha - k_{12}. \tag{6.21c}$$

Thus, when the same compartment is perturbed and observed, prior knowledge of any one rate constant (with the exception of $k_{12} = 0$ or $k_{21} = 0$) allows the other three to be identified uniquely. The most common prior assumption is that one of the compartments does not excrete material to be environment, so that either $k_{01} = 0$ or $k_{02} = 0$. The closed system, with k_{01} and k_{02} zero, is also uniquely identifiable from this experiment.

If, instead, compartment 1 (only) is perturbed and compartment 2 (only) observed, the equations in the rate constants are (6.14), (6.16) and (6.17). By inspection, it may be seen that, unless k_{21} can be identified from eqn (6.14), prior knowledge of any one rate constant will not allow the others to be identified.

Example 6.1. *An impulsive input is applied to compartment* 1 *of a general two-compartment system, for which it is known that* $k_{02} = 1$. *The normalized response in compartment* 1 *is* $y_1(t) = 0.5(e^{-3t} + e^{-t})$. *Determine* k_{02}, k_{12} *and* k_{21}.

Solution.

$$Y_1(s) = \frac{0.5}{s+3} + \frac{0.5}{s+1} = \frac{s+2}{s^2+4s+3}.$$

Substituting in eqns (6.15) to (6.17),

$$k_{02} + k_{12} = 2$$

$$(k_{01} + k_{21}) + (k_{02} + k_{12}) = 4$$

$$(k_{01} + k_{21})(k_{02} + k_{12}) - k_{12}k_{21} = 3.$$

With $$k_{02} = 1, \qquad k_{12} = 1.$$

Since $$(k_{01} + k_{21}) = (k_{02} + k_{12}) = 2,$$

$$4 - k_{21} = 3 \quad \text{so that} \quad k_{21} = 1.$$

Hence $$k_{02} = 1.$$

The apparent volume of distribution V_1 could also be determined if the measurement were of concentration and the input fraction b_1 and the normalization factor were known.

6.2.3 *Simultaneous observation of both compartments*

Returning to the general model of Fig. 6.2 with no prior knowledge of any rate constants, consider an experiment with compartment 1 (only) perturbed but with both compartments observed. The available equations in the rate constants are (6.14) to (6.17) and identifiability now hinges on whether $c_2 b_1$ is known. If it is, a_{21} $(=k_{21})$ is identifiable from eqn (6.14) and the remaining rate constants follow:

$$k_{01} \equiv -a_{11} - k_{21} = \gamma - \alpha - k_{21} \tag{6.22a}$$

$$k_{12} \equiv \frac{a_{12}a_{21}}{a_{21}} = \frac{1}{k_{21}}[-\delta - \alpha(\alpha - \gamma)] \tag{6.22b}$$

and $$k_{02} \equiv -a_{22} - k_{12} = \alpha - k_{12} \tag{6.22c}$$

so that the model is globally identifiable from this experiment provided c_2b_1 is known. The apparent volume of distribution V_1 is also identifiable from eqn (6.12a) provided the measurement is of concentration.

Example 6.2. *In a general two-compartment system, compartment* 1 *is forced by a rapidly administered input of* 1 *unit. The observed responses in compartments* 1 *and* 2 *are, respectively,*

$$y_1(t) = 0.8\,\mathrm{e}^{-10t} + 0.2\,\mathrm{e}^{-t}$$

and

$$y_2(t) = -0.8\,\mathrm{e}^{-10t} + 0.8\,\mathrm{e}^{-t}.$$

It is known that $c_2b_1 = 1$. Determine the rate constants.

Solution. Taking Laplace transforms,

$$Y_1(s) = \frac{0.8}{s+10} + \frac{0.2}{s+1} = \frac{s+2.8}{s^2+11s+10}$$

$$Y_2(s) = -\frac{0.8}{s+10} + \frac{0.8}{s+1} = \frac{7.2}{s^2+11s+10}.$$

With $c_2b_1 = 1$, we have, from eqn (6.14), $k_{21} = 7.2$. Then, since $\alpha = 2.8$, $\gamma = 11$, $\delta = 10$, we have, substituting into eqns (6.22),

$$k_{01} = 11 - 2.8 - 7.2 = 1$$

$$k_{12} = \frac{1}{7.2}[-10 + (2.8)(8.2)] = 1.8$$

$$k_{02} = 2.8 - 1.8 = 1.$$

6.2.4 *Perturbation of both compartments*

Suppose again that there is no prior knowledge of any rate constants, but let us now consider an experiment in which only compartment 1 is observed, while both compartments can be perturbed. If the two perturbations are applied at well-separated times, so that the two experimental responses are effectively separated, an additional numerator term from $G_{12}(s)$:

$$c_1b_2k_{12} = K_2 \tag{6.23}$$

is added to the eqns (6.11) and (6.12) available from $G_{11}(s)$. The rate constant k_{12} can be identified provided c_1b_2 is known; the other rate constants then follow and the model is globally identifiable. Again considering biomedical experiments, if compartment 1 contains the systemic blood, b_1 will usually be

1, but compartment 2 will not then contain the systemic blood and it is doubtful whether the input fraction b_2 will be known with good accuracy.

Doing experiments on the same system well separated in time is usually somewhat undesirable because the assumption that the system is stationary becomes more difficult to justify. Let us consider instead an experiment in which the two inputs are applied simultaneously. We must now consider the Laplace transform of the observation (since that is what is available to the experimenter) rather than the individual transfer functions $G_{11}(s)$ and $G_{12}(s)$. Thus

$$\begin{aligned} Y_1(s) &= G_{11}(s)U_1(s) + G_{12}(s)U_2(s) \\ &= \frac{1}{\Delta(s)}[c_1b_1(s - a_{22})U_1(s) + c_1b_2a_{12}U_2(s)] \end{aligned} \tag{6.24}$$

where the denominator, $\Delta(s)$, is still given by eqn (6.9).

If the inputs are of the same shape, so that $U_1(s) = M_1U(s)$ and $U_2(s) = M_2U(s)$, where M_1 and M_2 are known constants, then

$$Y_1(s) = \frac{U(s)}{\Delta(s)}[M_1c_1b_1s + M_1c_1b_1(-a_{22}) + M_2c_1b_2a_{12}]. \tag{6.25}$$

Again, c_1b_1 is known from the term in s in the numerator. If, in addition, b_2 is known and c_1 is available from c_1b_1 then $(-a_{22} + \mu a_{12})$, where μ is a known constant, can be identified from the constant term in the numerator. As before, though, there are only three equations in the four unknown rate constants and the model remains unidentifiable using two simultaneous inputs of the same shape.

Let us now investigate what happens when two inputs of different shape are applied simultaneously. It is impractical to consider all possible waveshapes, so the investigation will be confined to the two shapes most commonly encountered in biomedical applications—impulses and steps.

(i) **Experiment A.** $u_1(t)$ **an infusion;** $u_2(t)$ **impulsive.**

For this experiment, $U_1(s) = \dfrac{k_i}{s}$ and $U_2(s) = D_2$, with k_i and D_2 known. From eqn (6.24), the observation is given by

$$\begin{aligned} Y_1(s) &= \frac{1}{s\cdot\Delta(s)}[c_1b_1k_i(s - a_{22}) + c_1b_2a_{12}D_2s] \\ &= \frac{1}{s\cdot\Delta(s)}[s(c_1b_1k_i + c_1b_2D_2a_{12}) - c_1b_1k_ia_{22}]. \end{aligned} \tag{6.26}$$

The corresponding experimental response would be of the form:

$$Y_1(s) = \frac{K_5(s + \varepsilon)}{s(s^2 + \gamma s + \delta)} \tag{6.27}$$

where K_5, ε, γ and δ are measurable constants. Comparing coefficients,

$$c_1 b_1 k_i + c_1 b_2 D_2 a_{12} = K_5 \tag{6.28a}$$

$$-c_1 b_1 k_i a_{22} = K_5 \varepsilon \tag{6.28b}$$

together with the denominator eqns (6.11) as before. If the input fractions b_1 and b_2 are known, a_{12} $(=k_{12})$ follows from eqn (6.28a) *provided* c_1 *is known* and then k_{02} follows from eqn (6.28b) and k_{21} and k_{01} from eqns (6.11a) and (6.11b).

(ii) **Experiment B.** $u_1(t)$ **impulsive;** $u_2(t)$ **an infusion.**

For this experiment, $U_1(s) = D_1$ and $U_2(s) = \frac{k_i}{s}$, with D_1 and k_i known. From eqn (6.24), the observation is given by

$$Y_1(s) = \frac{1}{s \cdot \Delta(s)} [c_1 b_1 D_1 s(s - a_{22}) + c_1 b_2 a_{12} k_i]$$

$$= \frac{1}{s \cdot \Delta(s)} [c_1 b_1 D_1 s^2 - c_1 b_1 D_1 a_{22} s + c_1 b_2 a_{12} k_i]. \tag{6.29}$$

The corresponding experimental response would be of the form

$$Y_1(s) = \frac{K_6(s^2 + \mu s + \nu)}{s(s^2 + \gamma s + \delta)} \tag{6.30}$$

where K_6, μ, ν, γ and δ are measurable constants. Comparing coefficients,

$$c_1 b_1 D_1 = K_6 \tag{6.31a}$$

$$-a_{22} = \mu \tag{6.31b}$$

$$a_{12} = \nu \cdot \frac{b_1 D_1}{b_2 k_i} \tag{6.31c}$$

together with the denominator eqns (6.11). If the input fractions b_1 and b_2 are known, a_{12} $(=k_{12})$ is identifiable from eqn (6.31c); then k_{02} follows from eqn (6.31b) and k_{21} and k_{01} from eqns (6.11a) and (6.11b). The model is thus globally identifiable from this experiment regardless of whether the observation gain c_1 is known.

Example 6.3. *In a general two-compartment system, an input of* 1 *unit is rapidly administered to compartment* 1, *and starting at the same time, a constant input of* 0.1 *units per unit time is administered to compartment* 2. *It is known that both input fractions are unity (i.e.* $b_1 = b_2 = 1$). *Compartment* 1 *(only) is observed, and the response, normalized so that* $y_1(0) = 1$, *is given by*

$$y_1(t) = 0.78\,\mathrm{e}^{-10t} + 0.2\,\mathrm{e}^{-t} + 0.02.$$

Determine k_{01}, k_{02}, k_{12} *and* k_{21}.

Solution.

$$Y_1(s) = \frac{0.78}{s+10} + \frac{0.2}{s+1} + \frac{0.02}{s}$$

$$= \frac{s^2 + 3s + 0.2}{s(s^2 + 11s + 10)}$$

so $K_6 = 1, \quad \mu = 3, \quad v = 0.2, \quad \gamma = 11 \quad \text{and} \quad \delta = 10.$

From eqn (6.31c), $$k_{12} = \frac{(0.2)\cdot 1}{0.1} = 2.$$

From eqn (6.31b), $$k_{02} = 3 - 2 = 1.$$

Substituting in eqns (6.11a) and (6.11b),

$$k_{01} + k_{21} = 11 - 3 = 8$$

$$3(k_{01} + k_{21}) - 2k_{21} = 10$$

whence $$k_{01} = 1, \quad k_{21} = 7.$$

Summarizing the effect of waveshape when simultaneous inputs are applied, with compartment 1 (only) observed, input fractions b_1 and b_2 known and waveform amplitudes known, the general two-compartment model of Fig. 6.2 is:

(i). unidentifiable if the two inputs are of the same shape;
(ii). globally identifiable if $u_1(t)$ is a step and $u_2(t)$ is impulsive only if c_1 is known;
(iii). globally identifiable if $u_1(t)$ is impulsive and $u_2(t)$ is a step regardless of whether c_1 is known.

6.2.5 *Source compartment: a local identifiability problem*

Suppose now that $k_{12} = 0$ so that compartment 1 is a source compartment, as shown in Fig. 6.3. From eqns (6.8) and (6.9),

$$\mathbf{G}(s) = \begin{bmatrix} \dfrac{c_1 b_1}{s + k_{01} + k_{21}} & 0 \\ \dfrac{c_2 b_{21}}{(s + k_{01} + k_{21})(s + k_{02})} & \dfrac{c_2 b_2}{s + k_{02}} \end{bmatrix}. \tag{6.32}$$

Clearly, compartment 1 must be perturbed if all rate constants are to be estimated, while if only compartment 1 is observed, then the sum $(k_{01} + k_{21})$ can be identified, but no individual rate constant.

Consider instead therefore an experiment in which compartment 1 (only) is perturbed and compartment 2 (only) is observed. The experimental transfer function has the form

$$G_{21}(s) = \frac{K_3}{(s + \phi)(s + \psi)} \tag{6.33}$$

where K_3, ϕ and ψ are measurable constants. The denominator equations have two solutions:

either $\quad k_{01} + k_{21} = \phi, \qquad k_{02} = \psi \qquad$ (6.34a)

or $\quad k_{02} = \phi, \qquad k_{01} + k_{21} = \psi. \qquad$ (6.34b)

The numerator equation

$$c_2 b_1 k_{21} = K_3 \tag{6.35}$$

will allow k_{21} to be determined uniquely if $c_2 b_1$ is known, in which case k_{01} and k_{02} are locally identifiable with two solutions. If $c_2 b_1$ is not known, k_{02} and $(k_{01} + k_{21})$ are locally identifiable with two solutions, but k_{01} and k_{21} are not individually identifiable, locally or globally.

The unidirectional exchange model is very important in biomedicine for

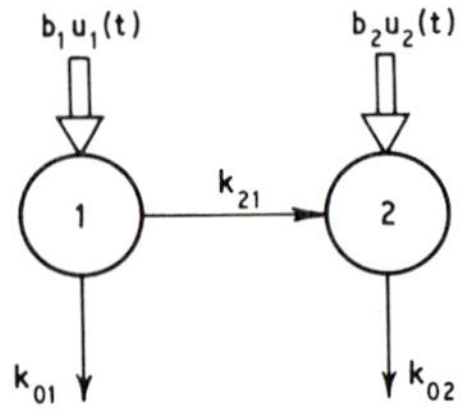

Figure 6.3 Two-compartment model with unidirectional exchange between compartments.

modelling the kinetics of a dose administered via an extravascular route and whose absorption kinetics are approximately first-order. The absorption compartment (no. 1) is not a well-defined physical entity and merely reflects that, for many substances, the complex processes involved in absorption of material behave overall in approximately first-order fashion. In such an application, k_{01} is usually taken as zero, and the discrepancy between quantity administered and the quantity appearing in the compartment (no. 2) containing the systemic blood is accounted for by the input fraction b_1 which is called the bioavailability fraction and is usually unknown. Measurement is possible only in compartment 2.

For this model, it is readily seen by putting $k_{01} = 0$ in eqns (6.34) that k_{21} and k_{02} are locally identifiable with two solutions. It would be possible to determine k_{21} (and hence k_{02}) uniquely from eqn (6.35) if $c_2 b_1$ were known, but, as pointed out above, the input fraction b_1 is often not known. Thus it is not possible to distinguish the absorption and elimination rate constants from this experiment, and prior knowledge as to which of absorption and elimination is the more rapid process has to be available for the ambiguity to be resolved.

In the absence of any such prior knowledge, it is interesting to see whether perturbation of compartment 2 as well as compartment 1 would resolve the ambiguity. If the inputs are well separated in time, then from eqn (6.32), k_{02} can be determined uniquely from $G_{22}(s)$ and with $k_{01} = 0$, k_{21} can then be determined uniquely from $G_{21}(s)$. Assuming b_2 is known (usually equal to 1), b_1 can also be estimated uniquely from the two responses and this is the usual method of determining the bioavailability fraction of a drug. If the two inputs are applied simultaneously, the Laplace transform of the observation is

$$Y_2(s) = \frac{c_2 b_1 k_{21} U_1(s)}{(s + k_{21})(s + k_{02})} + \frac{c_2 b_2 U_2(s)}{(s + k_{02})}. \tag{6.36}$$

It is readily confirmed that if the two inputs are of the same shape, so that $U_1(s) = M_1 U(s)$ and $U_2(s) = M_2 U(s)$, with M_1 and M_2 known constants, the system remains only locally identifiable unless b_1 is known. Instead, let the input to compartment 1 be impulsive, with $U_1(s) = D_1$ and the input to compartment 2 be a constant infusion, with $U_2(s) = k_i/s$ with D_1 and k_i known. Then

$$Y_2(s) = \frac{(c_2 b_1 k_{21} D_1 + c_2 b_2 k_i)s + c_2 b_2 k_i k_{21}}{s(s + k_{21})(s + k_{02})} \tag{6.37}$$

corresponding to an experimental response of the form

$$Y_2(s) = \frac{K_2(s + \alpha)}{s(s + \phi)(s + \psi)}. \tag{6.38}$$

Comparing coefficients,

$$c_2b_1k_{21}D_1 + c_2b_2k_i = K_2 \tag{6.39a}$$

$$c_2b_2k_ik_{21} = K_2\alpha \tag{6.39b}$$

and *either*
$$k_{21} = \phi, \qquad k_{02} = \psi \tag{6.39c}$$

or
$$k_{02} = \phi, \qquad k_{21} = \psi. \tag{6.39d}$$

Thus, if c_2b_2 is known, k_{21} can be estimated uniquely from eqn (6.39b) and k_{02} then follows. The bioavailability fraction b_1 can also be estimated uniquely from eqn (6.39a). The model is thus globally identifiable from this experiment provided c_2b_2 is known.

Example 6.4. *It is known that when a certain drug is administered via an intravascular route, its pharmacokinetics are adequately modelled by a compartmental model with one compartment, and that when it is administered via an extravascular route, the absorption is first-order. The model is thus as shown in Fig. 6.3, with* $k_{01} = 0$.

In an experiment, a dose of 1 *unit is rapidly administered to compartment* 1 *and, starting at the same time, a constant input of* 0.1 *units per unit time is infused into compartment* 2. *The measured response of compartment* 2 *is*

$$y_2(t) = 0.1 + 0.9\,\mathrm{e}^{-t} - \mathrm{e}^{-2t}.$$

If $c_2 = b_2 = 1$, *determine* k_{21}, k_{02} *and* b_1.

Solution.

$$Y_2(s) = \frac{0.1}{s} + \frac{0.9}{s+1} - \frac{1}{s+2}$$

$$= \frac{1.1s + 0.2}{s(s+1)(s+2)}.$$

Substituting into eqns (6.39a) and (6.39b),

$$b_1k_{21} + 0.1 = 1.1$$

$$0.1k_{21} = 0.2$$

whence
$$k_{21} = 2 \quad \text{and} \quad b_1 = 0.5.$$

Then from eqn (6.39d),
$$k_{02} = 1.$$

6.2.6 *Volumes of distribution*

If the same compartment (no. 1, say) is perturbed and observed, the input fraction b_1 is known and the measurement is of concentration, eqn (6.12a) allows the apparent volume of distribution V_1 to be identified uniquely. If, for example, $b_1 = 1$, $u_1(t) = D_1\,\delta(t)$ and $c_1 = 1/V_1$, then $V_1 = \dfrac{D_1}{y_1(0^+)}$, as for the one-compartment model. It is not possible to determine the apparent volume of distribution V_2 from this experiment. It is, however, possible to determine the steady-state volume of distribution $V_2(\infty)$ by considering the total amount of material in the body in the steady state during a constant infusion into compartment 1.

Let the material be infused at a constant rate k_i so that the differential equations for the compartmental quantities are

$$\frac{\mathrm{d}x_1}{\mathrm{d}t} = a_{11}x_1 + a_{12}x_2 + k_i$$

$$\frac{\mathrm{d}x_2}{\mathrm{d}t} = a_{21}x_1 + a_{22}x_2.$$

In the steady state $\left(\dfrac{\mathrm{d}x_1}{\mathrm{d}t} = \dfrac{\mathrm{d}x_2}{\mathrm{d}t} = 0\right)$,

$$x_2(\infty) = -\frac{a_{21}}{a_{22}}\,x_1(\infty).$$

The (total) steady-state volume of distribution is then defined as

$$V_\infty = \frac{x_1(\infty)}{C(\infty)} + \frac{x_2(\infty)}{C(\infty)}$$

where $C(\infty)$ is the concentration of material in compartment 1 at steady state. The first term on the right-hand side is V_1 as already found, and $\dfrac{x_2(\infty)}{C(\infty)}$ is taken as $V_2(\infty)$, which is related to V_1 by

$$V_2(\infty) = -\frac{a_{21}}{a_{22}}\,V_1 = \frac{k_{21}}{k_{02} + k_{12}}\,V_1. \tag{6.40}$$

Compartmental masses can also be determined if the density of the material is known. The definition is readily extended to a mammillary model with any number of compartments, provided it is the central compartment which is perturbed and observed.

6.3 Three-compartment systems

6.3.1 *The general three-compartment system*

For the general three-compartment system shown in Fig. 6.4, the transfer function matrix is, from Section 4.1,

$$\mathbf{G}(s) = \frac{1}{\Delta(s)}\begin{bmatrix} c_1b_1[(s-a_{22})(s-a_{33})-a_{23}a_{32}] & c_1b_2[a_{12}(s-a_{33})+a_{13}a_{32}] & c_1b_3[a_{13}(s-a_{22})+a_{12}a_{23}] \\ c_2b_1[a_{21}(s-a_{33})+a_{23}a_{31}] & c_2b_2[(s-a_{11})(s-a_{33})-a_{13}a_{31}] & c_2b_3[a_{23}(s-a_{11})+a_{13}a_{21}] \\ c_3b_1[a_{31}(s-a_{22})+a_{21}a_{32}] & c_3b_2[a_{32}(s-a_{11})+a_{12}a_{31}] & c_3b_3[(s-a_{11})(s-a_{22})-a_{12}a_{21}] \end{bmatrix} \tag{6.41}$$

where
$$\begin{aligned}\Delta(s) = s^3 &- (a_{11} + a_{22} + a_{33})s^2 \\ &+ (a_{11}a_{22} + a_{22}a_{33} + a_{33}a_{11} - a_{12}a_{21} - a_{13}a_{31} - a_{23}a_{32})s \\ &- a_{11}a_{22}a_{33} + a_{11}a_{23}a_{32} + a_{22}a_{13}a_{31} + a_{33}a_{12}a_{21} \\ &- a_{12}a_{23}a_{31} - a_{21}a_{32}a_{13}\end{aligned} \tag{6.42}$$

and
$$a_{11} = -(k_{01} + k_{21} + k_{31})$$
$$a_{12} = k_{12}, \text{ etc.}$$

For a single input, single observation experiment, if the perturbed and observed compartments are the same, the experimental transfer function will be of the form

$$G_{mm}(s) = \frac{K_1(s^2 + \alpha s + \beta)}{s^3 + \gamma s^2 + \delta s + \varepsilon} \tag{6.43}$$

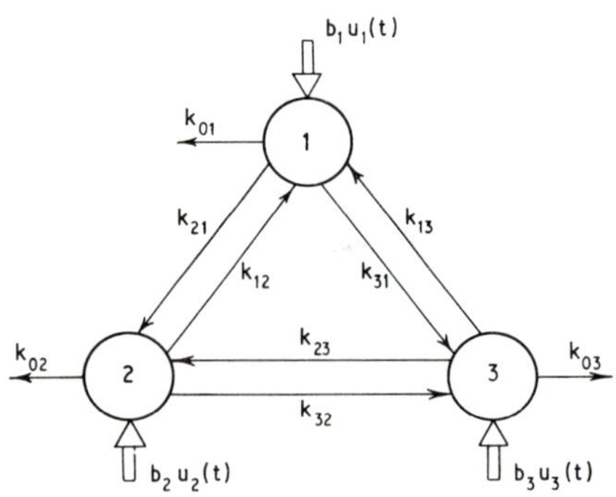

Figure 6.4 General three-compartment model.

while, for the general model, if the perturbed and observed compartments are different it will be of the form:

$$G_{mn}(s) = \frac{K_2(s + \zeta)}{s^3 + \gamma s^2 + \delta s + \varepsilon} \tag{6.44}$$

since, from the structure (Fig. 6.4), the compartments must be adjacent. (Here, K_1, α, β, γ, δ, ε, K_2 and ζ are measurable constants.)

In either case, there will be a maximum of five independent equations—two from the numerator (the term in s^2 in the numerator of $G_{mm}(s)$ involves $c_m b_m$ but not any of the rate constants) and three from the denominator. Thus from a single-input, single-output experiment, a maximum of five rate constants can be identified. For the general system, there are nine rate constants to be identified, so either two further transfer functions must be measured (and even then it is quite easy to obtain further equations which are not independent of those already available) or there must be prior knowledge of four rate constants for there to be five equations in the remaining five rate constants.

Let us now consider some simplified models of considerable practical interest, in which some of the inter-compartment links are assumed to be missing.

6.3.2 *The mammillary system: local identifiability due to symmetry*

For the mammillary system (Fig. 6.5), $a_{23} = a_{32} = 0$ by definition so that there are seven rate constants to be identified. Consider an SISO experiment in which only the central compartment is perturbed and observed. Intuitively, we would expect local identifiability to result since, from the symmetry of model and experiment, the peripheral compartments are indistinguishable. From eqns (6.41), (6.42) and (6.43),

$$G_{11}(s) = \frac{c_1 b_1 (s - a_{22})(s - a_{33})}{\Delta(s)} = \frac{K_1(s^2 + \alpha s + \beta)}{s^3 + \gamma s^2 + \delta s + \varepsilon} \tag{6.45}$$

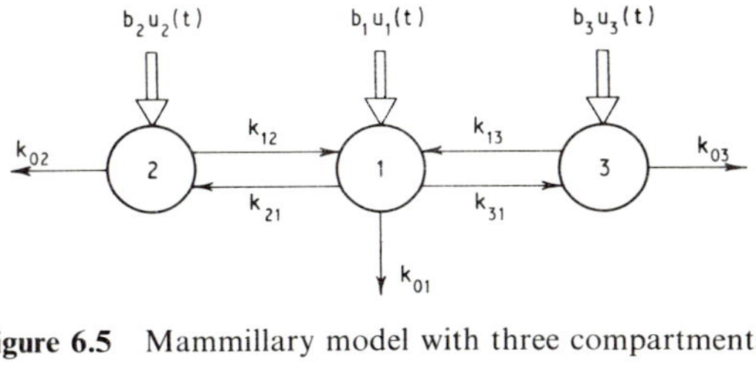

Figure 6.5 Mammillary model with three compartments.

where $$\Delta(s) = s^3 - (a_{11} + a_{22} + a_{33})s^2 + (a_{11}a_{22} + a_{22}a_{33} + a_{33}a_{11} - a_{12}a_{21} - a_{13}a_{31})s - a_{11}a_{22}a_{33} + a_{33}a_{12}a_{21} + a_{22}a_{13}a_{31}. \tag{6.46}$$

From the numerator, a_{22} and a_{33} are locally identifiable with two solutions while from the denominator, a_{11} is uniquely identifiable from the term in s^2 and $a_{12}a_{21}$ and $a_{13}a_{31}$ are locally identifiable with two solutions from the term in s and the constant term. No individual rate constant is identifiable, either globally or locally.

To achieve identifiability with an SISO experiment, prior knowledge of two further rate constants is necessary. By far the most common assumption is that excretion to the environment from two of the compartments is zero. We will consider two cases.

Case A. $k_{02} = k_{03} = 0$. (Excretion from the central compartment only.) Writing the elements of **A** in terms of rate constants, we have, from the numerator of $G_{11}(s)$,

$$k_{12} + k_{13} = \alpha \tag{6.47a}$$

$$k_{12}k_{13} = \beta \tag{6.47b}$$

which give two solutions for k_{12} and k_{13} which cannot be assigned uniquely. From the denominator of $G_{11}(s)$,

$$k_{01} + k_{21} + k_{31} + k_{12} + k_{13} = \gamma \tag{6.48a}$$

$$k_{13}k_{21} + k_{12}k_{31} + k_{01}(k_{12} + k_{13}) + k_{12}k_{13} = \delta \tag{6.48b}$$

$$k_{01}k_{12}k_{13} = \varepsilon. \tag{6.48c}$$

Rate constant k_{01} is found uniquely as $\frac{\varepsilon}{\beta}$ and there are then two solutions for k_{21} and k_{31} from

$$k_{21} + k_{31} = \gamma - \alpha - \frac{\varepsilon}{\beta} \tag{6.49a}$$

and $$k_{13}k_{21} + k_{12}k_{31} = \delta - \frac{\varepsilon}{\beta}\alpha - \beta. \tag{6.49b}$$

Reversal of k_{12} and k_{13} reverses k_{21} and k_{31} so that k_{12}, k_{13}, k_{21} and k_{31} are locally identifiable with two solutions. The additional relationship

$$c_1b_1 = K_1 \tag{6.50}$$

allows c_1b_1 to be determined uniquely (hence V_1 if b_1 is known and measurement is of concentration).

Case B. $k_{01} = k_{02} = 0$. (Excretion from one peripheral compartment only.)
From the numerator of $G_{11}(s)$,

$$k_{12} + (k_{13} + k_{03}) = \alpha \tag{6.51a}$$

$$k_{12}(k_{13} + k_{03}) = \beta \tag{6.51b}$$

giving two solutions for k_{12} and $(k_{13} + k_{03})$. The denominator equations are

$$k_{21} + k_{31} + k_{12} + k_{03} + k_{13} = \gamma \tag{6.52a}$$

$$(k_{21} + k_{31})(k_{12}) + k_{12}(k_{03} + k_{13}) + (k_{03} + k_{13})(k_{21} + k_{31}) - k_{12}k_{21} - k_{13}k_{31} = \delta \tag{6.52b}$$

$$(k_{21} + k_{31})k_{12}(k_{03} + k_{13}) - (k_{03} + k_{13})k_{12}k_{21} - k_{12}k_{13}k_{31} = \varepsilon. \tag{6.52c}$$

From eqns (6.51a) and (6.52a),

$$k_{21} + k_{31} = \gamma - \alpha. \tag{6.53a}$$

Substituting in eqns (6.52b) and (6.52c),

$$(\gamma - \alpha)\alpha + \beta - k_{12}k_{21} - k_{13}k_{31} = \delta \tag{6.53b}$$

$$(\gamma - \alpha)\beta - \beta k_{21} - k_{12}k_{13}k_{31} = \varepsilon \tag{6.53c}$$

which may be solved for $k_{13}k_{31}$ and k_{21} with two solutions, one for each value of k_{12}. Then k_{31} follows from eqn (6.53a), k_{13} from $k_{13}k_{31}$ and k_{03} from eqn (6.52a). All the rate constants are locally identifiable with two solutions from this experiment.

Example 6.5. *In a three-compartment mammillary system, the response in the central compartment to a rapidly administered input of* 1 *unit to the central compartment is*

$$y_1(t) = 0.7\,\mathrm{e}^{-5t} + 0.2\,\mathrm{e}^{-t} + 0.1\,\mathrm{e}^{-0.1t}$$

(*cf. Example 4.2.*) *Determine the unknown rate constants if there is prior knowledge* (*a*) *that excretion is from the central compartment only,* (*b*) *that excretion is from one of the peripheral compartments only.*

Solution.

$$Y_1(s) = \frac{0.7}{s+5} + \frac{0.2}{s+1} + \frac{0.1}{s+0.1}$$

$$= \frac{s^2 + 2.39s + 0.67}{s^3 + 6.1s^2 + 5.6s + 0.5}$$

so that $\alpha = 2.39$, $\beta = 0.67$, $\gamma = 6.1$, $\delta = 5.6$, $\varepsilon = 0.5$.

Case A. $k_{02} = k_{03} = 0$. Following the given method of solution,

$$k_{01} = \frac{\varepsilon}{\beta} = \frac{0.5}{0.67} = 0.75.$$

From eqn. (6.47a) and (6.47b), $k_{12} + k_{13} = 2.39$; $k_{12}k_{13} = 0.67$ whence

$$k_{12} = 0.325, \qquad k_{13} = 2.065$$

or

$$k_{12} = 2.065, \qquad k_{13} = 0.325.$$

Taking the first of these alternatives and substituting in eqns (6.49),

$$k_{21} + k_{31} = 6.1 - 2.39 - 0.75 = 2.96$$

$$2.065k_{21} + 0.325k_{31} = 5.6 - (0.75)(2.39) - 0.67 = 3.14$$

whence

$$k_{21} = \frac{\begin{vmatrix} 2.96 & 1 \\ 3.14 & 0.325 \end{vmatrix}}{\begin{vmatrix} 1 & 1 \\ 2.065 & 0.325 \end{vmatrix}} = \frac{2.18}{1.74} = 1.253$$

$$k_{31} = 2.96 - 1.253 = 1.707.$$

Taking the second of the alternatives reverses k_{21} and k_{31} so that the two sets of solutions are

$$k_{01} = 0.75,\ k_{12} = 0.325,\ k_{21} = 1.253,\ k_{13} = 2.065,\ k_{31} = 1.707,$$

$$k_{01} = 0.75,\ k_{12} = 2.065,\ k_{21} = 1.707,\ k_{13} = 0.325,\ k_{31} = 1.253.$$

Notice that neither set contains any negative rate constants, which would permit rejection of such a set.

Case B. $k_{01} = k_{02} = 0$.

From eqns (6.51),

$$k_{12} + (k_{13} + k_{03}) = 2.39; \qquad k_{12}(k_{13} + k_{03}) = 0.67$$

so that

$$k_{12} = 2.065, \qquad (k_{13} + k_{03}) = 0.325$$

or

$$k_{12} = 0.325, \qquad (k_{13} + k_{03}) = 2.065.$$

Taking the first of these alternatives and substituting in eqns (6.53b) and (6.53c),

$$k_{13}k_{31} + 2.065k_{21} = -5.6 + (3.71)(2.39) + 0.67 = 3.94$$

$$2.065k_{13}k_{31} + 0.325k_{21} = -0.5 + (3.71)(0.67) = 1.99$$

whence $k_{13}k_{31} = 0.409, \qquad k_{21} = 1.707.$

From eqn (6.53a), $k_{31} = 3.71 - k_{21} = 2$

so that
$$k_{13} = \frac{0.409}{2} = 0.205$$

and $k_{03} = 0.325 - 0.205 = 0.12.$

Taking the second of the alternatives gives $k_{13}k_{31} = 3.53$, $k_{21} = 1.253$

$$k_{31} = 3.71 - k_{21} = 2.46$$

$$k_{13} = \frac{3.53}{2.46} = 1.43$$

$$k_{03} = 2.065 - 1.43 = 0.635.$$

The two sets of solutions are

$$k_{12} = 2.065,\ k_{21} = 1.707,\ k_{31} = 2,\ k_{13} = 0.205,\ k_{03} = 0.12$$
$$k_{12} = 0.325,\ k_{21} = 1.253,\ k_{31} = 2.46,\ k_{13} = 1.43,\ k_{03} = 0.635.$$

Again, neither set contains any negative values.

6.3.3 *The mammillary system: global identifiability with a different SISO experiment*

Suppose now that one of the peripheral compartments (only) is perturbed and that the same compartment (only) is observed. (With an end compartment perturbed, such a model would often be described in the biomedical literature as a catenary model; as we have seen in Chapter 4, the three-compartment mammillary and catenary models have the same structure.)

Let compartment 2 be perturbed and observed. From eqns (6.41) and (6.43),

$$G_{22}(s) = \frac{c_2 b_2[(s - a_{11})(s - a_{33}) - a_{13}a_{31}]}{\Delta(s)} = \frac{K_2(s^2 + \alpha s + \beta)}{\Delta(s)}$$

where $\Delta(s)$ is as before, given by eqn (6.46). The numerator equations involving rate constants are thus

$$-(a_{11} + a_{33}) = \alpha \tag{6.54a}$$

$$a_{11}a_{33} - a_{13}a_{31} = \beta. \tag{6.54b}$$

The symmetry of the preceding section no longer exists, and it is instructive to consider the same prior knowledge assumptions as before to see whether the rate constants are now globally identifiable.

Case A. $k_{02} = k_{03} = 0$. (Excretion from the central compartment only.) From eqns (6.54),

$$k_{01} + k_{21} + k_{31} + k_{13} = \alpha \tag{6.55a}$$

$$k_{01}k_{13} + k_{21}k_{13} = \beta \tag{6.55b}$$

together with eqns (6.48) from the denominator. These can be solved uniquely for the rate constants, as follows:

$$k_{12} = \gamma - \alpha \tag{6.56a}$$

$$k_{12}(k_{13} + k_{01} + k_{31}) = \delta - \beta$$

so
$$k_{13} + k_{01} + k_{31} = \frac{\delta - \beta}{k_{12}}$$

$$\therefore \quad k_{21} = \alpha - \frac{\delta - \beta}{k_{12}} \tag{6.56b}$$

$$k_{01}k_{13} = \frac{\varepsilon}{k_{12}}$$

so
$$k_{13} = \left(\beta - \frac{\varepsilon}{k_{12}}\right)\frac{1}{k_{21}} \tag{6.56c}$$

$$k_{01} = \frac{\varepsilon}{k_{12}k_{13}} \tag{6.56d}$$

and
$$k_{31} = \gamma - (k_{12} + k_{21} + k_{13} + k_{01}). \tag{6.56e}$$

Case B. $k_{01} = k_{02} = 0$. (Excretion from the other peripheral compartment only.) The numerator equations are now

$$k_{21} + k_{31} + k_{03} + k_{13} = \alpha \tag{6.57a}$$

$$k_{21}k_{13} + k_{21}k_{03} + k_{31}k_{03} = \beta \tag{6.57b}$$

while the denominator equations are (6.52), as before.

These can be solved uniquely for the rate constants, as follows:

$$k_{12} = \gamma - \alpha \tag{6.58a}$$

$$k_{31}k_{12} + k_{03}k_{12} + k_{13}k_{12} = \delta - \beta$$

so $$k_{31} + k_{03} + k_{13} = \frac{\delta - \beta}{k_{12}}$$

$$\therefore \quad k_{21} = \alpha - \frac{\delta - \beta}{k_{12}} \tag{6.58b}$$

$$k_{31}k_{03} = \frac{\varepsilon}{k_{12}}$$

so $$k_{31} = \frac{\delta - \beta}{k_{12}} - \frac{1}{k_{21}}\left(\beta - \frac{\varepsilon}{k_{12}}\right) \tag{6.58c}$$

$$k_{03} = \frac{\varepsilon}{k_{12}k_{31}} \tag{6.58d}$$

and $$k_{13} = \gamma - (k_{21} + k_{31} + k_{12} + k_{03}). \tag{6.58e}$$

A unique solution also follows for the case when excretion is from the perturbed and observed compartment only.

Example 6.6. *A three-compartment mammillary system has excretion from one of the peripheral compartments only. When the other peripheral compartment is perturbed with a unit impulse, the response in the perturbed compartment is* $0.7\,\mathrm{e}^{-5t} + 0.2\,\mathrm{e}^{-t} + 0.1\,\mathrm{e}^{-0.1t}$. *Find the rate constants.*

Solution. Let the perturbed and observed compartment be no. 2; then $k_{01} = k_{02} = 0$. For the given perturbation and response,

$$Y_2(s) = \frac{s^2 + 2.39s + 0.67}{s^3 + 6.1s^2 + 5.6s + 0.5}$$

so that $\alpha = 2.39$, $\beta = 0.67$, $\gamma = 6.1$, $\delta = 5.6$ and $\varepsilon = 0.5$. Following the solution in eqns (6.58),

$$k_{12} = 6.1 - 2.39 = 3.71$$

$$k_{21} = 2.39 - \frac{5.6 - 0.67}{3.71} = 1.061$$

$$k_{31} = \frac{5.6 - 0.67}{3.71} - \frac{1}{1.061}\left(0.67 - \frac{0.5}{3.71}\right)$$

$$= 1.329 - 0.504 = 0.825$$

$$k_{03} = \frac{0.5}{(3.71)(0.825)} = 0.164$$

$$k_{13} = 6.1 - (3.71 + 1.061 + 0.825 + 0.164) = 0.340.$$

6.3.4 *Mammillary model with observation of two states: a redundant equation*

Let us return now to the situation where there is no prior knowledge of any of the seven rate constants of the mammillary model of Fig. 6.5, and consider an experiment in which the central compartment (only) is perturbed and compartments 1 and 2 are observed. The denominator of both transfer functions $G_{11}(s)$ and $G_{21}(s)$ gives

$$-(a_{11} + a_{22} + a_{33}) = \gamma \tag{6.59a}$$

$$a_{11}a_{22} + a_{22}a_{33} + a_{33}a_{11} - a_{12}a_{21} - a_{13}a_{31} = \delta \tag{6.59b}$$

$$-a_{11}a_{22}a_{33} + a_{33}a_{12}a_{21} + a_{22}a_{13}a_{31} = \varepsilon. \tag{6.59c}$$

The numerator of $G_{11}(s)$ gives

$$-(a_{22} + a_{33}) = \alpha \tag{6.59d}$$

$$a_{22}a_{33} = \beta \tag{6.59e}$$

(together with $c_1 b_1 = K_1$).

From eqn (6.41), the numerator of $G_{21}(s)$ is $c_2 b_1 a_{21}(s - a_{33})$, corresponding to a numerator in the experimental response of the form $K_2(s + \zeta)$, so that $G_{21}(s)$ provides two further equations:

$$c_2 b_1 a_{21} = K_2 \tag{6.59f}$$

$$-a_{33} = \zeta. \tag{6.59g}$$

There are now seven equations in the seven unknown elements of **A** which raises hopes that the system might be identifiable from this experiment. However, this does not prove to be the case, because one of the new equations is not independent. From eqns (6.59d) and (6.59g), $a_{22} = -\alpha + \zeta$ and from eqn (6.59e), $(\alpha - \zeta)\zeta = \beta$, so that there are only six independent equations in the seven unknowns.

Let us see what can be identified from this experiment. Assume that $c_2 b_1$ is known, so that a_{21} $(=k_{21})$ can be identified uniquely from eqn (6.59f). Then the following parameter combinations are identifiable: a_{33} from eqn (6.59g); a_{22} from eqn (6.59d) or eqn (6.59e); a_{11} from eqn (6.59a); $a_{12}a_{21}$ and $a_{13}a_{31}$ from eqns (6.59b) and (6.59c). From $a_{12}a_{21}$, a_{12} $(=k_{12})$ can be found and from a_{22} $(=-(k_{02} + k_{12}))$, k_{02} can be found.

Thus k_{21}, k_{12} and k_{02} can be identified uniquely. The four remaining rate constants are not identifiable individually, but from the above known parameter combinations, they are uniquely identifiable if there is prior knowledge of either k_{01} or k_{03}. Prior knowledge of k_{02} alone does not make the remaining rate constants identifiable since it is already identifiable.

Example 6.7. *A three-compartment mammillary system has the central compartment (no. 1) perturbed by a unit impulse. The responses in the central compartment and one of the peripheral compartments are:*

$$y_1(t) = 0.7\,\mathrm{e}^{-5t} + 0.2\,\mathrm{e}^{-t} + 0.1\,\mathrm{e}^{-0.1t}$$

$$y_2(t) = -0.408\,\mathrm{e}^{-5t} + 0.320\,\mathrm{e}^{-t} + 0.088\,\mathrm{e}^{-0.1t}$$

and it is known that $c_2 b_1 = 1$. Determine k_{21}, k_{12} and k_{02}.

If, in addition, it is known that $k_{01} = 0$, determine the remaining rate constants (k_{03}, k_{13} and k_{31}).

Solution. As before,

$$Y_1(s) = \frac{s^2 + 2.39s + 0.67}{s^3 + 6.1s^2 + 5.6s + 0.5}$$

so that $\alpha = 2.39$, $\beta = 0.67$, $\gamma = 6.1$, $\delta = 5.6$ and $\varepsilon = 0.5$.

$$Y_2(s) = -\frac{0.408}{s+5} + \frac{0.320}{s+1} + \frac{0.088}{s+0.1}$$

$$= \frac{1.71(s + 0.325)}{s^3 + 6.1s^2 + 5.6s + 0.5}.$$

Since $c_2 b_1 = 1$, $Y_2(s) = G_{22}(s)$, so that, from eqn (6.59f),

$$a_{21} = k_{21} = 1.71$$

$$-a_{33} = (k_{03} + k_{13}) = 0.325\ (=\zeta).$$

(Note that $(\alpha - \zeta)\zeta = (2.065)(0.325) = 0.67 = \beta$, the redundancy.)
From eqn (6.59d),

$$-a_{22} = \alpha + a_{33} = 2.39 - 0.325 = 2.065 = k_{02} + k_{12}.$$

(This could also be obtained from eqn (6.59e).) From eqn (6.59a),

$$-a_{11} = \gamma + a_{22} + a_{33} = 6.1 - 2.39 = 3.71.$$

Since $$k_{21} = 1.71, \qquad k_{01} + k_{31} = 2.$$

From eqns (6.59b) and (6.59c),

$$(3.71)(2.065) + (2.065)(0.325) + (0.325)(3.71) - k_{12}k_{21} - k_{13}k_{31} = 5.6$$

giving
$$k_{12}k_{21} + k_{13}k_{31} = 3.938$$

and
$$(3.71)(2.065)(0.325) - 0.325k_{12}k_{21} - 2.065k_{13}k_{31} = 0.5$$

giving
$$0.325k_{12}k_{21} + 2.065k_{13}k_{31} = 1.990.$$

Hence
$$k_{12}k_{21} = \frac{\begin{vmatrix} 3.938 & 1 \\ 1.990 & 2.065 \end{vmatrix}}{2.065 - 0.325} = 3.53$$

and
$$k_{13}k_{31} = 0.41.$$

Since
$$k_{21} = 1.71, \qquad k_{12} = \frac{3.53}{1.71} = 2.065$$

and since
$$a_{22} = -2.065, \qquad k_{02} = 0.$$

In the absence of prior knowledge of any rate constants therefore,

$$k_{21} = 1.71, \qquad k_{12} = 2.065, \qquad k_{02} = 0, \qquad (k_{03} + k_{13}) = 0.325,$$
$$(k_{01} + k_{31}) = 2 \quad \text{and} \quad k_{13}k_{31} = 0.41.$$

With the additional prior knowledge that $k_{01} = 0$,

$$k_{31} = 2,$$
$$k_{13}k_{31} = 0.41, \qquad \text{so} \qquad k_{13} = 0.205,$$

and
$$k_{03} = 0.325 - 0.205 = 0.12.$$

(If, instead, the prior knowledge had been that $k_{03} = 0$,

$$k_{13} = 0.325, \qquad k_{31} = \frac{0.41}{0.325} = 1.253$$

and
$$k_{01} = 2 - 1.253 \simeq 0.75.)$$

If there is no prior knowledge of any of the rate constants, then the system is uniquely identifiable from an experiment with the central compartment (only) perturbed, with known input fraction, and all three compartments observed, the peripheral compartments with known observation gains. There are now two redundant equations.

6.3.5 *A source compartment in a three-compartment model: an inadmissible solution*

The model shown in Fig. 6.6 represents a general two-compartment system (compartments 1 and 2) with an additional input to compartment 1 from a source compartment modelling an input of first-order form (for example, a drug administered via an extravascular route). In the model, it is assumed that $k_{31} = 0$, that there is no excretion from the source compartment ($k_{03} = 0$) and that compartment 3 cannot be observed. The Laplace transformed equations for x_1 and x_2 are:

$$\begin{bmatrix} sX_1(s) \\ sX_2(s) \end{bmatrix} = \begin{bmatrix} a_{11} & a_{12} \\ a_{21} & a_{22} \end{bmatrix} \begin{bmatrix} X_1(s) \\ X_2(s) \end{bmatrix} + \begin{bmatrix} b_1 U_1(s) + \dfrac{b_3 k_{13}}{s + k_{13}} U_3(s) \\ b_2 U_2(s) \end{bmatrix} \tag{6.60}$$

where $a_{11} = -(k_{01} + k_{21}), \quad a_{12} = k_{12}, \quad a_{21} = k_{21}$

and $a_{22} = -(k_{02} + k_{12})$.

If there is no input directly to compartments 1 and 2, and only compartment 1 is observed, the measured transfer function,

$$G_{13}(s) = \frac{c_1 b_1 k_{13}}{s + k_{13}} \frac{(s - a_{22})}{s^2 - (a_{11} + a_{22})s + a_{11}a_{22} - a_{12}a_{21}}. \tag{6.61}$$

The corresponding experimental transfer function is of the form

$$G_{13}(s) = \frac{K_1(s + \alpha)}{(s + \eta)(s^2 + \gamma s + \delta)} \tag{6.62}$$

corresponding to a unit impulse response of the form

$$y_1(t) = A\,e^{-\lambda t} + B\,e^{-\mu t} - (A + B)\,e^{-\nu t} \tag{6.63}$$

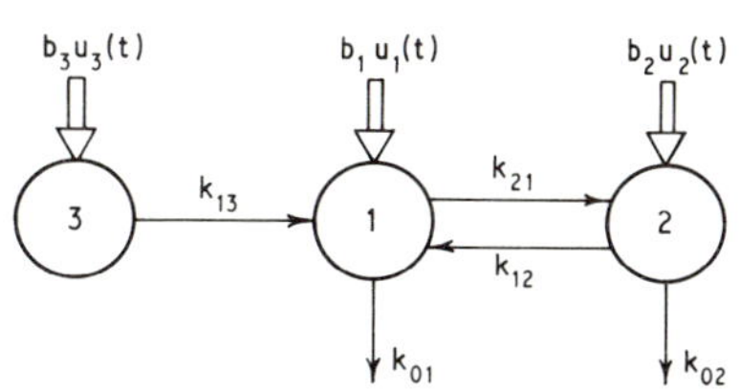

Figure 6.6 Three-compartment system including a source compartment.

where K_1, α, η, γ, δ, A, B, λ, μ and ν are constants. The equations in the rate constants are:

$$-a_{22} = \alpha \tag{6.64a}$$

$$-(a_{11} + a_{22}) = \gamma \tag{6.64b}$$

$$a_{11}a_{22} - a_{12}a_{21} = \delta \tag{6.64c}$$

$$c_1 b_3 k_{13} = K_1 \tag{6.64d}$$

$$k_{13} = \eta = \lambda \text{ or } \mu \text{ or } \nu. \tag{6.64e}$$

If $c_1 b_3$ is known, k_{13} can be found uniquely from eqn (6.64d). Since these equations are modelling an input via an extravascular route, for which the bioavailability fraction b_3 will not usually be known (so that eqn (6.60d) will not give k_{13}), the absorption rate k_{13} more usually can only be identified locally with three solutions. One or more of these solutions may be inadmissible, since it will result in negative rate constants (or combinations of rate constants)—see Example 6.8. With three possible solutions for k_{13}, there are also three solutions for γ and δ:

$$\text{if} \quad k_{13} = \lambda, \quad \gamma = \mu + \nu, \quad \delta = \mu\nu \tag{6.65a}$$

$$\text{if} \quad k_{13} = \mu, \quad \gamma = \lambda + \nu, \quad \delta = \lambda\nu \tag{6.65b}$$

$$\text{if} \quad k_{13} = \nu, \quad \gamma = \lambda + \mu, \quad \delta = \lambda\mu. \tag{6.65c}$$

The analysis now proceeds along similar lines to that given in Section 6.2.1; the rate constants k_{01}, k_{21}, k_{02} and k_{12} are not identifiable from this experiment.

If there is prior knowledge of one of these rate constants, the remaining rate constants are also identifiable—some locally and some globally. This will be illustrated with an example.

Example 6.8. *In the model shown in Fig. 6.6, $u_3(t) = \delta(t)$ but b_3 is unknown; $u_1(t) = u_2(t) = 0$. The response in compartment 1 is $y_1(t) = -0.408\,\mathrm{e}^{-5t} + 0.320\,\mathrm{e}^{-t} + 0.088\,\mathrm{e}^{-0.1t}$. Determine the remaining rate constants for the two cases where there is prior knowledge (i) that $k_{02} = 0$; (ii) that $k_{01} = 0$.*

Solution.

$$Y_1(s) = \frac{1.71(s + 0.325)}{s^3 + 6.1s^2 + 5.6s + 0.5}.$$

Case (i). $k_{02} = 0$. From eqn (6.64a),

$$-a_{22} = k_{12} = 0.325.$$

Since $-a_{11} = (k_{01} + k_{21})$, eqns (6.64a), (6.64b) and (6.64c) give

$$k_{01} = \frac{\delta}{\alpha} \quad \text{and} \quad k_{21} = \gamma - \alpha - k_{01}.$$

There are three possible values for k_{13}.

(a). $k_{13} = 5$. Then the denominator of $Y_1(s) = (s+5)(s^2 + 1.1s + 0.1)$ so

$$\gamma = 1.1, \qquad \delta = 0.1$$

so that $$k_{01} = \frac{0.1}{0.325} = 0.308$$

and $$k_{21} = 1.1 - 0.325 - 0.308 = 0.467.$$

(b). $k_{13} = 1$. The denominator of $Y_1(s) = (s+1)(s^2 + 5.1s + 0.5)$ so $\gamma = 5.1$, $\delta = 0.5$

so that $$k_{01} = \frac{0.5}{0.325} = 1.54$$

and $$k_{21} = 5.1 - 0.325 - 1.54 = 3.235.$$

(c). $k_{13} = 0.1$. The denominator of $Y_1(s) = (s + 0.1)(s^2 + 6s + 5)$ so $\gamma = 6$, $\delta = 5$

so that $$k_{01} = \frac{5}{0.325} = 15.4$$

and $$k_{21} = 6 - 0.325 - 15.4$$

which is negative, so that this case is inadmissible.

The two possible sets of values for the rate constants are thus

$$k_{13} = 5, \quad k_{12} = 0.325, \quad k_{21} = 0.467, \quad k_{01} = 0.308$$

and $$k_{13} = 1, \quad k_{12} = 0.325, \quad k_{21} = 3.235, \quad k_{01} = 1.54.$$

Case (ii). $k_{01} = 0$. From eqns (6.64a) and (6.64b), $k_{21} = \gamma - \alpha$.

From eqn (6.64c), $$k_{02} = \frac{\delta}{k_{21}}$$

and from eqn (6.64a), $$k_{12} = \alpha - k_{02}.$$

Again considering the three possible values for k_{13}:

(a). $k_{13} = 5$

$$k_{21} = 1.1 - 0.325 = 0.775$$

$$k_{02} = \frac{0.1}{0.775} = 0.129$$

$$k_{12} = 0.325 - 0.129 = 0.196.$$

(b) $k_{13} = 1$

$$k_{21} = 5.1 - 0.325 = 4.775$$

$$k_{02} = \frac{0.5}{4.775} = 0.209$$

$$k_{12} = 0.325 - 0.209 = 0.116.$$

(c) $k_{13} = 0.1$

$$k_{21} = 6 - 0.325 = 5.675$$

$$k_{02} = \frac{5}{5.675} = 0.881$$

$$k_{12} = 0.325 - 0.881$$

which is inadmissible.

The two possible sets of values for the rate constants are thus

$$k_{13} = 5, \quad k_{12} = 0.196, \quad k_{21} = 0.775, \quad k_{02} = 0.129$$

and

$$k_{13} = 1, \quad k_{12} = 0.116, \quad k_{21} = 4.775, \quad k_{02} = 0.209.$$

We have seen in both cases of the above Example that it is inadmissible for k_{13} to be the smallest of the decay constants, since one of the other rate constants then turns out to be negative. Inadmissible cases for this model have been tabulated by Chau (1977), who gives the impulse response in the form $y_1(t) = A\,e^{-\alpha t} + B\,e^{-\beta t} + E\,e^{-\gamma t}$, where $A + B + E = 0$, with the decay constants ordered $\gamma > \beta > \alpha$. Then $A > 0$, otherwise the impulse response will be negative for large values of time. The condition $B > 0$, $E > 0$ is always inadmissible since $A + B + E = 0$, and the following combinations of B and E are also inadmissible:

if $k_{13} = \alpha$, $B < 0$, $E > 0$ and $B > 0$, $E < 0$

if $k_{13} = \beta$, $B < 0$, $E > 0$

if $k_{13} = \gamma$, $B < 0$, $E > 0$ and $B < 0$, $E < 0$.

In Example 6.8, using Chau's terminology, $A = 0.088$, $B = 0.320$ and $E = -0.408$, so since $B > 0$ and $E < 0$, $k_{13} = \alpha$ $(=0.1)$ is inadmissible. From the list of inadmissible combinations of B and E, it is seen that there are no admissible values of k_{13} if $B < 0$, $E > 0$ and, as seen above, if $B > 0$, $E > 0$. If $B > 0$, $E < 0$, k_{13} can only be β or γ, while if $B < 0$, $E < 0$, k_{13} can only be α or β.

6.3.6 *Catenary model with unidirectional exchange: the effect of input and observation gains on identifiability*

Several of the examples described in this chapter have illustrated how knowledge of input and observation gains can affect identifiability when input and observed compartments are different. This point is particularly well illustrated when considering the three-compartment catenary model with unidirectional exchange, shown in Fig. 6.7. If the input is to compartment 1 only, the transfer functions between input and observations of the three compartments are:

$$G_{11}(s) = \frac{c_1 b_1}{s + k_{01} + k_{21}} \tag{6.66a}$$

$$G_{21}(s) = \frac{c_2 b_1 k_{21}}{(s + k_{01} + k_{21})(s + k_{02} + k_{32})} \tag{6.66b}$$

$$G_{31}(s) = \frac{c_3 b_1 k_{21} k_{32}}{(s + k_{01} + k_{21})(s + k_{02} + k_{32})(s + k_{03})}. \tag{6.66c}$$

If all the compartments can be observed and c_2, c_3 and b_1 are known, then all the rate constants can be identified uniquely from successive denominator and numerator terms, but if c_2, c_3 and b_1 are not known, only parameter combinations $(k_{01} + k_{21})$, $(k_{02} + k_{32})$ and k_{03} can be identified. If, on the

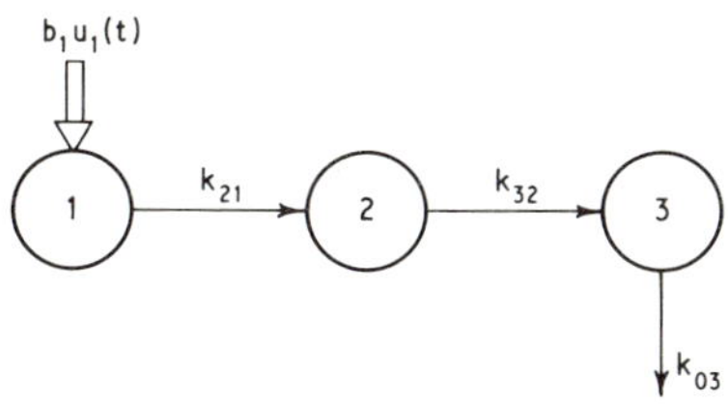

Figure 6.7 Catenary three-compartment model with unidirectional exchange between compartments.

other hand, only compartment 3 is observed, the same parameter combinations are locally identifiable with six solutions, corresponding to the number of ways in which they can be assigned to the three roots of the experimental transfer function; the product $k_{21}k_{32}$ can also be identified (uniquely) if c_3b_1 is known.

6.4 Systems with more than three compartments

Laplace transform identifiability analysis becomes very laborious for most systems with more than three compartments. Redundancies are often difficult to spot and it is usually far from clear how to proceed with a trial solution. This is not true for all systems, though, and some results have been obtained for two systems of considerable practical importance—mammillary systems and catenary systems (Cobelli, Lepschy and Romanin-Jacur, 1979b).

As mentioned earlier, these systems are no longer structurally identical when the number of compartments exceeds three. The results are for closed or almost closed (excretion from one compartment only) systems in which only one compartment is perturbed and the same compartment (only) is observed. The approach taken is to derive iterative solutions for the two new rate constants as each new compartment is added to the system. Their results are summarized below.

(*a*) *Mammillary system with p compartments.* If the perturbed and observed compartment is the central compartment (no. 1), then, for the closed system, all the rate constants are locally identifiable with $(p-1)!$ solutions, the number of solutions corresponding to arbitrary assignment of the rate constant pairs to the $(p-1)$ peripheral compartments. For the almost closed system, the rate constants are again locally identifiable with $(p-1)!$ solutions with the exception that if the excretion is from the central compartment, then k_{01} is uniquely identifiable.

If the perturbed and observed compartment is one of the peripheral compartments, then the closed or almost closed system is locally identifiable with $(p-2)!$ solutions (again with the exception that if the non-zero elimination rate constant is from the perturbed and observed compartment, then that rate constant is uniquely identifiable).

(*b*) *Catenary system with p compartments.* If the perturbed and observed compartment is one of the extremal compartments (1 or p), the rate constants are uniquely identifiable for the closed or almost closed system. If the perturbed and observed compartment is intermediate, the rate constants are locally identifiable, the number of solutions increasing as the distance from the nearer extremal compartment increases.

In the absence of prior knowledge of rate constants, neither mammillary nor catenary systems are identifiable from this type of experiment if there is excretion from more than one compartment.

7 Parameter Bounds for Unidentifiable Compartmental Models

7.1 How bounds arise

We have seen in Chapter 6 that many compartmental models of practical importance are unidentifiable from a particular single-input, single-output experiment, but that they can often be made identifiable if it is possible to perturb more than one input or to observe more than one compartment, or if there is some prior knowledge of the rate constants. In many experimental situations, though, it is not possible to perturb more than one input or to observe more than one state, nor does reliable prior knowledge exist of any rate constants. Then, such important models as the general two-compartment model and the three-compartment mammillary model are unidentifiable.

However, for a compartmental system, the elements of **A** are constrained. Specifically,

$$a_{ij} = k_{ij} \geqslant 0, \qquad i \neq j \tag{7.1}$$

$$a_{ii} = -\left(k_{0i} + \sum_{\substack{j=1 \\ j \neq i}}^{p} k_{ji}\right) \leqslant 0. \tag{7.2}$$

The latter inequality can be expressed as

$$a_{ii} \leqslant -\sum_{\substack{j=1 \\ j \neq i}}^{p} k_{ji} \tag{7.3}$$

which is a tighter constraint if there is any information that any of the k_{ji} is non-zero. These limitations on the possible values of the elements of **A** mean that the rate constants in an unidentifiable model must lie within a finite range rather than the range 0 to ∞ expected from the definition of a rate constant.

The idea that rate constant values are bounded was originated by Berman and Schoenfeld (1956). Using a Laplace transform approach, Skinner *et al.* (1959) derived bounds for certain parameters of a mammillary three-compartment model with simultaneous impulsive perturbation of the central compartment and one of the peripheral compartments. Little then seems to have appeared on bounds for many years. Although Rubinow and Winzer (1971) showed that many compartmental model parameters are bounded, and obtained the possible ranges using a normal-mode approach, the emphasis in their paper was on determining the set of coefficient matrices appearing in the compartmental equations which are compatible with the part of the solution that is observed. It was not until 1977 that a paper specifically concerned with bounds appeared, dealing with the general two-compartment model with impulsive input and with first-order input and using a modal approach (Chau, 1977). The lack of physiology attached to many linear compartmental models in pharmacokinetics provided the motivation for Chau's work. DiStefano and his co-workers have derived bounds for the thyroid kinetic model of Fig. 4.22 by writing down all known parameter combinations and inequalities which apply. They determined bounds for thyroid hormone kinetic studies in sheep (Wilson *et al.*, 1977) and more recently refined and extended the analysis and applied it in a series of physiological experiments in the rat (DiStefano *et al.*, 1982).

In this chapter, we will consider the general two-compartment model with impulsive input, the same model with first-order input and the three-compartment mammillary model. In each case, Laplace transforms will be used to give the bounds in terms of the amplitudes and decay constants of a multi-exponential fit, so allowing direct comparison with the identifiability analysis of Chapter 6.

7.2 The general two-compartmental model (Fig. 7.1)

Consider a single-input, single-ouput experiment with compartment 1 perturbed and observed (cf. Section 6.2.1). From eqns (6.10), the measured transfer function

$$G_{11}(s) = \frac{c_1 b_1 (s - a_{22})}{s^2 - (a_{11} + a_{22})s + a_{11}a_{22} - a_{12}a_{21}} \tag{7.4}$$

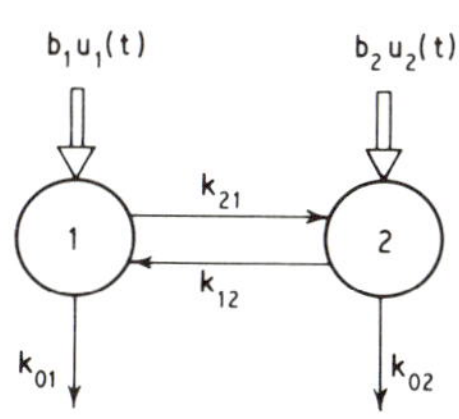

Figure 7.1 General two-compartment model.

and the parameter combinations a_{22}, a_{11} and $a_{12}a_{21}$ may be uniquely identified but no individual rate constant. From eqns (7.1) and (7.3), the identifiable quantities are bounded by

$$a_{11} \leqslant -k_{21} \tag{7.5}$$

$$a_{22} \leqslant -k_{12} \tag{7.6}$$

$$a_{12}a_{21} = k_{12}k_{21} \geqslant 0 \tag{7.7}$$

Let $k_{12}k_{21} = g$; the bounds on k_{21} can then be found as follows:

$$k_{21} = \frac{g}{-a_{22} - k_{02}} \geqslant -\frac{g}{a_{22}}$$

so that, *in terms of measurable quantities only*,

$$-\frac{g}{a_{22}} \leqslant k_{21} \leqslant -a_{11}. \tag{7.8}$$

Similarly,

$$-\frac{g}{a_{11}} \leqslant k_{12} \leqslant -a_{22}. \tag{7.9}$$

Further, since $k_{01} = -a_{11} - k_{21}$,

$$0 \leqslant k_{01} \leqslant -a_{11} + \frac{g}{a_{22}} \tag{7.10}$$

and similarly,

$$0 \leqslant k_{02} \leqslant -a_{22} + \frac{g}{a_{11}}. \tag{7.11}$$

The bounds are more usefully expressed in terms of the parameters describing a curve fit to a particular response. If the input is impulsive, for example, the measured response in the same compartment is:

$$y_1(t) = A\,\mathrm{e}^{-\alpha t} + B\,\mathrm{e}^{-\beta t} \tag{7.12}$$

where A, B, α and β are constants.

$$Y_1(s) = \frac{A}{s+\alpha} + \frac{B}{s+\beta}$$

$$= (A+B)\frac{s + \dfrac{A\beta + B\alpha}{A+B}}{s^2 + (\alpha+\beta)s + \alpha\beta}. \tag{7.13}$$

Comparing eqns (7.4) and (7.13),

$$-a_{22} = \frac{A\beta + B\alpha}{A+B}. \tag{7.14}$$

Since $-(a_{11} + a_{22}) = \alpha + \beta$,

$$-a_{11} = \frac{A\alpha + B\beta}{A + B} \tag{7.15}$$

and since $a_{11}a_{22} - k_{12}k_{21} = a_{11}a_{22} - g = \alpha\beta$,

$$\begin{aligned} g &= \frac{(A\alpha + B\beta)(A\beta + B\alpha) - (A + B)^2\alpha\beta}{(A + B)^2} \\ &= \frac{(A^2 + B^2)\alpha\beta + AB(\alpha^2 + \beta^2) - (A + B)^2\alpha\beta}{(A + B)^2} \\ &= \frac{AB(\alpha - \beta)^2}{(A + B)^2}. \end{aligned} \tag{7.16}$$

Thus, the bounds may be written in terms of A, B, α and β as follows:

From eqn (7.8) $$\frac{AB(\alpha - \beta)^2}{(A + B)(A\beta + B\alpha)} \leqslant k_{21} \leqslant \frac{A\alpha + B\beta}{A + B}. \tag{7.17}$$

From eqn (7.9) $$\frac{AB(\alpha - \beta)^2}{(A + B)(A\alpha + B\beta)} \leqslant k_{12} \leqslant \frac{A\beta + B\alpha}{A + B}. \tag{7.18}$$

From eqn (7.10)

$$0 \leqslant k_{01} \leqslant \frac{A\alpha + B\beta}{A + B} - \frac{AB(\alpha - \beta)^2}{(A + B)(A\beta + B\alpha)}.$$

$$\begin{aligned} \text{Upper limit} &= \frac{(A\alpha + B\beta)(A\beta + B\alpha) - AB(\alpha - \beta)^2}{(A + B)(A\beta + B\alpha)} \\ &= \frac{(A + B)^2\alpha\beta}{(A + B)(A\beta + B\alpha)} = \frac{(A + B)\alpha\beta}{A\beta + B\alpha} \end{aligned}$$

so

$$0 \leqslant k_{01} \leqslant \frac{(A + B)\alpha\beta}{A\beta + B\alpha}. \tag{7.19}$$

Similarly, from eqn (7.11)

$$0 \leqslant k_{02} \leqslant \frac{(A + B)\alpha\beta}{A\alpha + B\beta}. \tag{7.20}$$

Example 7.1. *An input of* 1 *unit is rapidly administered to one compartment of a general two-compartment system. Determine bounds on the rate constants when the normalized response of the same compartment is as follows:*

(i). $y_1(t) = 0.5(e^{-3t} + e^{-t})$ (cf. Example 6.1)
(ii). $y_1(t) = 0.9\,e^{-10t} + 0.1\,e^{-t}$.

Solution. Case (i). For the given impulse response, $A = B = 0.5$, $\alpha = 3$ and $\beta = 1$ so, substituting into eqns (7.17) to (7.20),

$$0.5 \leqslant k_{21} \leqslant 2$$

$$0.5 \leqslant k_{12} \leqslant 2$$

$$0 \leqslant k_{01} \leqslant 1.5$$

$$0 \leqslant k_{02} \leqslant 1.5.$$

Note that the bounds include the values $k_{21} = k_{12} = k_{01} = k_{02} = 1$ found in Example 6.1.

Case (ii). For the given impulse response, $A = 0.9$, $B = 0.1$, $\alpha = 10$, $\beta = 1$. Thus

$$3.84 \leqslant k_{21} \leqslant 9.1$$

$$0.80 \leqslant k_{12} \leqslant 1.9$$

$$0 \leqslant k_{01} \leqslant 5.26$$

$$0 \leqslant k_{02} \leqslant 1.10.$$

Before leaving this example, it is worthwhile noting that if there is prior knowledge that one of the rate constants has a value equal to one of the bounds, then the other rate constants also have values equal to one of their bounds. For example, if it were known in Case (ii) that there is no excretion from compartment 2, so that k_{02} is at its lower bound, then, substituting in eqns (6.18),

$$k_{12} = 1.9 \quad \text{(its upper bound)}$$

$$k_{21} = 3.84 \text{ (its lower bound)}$$

and

$$k_{01} = 5.26 \text{ (its upper bound)}.$$

This is a general result for a two-compartment system, from the form of the equations for the bounds.

7.3. Three-compartment model including a source compartment (Fig. 7.2)

Consider a single-input, single-output experiment with compartment 3 perturbed and compartment 1 observed. As we have noted in Section 6.3.5, this model represents a general two-compartment system (compartments 1

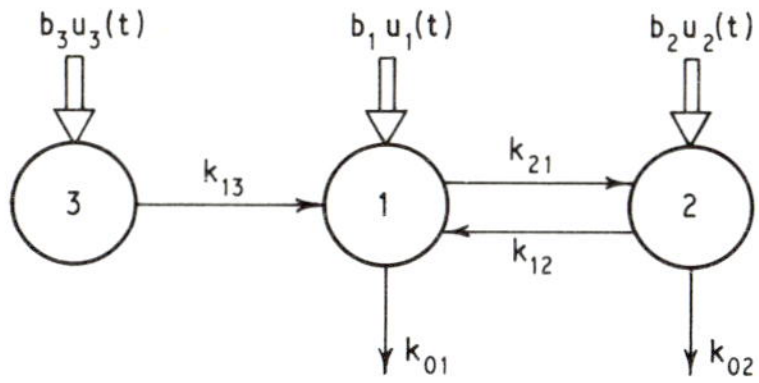

Figure 7.2 Three-compartment system including a source compartment.

and 2) with an input to compartment 1 from a source compartment, modelling an input of first-order form. The measured transfer function is, from eqn (6.61),

$$G_{13}(s) = \frac{c_1 b_3 k_{13}}{s + k_{13}} \frac{(s - a_{22})}{s^2 - (a_{11} + a_{22})s + a_{11}a_{22} - a_{12}a_{21}} \tag{7.21}$$

where a_{11}, a_{22} and $a_{12}a_{21}$ are bounded as in eqns (7.5)–(7.7). There is the additional constraint

$$k_{13} \geqslant 0. \tag{7.22}$$

With the input to compartment 3 of impulsive form (to achieve the first-order input to compartment 1),

$$y_1(t) = A\,\mathrm{e}^{-\alpha t} + B\,\mathrm{e}^{-\beta t} - (A + B)\,\mathrm{e}^{-k_{13}t} \tag{7.23}$$

where A, B, α and β are constants. Thus

$$Y_1(s) = \frac{A}{s + \alpha} + \frac{B}{s + \beta} - \frac{A + B}{s + k_{13}}$$

$$= \frac{\{[A(\beta + k_{13}) + B(\alpha + k_{13}) - (A + B)(\alpha + \beta)]s + A\beta k_{13} + B\alpha k_{13} - (A + B)\alpha\beta\}}{(s + k_{13})(s^2 + (\alpha + \beta)s + \alpha\beta)}. \tag{7.24}$$

Comparing the numerators of eqns (7.21) and (7.24),

$$A(\beta + k_{13}) + B(\alpha + k_{13}) - (A + B)(\alpha + \beta) = c_1 b_3 k_{13} \tag{7.25}$$

and

$$-a_{22} = \frac{A\beta k_{13} + B\alpha k_{13} - (A + B)\alpha\beta}{A(\beta + k_{13}) + B(\alpha + k_{13}) - (A + B)(\alpha + \beta)}$$

$$= \frac{A\beta(k_{13} - \alpha) + B\alpha(k_{13} - \beta)}{A(k_{13} - \alpha) + B(k_{13} - \beta)} \tag{7.26}$$

Equation (7.25) only proves useful if $c_1 b_3$ is known. In biomedical applications at least, it is unlikely that the input fraction will be known in this case. The form of eqn (7.26) is similar to that of eqn (7.14) and can be made even more similar in appearance if we define

$$L_1 \triangleq A(k_{13} - \alpha) \tag{7.27}$$

$$L_2 \triangleq B(k_{13} - \beta). \tag{7.28}$$

Then

$$-a_{22} = \frac{L_1\beta + L_2\alpha}{L_1 + L_2} \tag{7.29}$$

and since $-(a_{11} + a_{22}) = \alpha + \beta$ and $g = k_{12}k_{21} = a_{11}a_{22} - \alpha\beta$ as before,

$$-a_{11} = \frac{L_1\alpha + L_2\beta}{L_1 + L_2} \tag{7.30}$$

and

$$g = \frac{L_1 L_2(\alpha - \beta)^2}{(L_1 + L_2)^2}. \tag{7.31}$$

The bounds now follow by analogy to Section 7.2, with L_1 and L_2 replacing A and B in eqns (7.17) to (7.20). In the absence of prior knowledge of c_1b_3, the rate constant k_{13} is (in theory) locally identifiable with three solutions, so that three different cases have to be examined.

Example 7.2. *In the model of Fig. 7.2, an impulsive input of* 1 *unit is administered to compartment* 3, *and the impulse response in compartment* 1 *is* (*cf. Example 6.8*),

$$y_1(t) = -0.408\,\mathrm{e}^{-5t} + 0.320\,\mathrm{e}^{-t} + 0.088\,\mathrm{e}^{-0.1t}.$$

Determine bounds for the rate constants k_{21}, k_{12}, k_{01} *and* k_{02}.

Solution. Examining the three cases $k_{13} = 5$, 1 and 0.1 in turn, as in Example 6.8:

Case (i). $k_{13} = 5$ $\quad A = 0.320 \quad B = 0.088$

$\alpha = 1 \quad \beta = 0.1.$

From the definitions, $L_1 = 0.320(5 - 1) = 1.280$

$L_2 = 0.088(5 - 0.1) = 0.431.$

Substituting into eqns (7.17) to (7.20) replacing A by L_1 and B by L_2,

$$0.467 \leqslant k_{21} \leqslant 0.773$$

$$0.198 \leqslant k_{12} \leqslant 0.325$$

$$0 \leqslant k_{01} \leqslant 0.308$$

$$0 \leqslant k_{02} \leqslant 0.129.$$

Case (ii). With $k_{13} = 1$, $L_1 = 1.632$, $L_2 = 0.079$ and the rate constant bounds are:

$$3.235 \leqslant k_{21} \leqslant 4.775$$

$$0.221 \leqslant k_{12} \leqslant 0.325$$

$$0 \leqslant k_{01} \leqslant 1.54$$

$$0 \leqslant k_{02} \leqslant 0.105.$$

Case (iii). As we have already seen in Section 6.3.5, the case with $k_{13} = 0.1$ is inadmissible with the given form of impulse response. If the procedure for determining the rate constant bounds is followed, it is found that L_1 is positive and L_2 negative, which is inadmissible since from eqn (7.31),

$$g = \frac{L_1 L_2 (\alpha - \beta)^2}{(L_1 + L_2)^2}$$

and g $(= k_{12}k_{21})$ cannot be negative.

Comparison with the values found in Example 6.8 confirms that if there is prior knowledge that one of the rate constants equals one of its bounds, the others also take values equal to one of their bounds.

7.4 Three-compartment mammillary model (Fig. 7.3)

Consider a single-input, single-output experiment, with the central compartment perturbed and observed (cf. Section 6.3.2). The measured transfer function is, from eqns (6.45) and (6.46),

$$G_{11}(s) = \frac{c_1 b_1 (s - a_{22})(s - a_{33})}{\Delta(s)} \tag{7.32}$$

where
$$\begin{aligned}\Delta(s) = s^3 &- (a_{11} + a_{22} + a_{33})s^2 \\ &+ (a_{11}a_{22} + a_{22}a_{33} + a_{33}a_{11} - a_{12}a_{21} - a_{13}a_{31})s \\ &- a_{11}a_{22}a_{33} + a_{33}a_{12}a_{21} + a_{22}a_{13}a_{31}.\end{aligned} \tag{7.33}$$

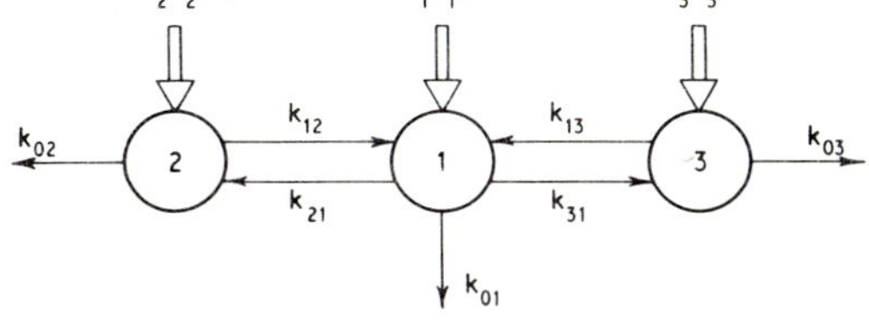

Figure 7.3 Mammillary model with three compartments.

From the numerator, a_{22} and a_{33} are locally identifiable with two solutions, then from the denominator, a_{11} is uniquely identifiable and $a_{12}a_{21}$ and $a_{13}a_{31}$ are locally identifiable with two solutions. The peripheral compartments are indistinguishable from this experiment. The analysis will now proceed as before, with the bounds on the rate constants being established in terms of the known parameter combinations listed above.

The relevant inequalities are, from eqn (7.3),

$$a_{11} \leqslant -(k_{21} + k_{31}) \tag{7.34}$$

$$a_{22} \leqslant -k_{12} \tag{7.35}$$

$$a_{33} \leqslant -k_{13} \tag{7.36}$$

and, from eqn (7.1),

$$a_{12}a_{21} = k_{12}k_{21} = g_1 \text{ (say) } \geqslant 0 \tag{7.37}$$

and

$$a_{13}a_{31} = k_{13}k_{31} = g_2 \text{ (say) } \geqslant 0. \tag{7.38}$$

The lower bounds on k_{21} and k_{31} are readily established as

$$k_{21} = \frac{g_1}{-a_{22} - k_{02}} \geqslant -\frac{g_1}{a_{22}} \tag{7.39}$$

and

$$k_{31} = \frac{g_2}{-a_{33} - k_{03}} \geqslant -\frac{g_2}{a_{33}}. \tag{7.40}$$

The upper bounds are then established using eqn (7.34):

$$-k_{21} \geqslant a_{11} + k_{31}$$

so

$$-k_{21} \geqslant a_{11} - \frac{g_2}{a_{33}}. \tag{7.41}$$

Similarly,

$$-k_{31} \geqslant a_{11} - \frac{g_1}{a_{22}}. \tag{7.42}$$

Thus the bounds on k_{21} and k_{31} are:

$$-\frac{g_1}{a_{22}} \leqslant k_{21} \leqslant -a_{11} + \frac{g_2}{a_{33}} \tag{7.43}$$

$$-\frac{g_2}{a_{33}} \leqslant k_{31} \leqslant -a_{11} + \frac{g_1}{a_{22}}. \tag{7.44}$$

The upper bounds on k_{12} and k_{13} are given by eqns (7.35) and (7.36). The lower bounds are established by proceeding along similar lines to above:

$$k_{12} = \frac{g_1}{k_{21}} \geqslant \frac{g_1}{-a_{11} + g_2/a_{33}}$$

and
$$k_{13} = \frac{g_2}{k_{31}} \geqslant \frac{g_2}{-a_{11} + g_1/a_{22}}$$

so the bounds on k_{12} and k_{13} are:

$$\frac{g_1}{-a_{11} + g_2/a_{33}} \leqslant k_{12} \leqslant -a_{22} \tag{7.45}$$

and
$$\frac{g_2}{-a_{11} + g_1/a_{22}} \leqslant k_{13} \leqslant -a_{33}. \tag{7.46}$$

Finally, since $k_{02} = -a_{22} - k_{12}$, $k_{03} = -a_{33} - k_{13}$ and $k_{01} = -a_{11} - k_{21} - k_{31}$, the bounds for the excretion rate constants are:

$$0 \leqslant k_{02} \leqslant \frac{g_1}{a_{11} - g_2/a_{33}} - a_{22} \tag{7.47}$$

$$0 \leqslant k_{03} \leqslant \frac{g_2}{a_{11} - g_1/a_{22}} - a_{33} \tag{7.48}$$

and
$$0 \leqslant k_{01} \leqslant \frac{g_1}{a_{22}} + \frac{g_1}{a_{33}} - a_{11}. \tag{7.49}$$

As before, the bounds are more usefully expressed in terms of parameters describing a curve fit to a particular response. For example, if the input is a unit impulse,

$$y_1(t) = A\,\mathrm{e}^{-\alpha t} + B\,\mathrm{e}^{-\beta t} + C\,\mathrm{e}^{-\gamma t} \tag{7.50}$$

where A, B, C, α, β and γ are constants,

$$Y_1(s) = \frac{A}{s+\alpha} + \frac{B}{s+\beta} + \frac{C}{s+\gamma}$$

$$= (A+B+C)\frac{s^2 + \dfrac{A(\beta+\gamma)+B(\gamma+\alpha)+C(\alpha+\beta)}{A+B+C}s + \dfrac{A\beta\gamma + B\gamma\alpha + C\alpha\beta}{A+B+C}}{s^3 + (a+\beta+\gamma)s^2 - (\alpha\beta+\beta\gamma+\gamma\alpha)s + \alpha\beta\gamma}. \tag{7.51}$$

By comparison with eqn (7.32),

$$-(a_{22}+a_{33})=\frac{A(\beta+\gamma)+B(\gamma+\alpha)+C(\alpha+\beta)}{A+B+C}=g_3 \text{ (say)} \quad (7.52)$$

and

$$a_{22}a_{33}=\frac{A\beta\gamma+B\gamma\alpha+C\alpha\beta}{A+B+C}=g_4 \text{ (say)} \quad (7.53)$$

so that a_{22} and a_{33} are the roots of a quadratic:

$$a_{22}, a_{33}=\tfrac{1}{2}[-g_3\pm(g_3^2-4g_4)^{1/2}]. \quad (7.54)$$

Comparing the denominators (eqns (7.33) and (7.51)),

$$-(a_{11}+a_{22}+a_{33})=\alpha+\beta+\gamma \quad (7.55)$$

$$a_{11}a_{22}+a_{22}a_{33}+a_{33}a_{11}-a_{12}a_{21}-a_{13}a_{31}=\alpha\beta+\beta\gamma+\gamma\alpha \quad (7.56)$$

$$-a_{11}a_{22}a_{33}+a_{12}a_{21}a_{33}+a_{13}a_{31}a_{22}=\alpha\beta\gamma. \quad (7.57)$$

From eqn (7.55)

$$\begin{aligned}-a_{11}&=a_{22}+a_{33}+\alpha+\beta+\gamma\\&=-g_3+\alpha+\beta+\gamma. \quad (7.58)\end{aligned}$$

Remaining terms g_1 and g_2 are found by simultaneous solution of eqns (7.56) and (7.57):

$$g_1+g_2=a_{11}a_{22}+a_{22}a_{33}+a_{33}a_{11}-(\alpha\beta+\beta\gamma+\gamma\alpha)$$

$$a_{33}g_1+a_{22}g_2=a_{11}a_{22}a_{33}+\alpha\beta\gamma$$

whence

$$\begin{aligned}g_1&=\frac{a_{11}a_{22}^2+a_{22}^2a_{33}+a_{33}a_{11}a_{22}-a_{22}(\alpha\beta+\beta\gamma+\gamma\alpha)-a_{11}a_{22}a_{33}-\alpha\beta\gamma}{a_{22}-a_{33}}\\&=\frac{a_{22}(\alpha\beta+\beta\gamma+\gamma\alpha)-a_{22}^2(a_{11}+a_{33})+\alpha\beta\gamma}{a_{33}-a_{22}} \quad (7.59)\end{aligned}$$

and

$$g_2=\frac{a_{33}(\alpha\beta+\beta\gamma+\gamma\alpha)-a_{33}^2(a_{11}+a_{22})+\alpha\beta\gamma}{a_{22}-a_{33}}. \quad (7.60)$$

Example 7.3. *In a three-compartment mammillary system, the normalized response in compartment* 1 *to a rapidly-administered input of* 1 *unit to compartment* 1 *is*

$$y_1(t)=0.7\,\mathrm{e}^{-5t}+0.2\,\mathrm{e}^{-t}+0.1\,\mathrm{e}^{-0.1t}$$

(*cf.* *Example* 6.5). *Find the bounds on the rate constants.*

Solution. $A = 0.7$, $B = 0.2$, $C = 0.1$, $\alpha = 5$, $\beta = 1$ and $\gamma = 0.1$.

From eqns (7.52) and (7.53),

$$g_3 = 2.39, \quad g_4 = 0.67.$$

From eqn (7.54),

$$\begin{aligned} a_{22}, a_{33} &= \tfrac{1}{2}[-2.39 \pm ((2.39)^2 - (4)(0.67))^{1/2}] \\ &= -0.325, \quad -2.065. \end{aligned}$$

From eqn (7.58), $$a_{11} = -3.71$$

From eqn (7.59),

$$g_1 = \frac{-(0.325)[(5)(1)+(1)(0.1)+(0.1)(5)]+(0.106)(3.71+2.065)+(5)(1)(0.1)}{-2.065+0.325}$$

$$=0.407.$$

From eqn (7.60)

$$g_2 = \frac{-(2.065)[(5)(1)+(1)(0.1)+(0.1)(5)]+(4.26)(3.71+2.065)+(5)(1)(0.1)}{2.065-0.325}$$

$$=3.52.$$

Substituting into eqns (7.43) to (7.49), the bounds on the rate constants are:

$$1.253 \leqslant k_{21} \leqslant 2.01$$

$$1.707 \leqslant k_{31} \leqslant 2.46$$

$$0.203 \leqslant k_{12} \leqslant 0.325$$

$$1.0 \leqslant k_{13} \leqslant 2.065$$

$$0 \leqslant k_{02} \leqslant 0.122$$

$$0 \leqslant k_{03} \leqslant 1.065$$

$$0 \leqslant k_{01} \leqslant 0.75.$$

Remember that compartments 2 and 3 cannot be distinguished from this experiment, so that a second set of bounds exists with k_{21} and k_{31}, k_{12} and k_{13} and k_{02} and k_{03} interchanged, but k_{01} unchanged.

The approach is equally applicable when there is some prior knowledge of the rate constants available, but not enough for the system to be identifiable. This will be illustrated with a further example.

Example 7.4. *In experiments on rats reported by DiStefano et al. (1982), measurements were taken of concentration of triiodothyronine* (T_3) *in the systemic plasma following rapid intravenous injection of* T_3. *For one particular rat, the impulse response was*

$$y_1(t) = 4.59\,\mathrm{e}^{-1.74t} + 0.32\,\mathrm{e}^{-0.031t} + 0.10\,\mathrm{e}^{-0.0017t}.$$

The model assumed was the three-compartment mammillary model (with input to, and observation of, the central compartment), with prior knowledge that there was no excretion from the central compartment ($k_{01} = 0$) *and a prior assumption that* $a_{22} < a_{33}$. *Find the bounds on the remaining six rate constants.*

Solution. As we have already seen in Section 6.3.2, the system remains unidentifiable from this experiment; prior knowledge of two rate constants is necessary for there to be any chance of identifying the remaining five. Bounds can still be established, though. Substituting in eqn (7.54), with the constraint that $a_{22} < a_{33}$,

$$a_{22} = -0.17, \quad a_{33} = -0.0079.$$

Then, substituting in eqns (7.58), (7.59) and (7.60),

$$a_{11} = -1.60, \quad g_1 = 0.22 \quad \text{and} \quad g_2 = 0.0016.$$

The bounds on the rate constants are then given by substitution into eqns (7.43) to (7.48):

$$1.29 \leqslant k_{21} \leqslant 1.40$$
$$0.20 \leqslant k_{31} \leqslant 0.31$$
$$0.158 \leqslant k_{12} \leqslant 0.17$$
$$0.0052 \leqslant k_{13} \leqslant 0.0079$$
$$0 \leqslant k_{02} \leqslant 0.012$$
$$0 \leqslant k_{03} \leqslant 0.0027.$$

7.5 The width of the bounds

In this chapter, bounds for the possible values of rate constants of three unidentifiable compartmental systems of considerable practical interest have been established. Bounds will always be somewhat less satisfactory than unique values, but they can nevertheless prove useful.

At worst, if the bounds are fairly wide (as in the two-exponential curves considered in Example 7.1) they may still provide grounds for acceptance or rejection of a postulated model. For example, in case (ii) of that example, if it were known that k_{21} could not possibly be as large as 3.84, there would be grounds for rejection of the general two-compartment model. At best, if the bounds are narrow, as in Example 7.4, they are virtually as good as having a unique value for each rate constant.

The relationship between width of bounds and the curve parameters is relatively complicated, particularly for the three-compartment model. Changing one of the curve parameters usually narrows some of the bounds and widens others. Further work needs to be done to relate values of the curve parameters to the widths of the bounds.

8 Some Practical Problems of Compartmental Modelling

8.1 Introduction

In this chapter, a number of practical problems which arise in describing systems by linear, time-invariant compartmental models will be discussed. In Section 8.2, problems due to imperfect input–output data will be described. The imperfections—noise, limited length of records and spacing (in time) of sampled observations—all have an effect on the accuracy with which parameters can be estimated and the problems will be lumped together under the title of numerical identifiability, the adjective being chosen to distinguish these problems from deterministic identifiability considered in Chapter 6, where perfect records were assumed. In Section 8.3, some recent work on optimal experiment design applied to compartmental modelling will be described. Finally, in Section 8.4, a different type of problem, that of choosing between models which are input–output equivalent, will be considered.

The relative importance of numerical identifiability and model selection depends on the purposes for which the model is intended. A wide range of purposes exists for mathematical modelling, from purely explanatory models which merely seek a conformation to input–output data (often referred to as black-box models) through to models which mirror the subsystems comprising the actual system as closely as possible and from which those parameters which cannot be measured directly can be estimated.

In black-box modelling (which is the approach used in many pharmacokinetic applications of compartmental models) model selection can present a considerable problem. For example, if an impulse response consists of two exponentials, indicating an open two-compartment model, prior knowledge of one of the rate constants must be assumed for the other three to be identified. This is usually done by assuming that excretion from one of the compartments is negligible, but the experimenter often has no basis on which to decide whether the excretion should be from the perturbed compartment or the unperturbed compartment. The accuracy with which compartmental parameters can be estimated is not a problem for this type of application, since its objective is to reflect input–output behaviour.

On the other hand, for models for which the structure has been postulated beforehand, as in most biomedical and ecosystem applications of compartmental modelling, imperfections in the data pose much more of a problem since they severely affect the accuracy with which some (but not necessarily all) of the model parameters can be estimated. The model now attempts to reflect the behaviour of the subsystems as well as the overall system, and some physical or physiological quantities estimated from the model may be considerably in error.

8.2 Numerical identifiability

8.2.1 *The nature of the experimental data*

As we have seen in Chapters 3 and 4, the impulse (and hence step) responses of linear, time-invariant compartmental models are multi-exponential. For a model with p compartments, none of which is a sink, there are generally p exponentials for an open system and $(p - 1)$ exponentials for a closed system. In biomedical applications, the inputs are usually impulses or steps and very often there is only one impulse response or step response to which the experimenter has to fit exponentials from which the compartmental model parameters (the rate constants) may be estimated. The curve fitting will normally be done on a digital computer, so that the experimental responses will be required in discrete (sampled) form; in most biomedical applications, they already consist of sets of samples.

It is therefore very pertinent to enquire how many exponentials can realistically be fitted to impulse or step response data. As we will see by the time we get to Section 8.2.7, it is rarely possible in practice to fit more than three exponentials. Just occasionally, four exponentials can be fitted if the responses are available over a very long time, are comparatively noise-free and the eigenvalues are well spaced. This limits the complexity of a compartmental model, since if the rate constants are being estimated on the basis of such single-input, single-output data alone, with no other prior knowledge, it is rarely possible to fit a model with more than three compartments (four if the model is closed). To fit a model with more compartments requires either prior knowledge of some of the parameters or the ability to isolate parts of the overall system and to perform experiments on these to determine their parameters. If neither of these is possible, it is necessary to reduce the model systematically, preserving the isomorphism with the system as far as possible. The usual technique adopted is to treat very rapid dynamic phenomena as pure gains and to ignore small amplitude slow phenomena.

In many applications in compartmental modelling, particularly in biomedicine, some limited model validation is possible through the area under the (impulse response) curve (AUC) and, if the perturbed and observed compartments are the same, the zero-time intercept ($x_1(0^+)$). For example, if an experiment is concerned with the kinetics of bilirubin, such as that described in Application Example 1 in Chapter 11, and the observation is of plasma clearance following rapid intravenous injection of labelled bilirubin, then the unconjugated bilirubin production rate (u.b.p.) is inversely proportional to AUC while the initial (or apparent) volume of distribution V_1 is inversely proportional to $x_1(0^+)$. Independent measurements of u.b.p. and V_1 are available to compare with those found from the compartmental modelling. In most of the examples in the remainder of this chapter, values of AUC and $x_1(0^+)$ will be quoted.

Most published examples showing the effects of measurement accuracy and noise on the estimation and decay constants of multi-exponential curves have been confined to examining specific curves; theoretical analysis has been largely absent. Although the asymptotic properties, i.e. the properties as the data sample becomes very large, of nonlinear least-squares estimators were discussed as long ago as 1969 by Jennrich, only very recently has attention been focussed on limited samples (Van den Bos, 1980, 1981). This theoretical work will be discussed briefly in Section 8.2.6.

8.2.2 *Measurement accuracy*

A sobering demonstration of the limitations of fitting exponentials to data was provided more than twenty-five years ago by Lanczos (1957, Chapter IV, Section 23). He considered records with a regular sampling interval h, and assumed a linear relationship between $(m + 1)$ successive values of the function $f(t)$:

$$c_0 f(t) + c_1 f(t + h) + c_2 f(t + 2h) + \ldots + c_m f(t + mh) = 0. \quad (8.1)$$

For a multi-exponential fit,

$$f(t) = A_1\,\mathrm{e}^{\lambda_1 t} + A_2\,\mathrm{e}^{\lambda_1 t} + \ldots + A_m\,\mathrm{e}^{\lambda_m t} \quad (8.2)$$

and by putting $\xi_i = \mathrm{e}^{\lambda_i h}$, an algebraic equation is obtained for the determination of the ξ_i:

$$c_0 + c_1\xi + c_2\xi^2 + \ldots + c_m\xi^m = 0. \quad (8.3)$$

Since both amplitudes and decay constants are to be calculated, the smallest number of function values needed is $2m$, and the c_i are then obtained by solving the set of linear equations:

$$
\begin{aligned}
y_1 c_0 + y_2 c_1 + \ldots + y_m c_{m-1} + y_{m+1} &= 0 \\
y_2 c_0 + y_3 c_1 + \ldots + y_{m+1} c_{m-1} + y_{m+2} &= 0 \\
\cdots\cdots\cdots\cdots\cdots\cdots\cdots\cdots & \\
y_m c_0 + y_{m+1} c_1 + \ldots + y_{2m-1} c_{m-1} + y_{2m} &= 0
\end{aligned}
\tag{8.4}
$$

where $y_1 \ldots y_{2m}$ are the values at successive sampling instants.

As the first example, Lanczos took a set of data obtained by sampling the function:

$$x_1(t) = 1.5576\,\mathrm{e}^{-5t} + 0.8607\,\mathrm{e}^{-3t} + 0.0951\,\mathrm{e}^{-t} \tag{8.5}$$

at regular intervals of 0.05 from $t = 0$ to $t = 1.15$ (24 samples in all). For the function used, $x_1(0^+) = 2.5134$ and AUC $= 0.6935$. The Laplace transform is

$$X_1(s) = \frac{2.5134(s + 3.725)(s + 1.111)}{(s + 5)(s + 3)(s + 1)}. \tag{8.6}$$

Note that there is quite close pole-zero cancellation.

To simulate the problems of measurement accuracy, the values used were limited to two decimal places, ranging from $y_1 = 2.51$ to $y_{24} = 0.06$. In the first of his demonstrations, Lanczos tried to fit a three-exponential curve to these data and as there are more data points than parameters to be estimated, the sum of four consecutive ordinates was used for the y's in eqns (8.4). He showed that, to within the limitations of the measurement accuracy, two of the eqns (8.4) are redundant, and reached the following very important conclusion. "It would be idle to hope that some other modified mathematical procedure could give better results, since the difficulty lies not with the manner of the evaluation but with the extraordinary sensitivity of the exponents and amplitudes to very small changes of the data which no amount of least-square or other form of statistics could remedy. The only remedy would be an increase of accuracy to limits which are far beyond the possibilities of our present measuring devices."

Lanczos then revised the problem and tried to fit a two-exponential curve to the data. Redundancy no longer occurred and the function fitted was

$$x_1(t) = 2.202\,\mathrm{e}^{-4.45t} + 0.305\,\mathrm{e}^{-1.58t} \tag{8.7}$$

(for which $x_1(0^+) = 2.507$ and AUC $= 0.6879$, both very similar to the true values). The Laplace transform of this response is

$$X_1(s) = \frac{2.507(s + 1.929)}{(s + 4.45)(s + 1.58)}.$$

Note that the zero and the poles are between those in eqn (8.6

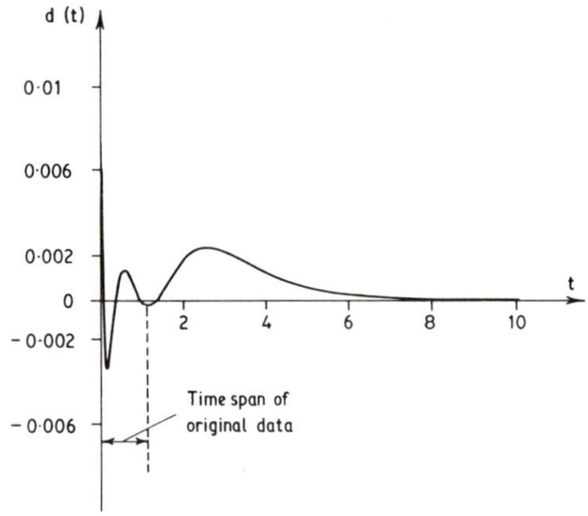

Figure 8.1 Difference $d_1(t)$, eqn (8.9), between original function (8.5) and two-exponential fit (8.7) of Lanczos.

The difference:

$$d_1(t) = 1.5576\,\mathrm{e}^{-5t} + 0.8607\,\mathrm{e}^{-3t} + 0.0951\,\mathrm{e}^{-t} - 2.202\,\mathrm{e}^{-4.45t} - 0.305\,\mathrm{e}^{-1.58t} \tag{8.9}$$

between the responses of eqns (8.5) and (8.7) is plotted against time in Fig. 8.1, which is extended some way beyond the time of the last sample to check that no large differences occur outside the original data range. It may be seen that the differences are very small, the maximum at the sampling instants being 0.0064.

Lanczos made a further attempt to fit a three-exponential curve to the original data by providing good estimates of the eigenvalues (-6, -2.7 and -1.2, compared with the true values of -5, -3 and -1) and using an iterative procedure to calculate first the λ's and then the A's. He obtained the following fit:

$$x_1(t) = 1.68\,\mathrm{e}^{-4.96t} + 0.79\,\mathrm{e}^{-2.73t} + 0.041\,\mathrm{e}^{-0.50t} \tag{8.10}$$

with Laplace transform:

$$X_1(s) = \frac{2.511(s + 3.496)(s + 0.548)}{(s + 4.96)(s + 2.73)(s + 0.50)}. \tag{8.11}$$

For this fit, $x_1(0^+) = 2.511$ (very close to its true value) and AUC $= 0.7101$ which is rather greater than the true value of 0.6935. The reason why is indicated in Fig. 8.2 which shows the difference:

$$d_1(t) = 1.5576\,\mathrm{e}^{-5t} + 0.8607\,\mathrm{e}^{-3t} + 0.0951\,\mathrm{e}^{-t} - 1.68\,\mathrm{e}^{-4.96t} - 0.79\,\mathrm{e}^{-2.73t} - 0.041\,\mathrm{e}^{-0.50t} \tag{8.12}$$

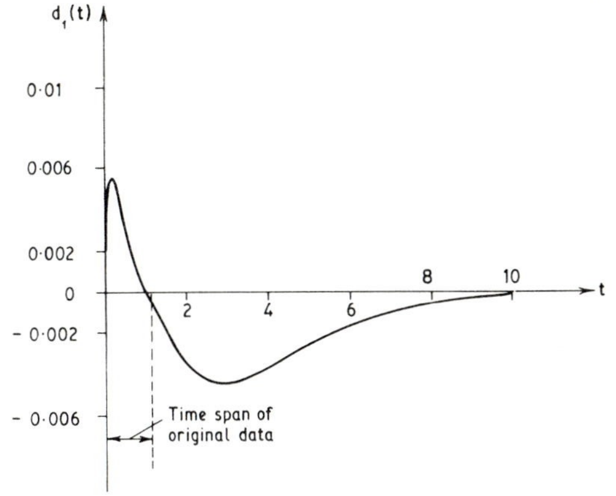

Figure 8.2 Difference $d_1(t)$, eqn (8.12), between original function (8.5) and three-exponential fit (8.10) of Lanczos.

between the responses of eqns (8.5) and (8.10). The agreement within the time range of the original data is closer than previously, but there is a larger magnitude difference outside the time range.

Rate constants for compartmental models corresponding to the responses of eqns (8.5), (8.10) and (8.7) are shown in Fig. 8.3. For the three-exponential responses, parameters have been determined for mammillary

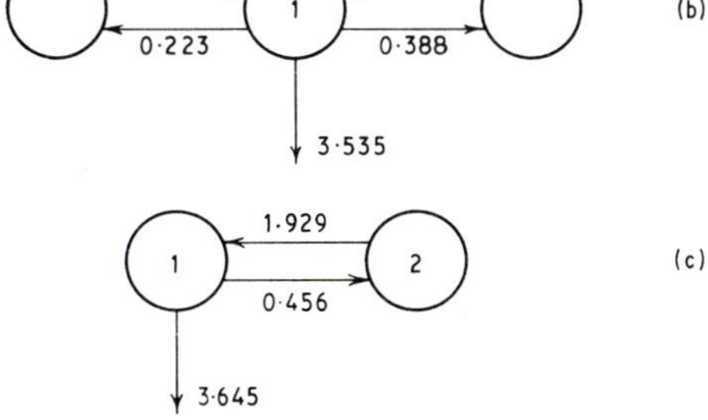

Figure 8.3 Compartmental models corresponding to original function and curve fits of Lanczos. (a) Mammillary three-compartment model, original function (8.5); (b) Mammillary three-compartment model, 3-exponential fit (8.10); (c) Two-compartment model, 2-exponential fit (8.7).

3-compartment models with excretion from the central (perturbed and observed) compartment only, while for the two-exponential response of eqn (8.7), parameters have been determined for a 2-compartment model with excretion only from the perturbed and observed compartment. (The choice of structure is arbitrary in all cases.) With the exception of the elimination rate constant, the parameters of the three-compartment models (Fig. 8.3(a) and (b)) are markedly different, despite the similarity of the impulse responses. The impulse response curves are very insensitive to large changes in some of the rate constants; some of the rate constants are very sensitive to small changes in the impulse responses. The elimination rate constant for the two-compartment model of Fig. 8.3(c) is similar to those of the three-compartment models.

8.2.3 *The effects of noise*

The diminishing difference in sum of squared error as noise increases, between models of different order has been illustrated by Finkelstein and Carson (1979) who used a standard nonlinear least-squares curve-fitting algorithm on a digital computer. Their data were constructed by sampling the response:

$$x_1(t) = e^{-0.01t} + e^{-0.005t} + e^{-0.001t} \tag{8.13}$$

at regular intervals of $t = 30$ from $t = 0$ to $t = 480$ (17 samples in all). For the above response, $x_1(0^+) = 3$ and AUC $= 1300$. The resulting data sequence is thus somewhat similar to that used by Lanczos (remembering the difference in scale of the decay constants and of the sampling interval), the main differences being that the decay constants in eqn (8.13) are more spread out than those in eqn (8.5), while in eqn (8.13), the amplitudes of the exponentials are all the same. The Laplace transform of eqn (8.13) is

$$X_1(s) = \frac{3(s + 0.00798)(s + 0.00272)}{(s + 0.01)(s + 0.005)(s + 0.001)} \tag{8.14}$$

and there is no particularly close pole-zero cancellation.

Using data correct to several decimal places, the computer program was able, in the noise-free case, to estimate amplitudes and exponents very close to the true values, but when 6% of noise was added to the data (the percentage referring to the root mean square percentage errors in the experimental sample of data points), a marked deterioration in the sum of squares occurred, the estimated curve being

$$x_1(t) = 0.956\, e^{-0.0115t} + 0.586\, e^{-0.00271t} + 1.377\, e^{-0.00223t} \tag{8.15}$$

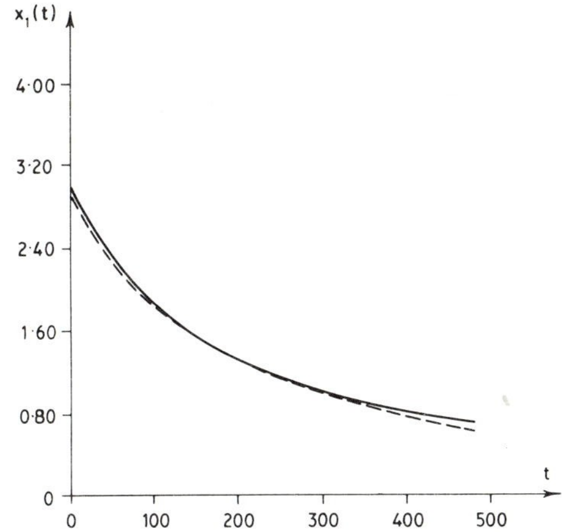

Figure 8.4 Unit impulse responses of models of Finkelstein and Carson. (———) Original function (8.13); (– – – –) 3-exponential fit (8.15).

for which $x_1(0^+) = 2.919$ and AUC = 917. The responses of eqns (8.13) and (8.15) are compared in Fig. 8.4 and the difference $d_1(t)$ given by

$$d_1(t) = e^{-0.01t} + e^{-0.005t} + e^{-0.001t} - 0.956\, e^{-0.0115t} - 0.586\, e^{-0.00271t} - 1.377\, e^{-0.00223t} \tag{8.16}$$

is shown in Fig. 8.5, which is continued well beyond the time range of the

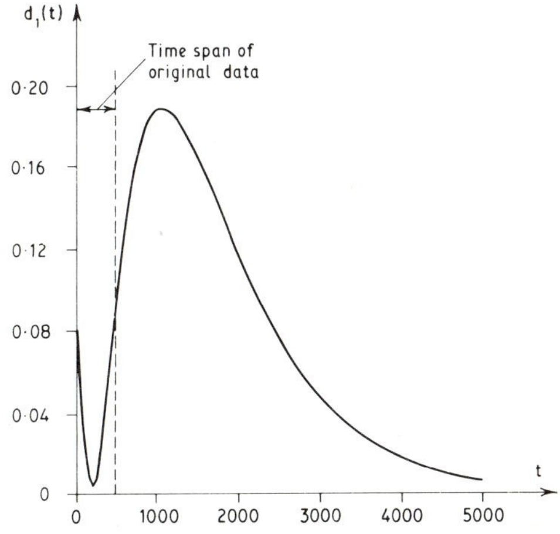

Figure 8.5 Difference $d_1(t)$, eqn (8.16), between original function (8.13) and three-exponential fit (8.15) of Finkelstein and Carson.

original data. The agreement in Fig. 8.4 seems very close, but there is a long-persisting error which is seen in Fig. 8.5 and which is the reason for the large discrepancy in AUC.

At least part of the problems with fitting this set of data is due to the set being rather short, in fact less than half the longest time constant (1000), but this is quite a common feature of many biomedical applications. Equation (8.15) immediately indicates that there are problems in the curve fitting, because two of the decay constants are very close to each other. This is also seen in the approximate pole-zero cancellation in the Laplace transform of eqn (8.15):

$$X_1(s) = \frac{2.919(s + 0.00855)(s + 0.00255)}{(s + 0.0115)(s + 0.00271)(s + 0.00223)}. \tag{8.17}$$

Rate constants for the three-compartment mammillary model with excretion from the central (perturbed and observed) compartment only and with impulse responses given by eqns (8.13) and (8.15) are shown in Fig. 8.6(a) and (b). Despite the closeness of the impulse responses within the time range of the data, there are substantial differences in some of the rate constants.

Finkelstein and Carson also fitted two-exponential curves to the data generated from eqn (8.13). In the noise-free case, the estimated curve was

$$x_1(t) = 1.203\,\mathrm{e}^{-0.0112t} + 1.808\,\mathrm{e}^{-0.00202t} \tag{8.18}$$

for which $x_1(0^+) = 3.011$, AUC = 1002 and

$$X_1(s) = \frac{3.011(s + 0.00753)}{(s + 0.0112)(s + 0.00202)}. \tag{8.19}$$

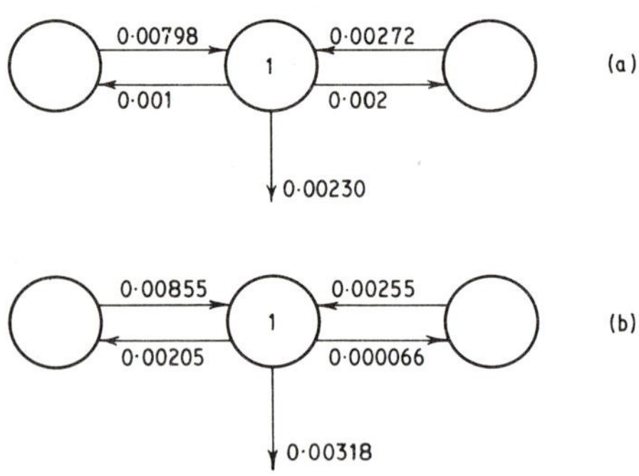

al ...ental models corresponding to original function and curve fit of ...on. (a) Mammillary three-compartment model, original function (8.13); ...e-compartment model, 3-exponential fit (8.15).

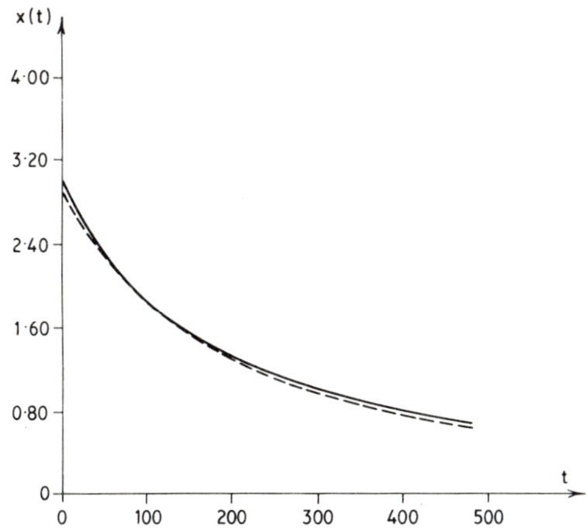

Figure 8.7 Unit impulse responses of models of Finkelstein and Carson. (———) 2-exponential fit (8.18) (noise-free data); (----) 2-exponential fit (8.20) (data with 6% noise added).

For the data set with 6% noise added, the estimated curve was

$$x_1(t) = 1.349\,\mathrm{e}^{-0.00861t} + 1.557\,\mathrm{e}^{-0.00188t} \tag{8.20}$$

for which $x_1(0^+) = 2.906$, AUC = 985 and

$$X_1(s) = \frac{2.906(s + 0.00549)}{(s + 0.00861)(s + 0.00188)}. \tag{8.21}$$

The responses of eqns (8.18) and (8.20) are compared in Fig. 8.7 and are again seen to be very similar over the time range of the data. The difference:

$$d_1(t) = 1.203\,\mathrm{e}^{-0.0112t} + 1.808\,\mathrm{e}^{-0.00202t} - 1.349\,\mathrm{e}^{-0.00861t} - 1.557\,\mathrm{e}^{-0.00188t} \tag{8.22}$$

is shown in Fig. 8.8 over an extended time range, from which it may be seen that the agreement between the two curves is closer outside the time span of the data than in the comparison between the three-exponential fits shown in Fig. 8.5. Rate constants for two-compartment models with excretion from the perturbed and observed compartment only and with impulse responses given by eqns (8.18) and (8.20) are shown in Fig. 8.9(a) and (b). The parameters change a good deal less than those of the three-compartment models of Fig. 8.6.

The sums of squared error computed by Finkelstein and Carson indicated that, in the noise-free case, the three-exponential fit was very much better

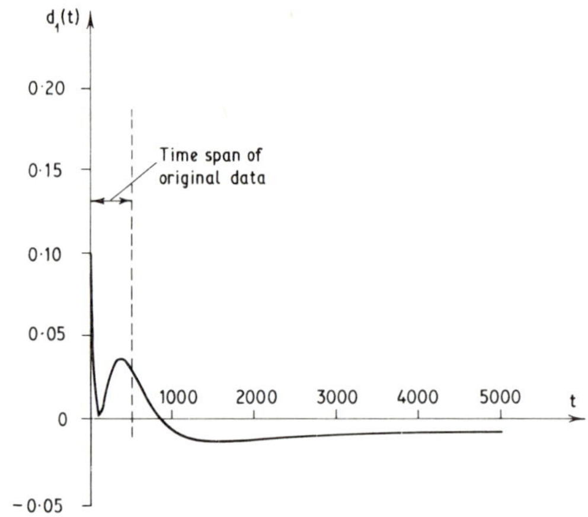

Figure 8.8 Difference $d_1(t)$, eqn (8.22), between two-exponential fits (8.18) and (8.20) of Finkelstein and Carson.

than the two-exponential fit, whereas with 6% noise added, the two fits were comparable. Such a noise level would not be uncommon in many physiological situations.

The presence of multiple minima in the criterion function (sum of squared error) being minimized can prove a considerable problem, particularly when three or more exponentials are being fitted, and it is essential in nonlinear curve fitting to use initial estimates of the parameters as close as possible to the true values. Good initial estimates can usually be obtained in the case of multi-exponential curves by plotting the data on semilogarithmic paper and curve peeling. Instructions for using most nonlinear curve fitting algorithms

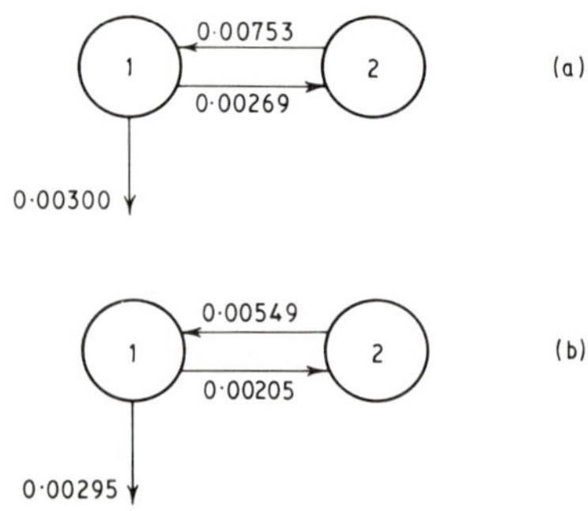

Figure 8.9 Two-compartment models corresponding to curve fits of Finkelstein and Carson. (a) 2-exponential fit (8.18); (b) 2-exponential fit (8.20).

recommend using several initial estimates in order to improve the chances of reaching the global minimum rather than stopping in a local minimum possibly well away from the true parameter values. Convergence, in a small number of iterations, to the same minimum from a wide selection of initial estimates is a useful indication that the model is well posed for the available data.

8.2.4 *Limited time range of data*

Even when response data are noise-free, erroneous results may be obtained if the responses are available over only a limited range of time. In a typical measurement of plasma clearance, for example, the first sample is usually taken several minutes after the injection of the tracer and the last several hours (but usually not days) after injection.

We will now examine three possible sources of error which may occur when the time range of the data is limited.

(a) Neglecting a fast transient. Consider a system whose impulse response is accurately described by

$$x_1(t) = 0.9\,e^{-10t} + 0.1\,e^{-t} \tag{8.23}$$

with $x_1(0^+) = 1$, AUC $= 0.19$ and

$$X_1(s) = \frac{s + 1.9}{(s + 10)(s + 1)}. \tag{8.24}$$

Suppose that measurement only starts some time after the impulsive input is applied. If measurement starts at $t = 0.5$, the first term in eqn (8.23) is only 10% of the second term (and gets smaller as t increases) so that the first term could well be omitted in the curve fitting and the curve

$$x_1(t) = 0.1\,e^{-t} \tag{8.25}$$

fitted to the data. The two fits are compared in Fig. 8.10. Neglecting the fast transient means that the zero time intercept is 0.1 instead of 1, so that the initial volume of distribution would be over-estimated by a factor of 10, and that AUC is 0.1, just over half its true value.

The difficulty of estimating fast transients when not measuring at small values of time has also been pointed out by Shipley and Clark (1972) who compared the following three curves:

Curve A

$$x_1(t) = 0.4\,e^{-0.5t} + 0.3\,e^{-0.2t} + 0.2\,e^{-0.1t} + 0.1\,e^{-0.01t} \tag{8.26}$$

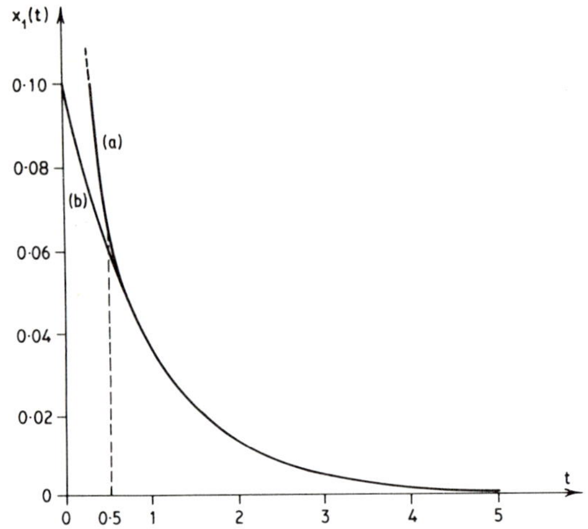

Figure 8.10 Effect of neglecting a fast transient. Impulse responses of (a) eqn (8.23); (b) eqn (8.25).

with $x_1(0^+) = 1$, AUC = 14.3 and

$$X_1(s) = \frac{(s + 0.362)(s + 0.138)(s + 0.0285)}{(s + 0.5)(s + 0.2)(s + 0.1)(s + 0.01)}. \tag{8.27}$$

Curve B

$$x_1(t) = 0.8\,\mathrm{e}^{-0.2t} + 0.1\,\mathrm{e}^{-0.1t} + 0.1\,\mathrm{e}^{-0.01t} \tag{8.28}$$

with $x_1(0^+) = 1$, AUC = 15.0 and

$$X_1(s) = \frac{(s + 0.112)(s + 0.0267)}{(s + 0.2)(s + 0.1)(s + 0.01)}. \tag{8.29}$$

Curve C

$$x_1(t) = 0.9\,\mathrm{e}^{-0.2t} + 0.1\,\mathrm{e}^{-0.01t} \tag{8.30}$$

with $x_1(0^+) = 1$, AUC = 14.5 and

$$X_1(s) = \frac{s + 0.029}{(s + 0.2)(s + 0.01)}. \tag{8.31}$$

Curves A and B are compared in Fig. 8.11 and curves A and C in Fig. 8.12. Shipley and Clark (1972) note that the curves are similar for $t > 10$ and that it would be difficult to distinguish between them if data were only available

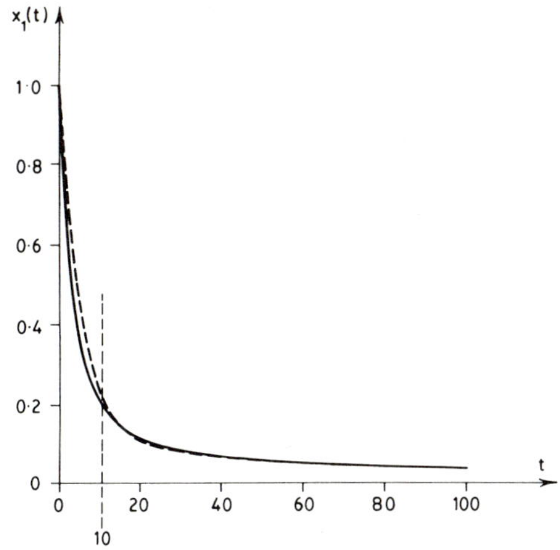

Figure 8.11 Unit impulse responses of models of Shipley and Clark. (———) 4-exponential curve (8.26); (– – – –) 3-exponential curve (8.28).

after $t = 10$. Rate constants for catenary models with compartment 1 perturbed and observed and excretion only from the other extreme compartment are shown in Fig. 8.13(a), (b) and (c) (corresponding respectively to the responses of eqns (8.26), (8.28) and (8.30)). The rate constants for transfer to and from compartment 1 are not very similar between

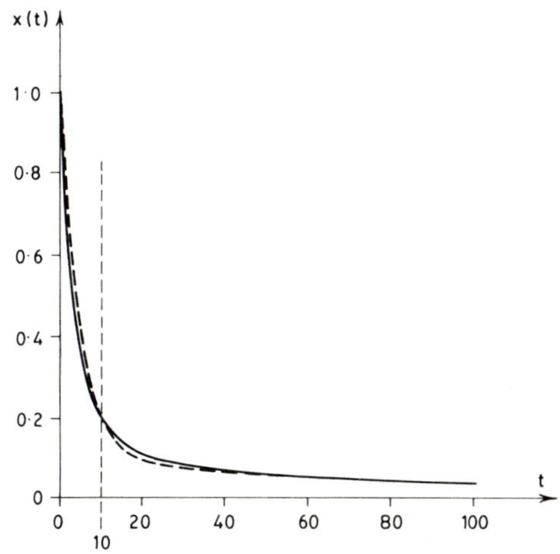

Figure 8.12 Unit impulse responses of models of Shipley and Clark. (———) 4-exponential curve (8.26); (– – – –) 2-exponential curve (8.30).

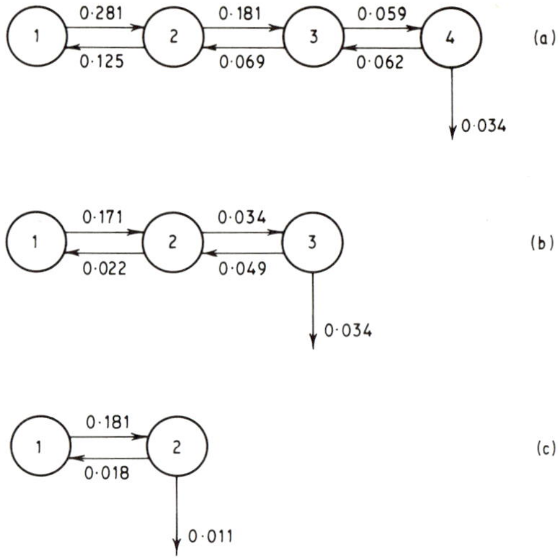

Figure 8.13 Compartmental models (catenary structure) corresponding to curve fits of Shipley and Clark. (a) 4-exponential curve (8.26); (b) 3-exponential curve (8.28); (c) 2-exponential curve (8.30).

the three models, even though, over the time range considered ($t > 10$), the responses are similar.

(b) Neglecting a small time delay. Consider again the response of eqn (8.23) but suppose that the system has a small time delay so that the true response is

$$x_1(t) = 0, \qquad 0 \leqslant t < 0.1 \tag{8.32a}$$

$$= 0.9\,\mathrm{e}^{-10(t-0.1)} + 0.1\,\mathrm{e}^{-(t-0.1)}, \qquad t > 0.1. \tag{8.32b}$$

If measurement does not begin until $t = 0.1$ (or later), the response is indistinguishable from

$$x_1(t) = 0.9(\mathrm{e}^{1})\,\mathrm{e}^{-10t} + 0.1(\mathrm{e}^{0.1})\,\mathrm{e}^{-t}$$

$$= 2.45\,\mathrm{e}^{-10t} + 0.11\,\mathrm{e}^{-t}, \qquad t > 0. \tag{8.33}$$

If this response is extrapolated back to zero, as in Fig. 8.14, the estimated zero-time intercept is 2.56, compared with the value of 1 for the response with no time delay (eqn 8.23)), so that the initial volume of distribution would be underestimated by a factor of 0.39. The AUC is 0.355, nearly double its true value.

Such an error could arise through incorrect timing of an impulsive input.

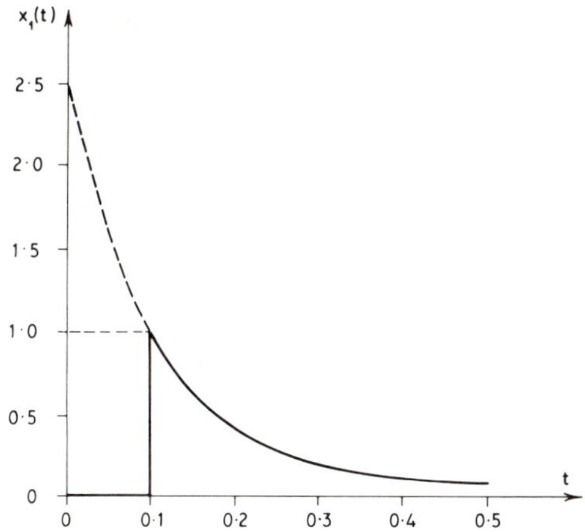

Figure 8.14 Effect of neglecting a small time delay. (———) Curve described by eqns (8.32); (– – – –) Curve of eqn (8.33) extrapolated back to $t = 0$.

(c) **Neglecting a small slow transient.** Premature termination of a response curve can lead to a small-amplitude slow transient being missed. This will be illustrated by considering the kinetics of bilirubin.

By taking response data over a period of at least 24 hours after rapid intravenous injection of radioactively-labelled bilirubin, Berk *et al.* (1969) found that the response curve was tri-exponential, a finding confirmed by Owens, Jones and Carson (1977) who took measurements up to 48 hours after injection. In another study, taking values over only a four-hour period after injection, Cobelli, Frezza and Tiribelli (1975) found that only two exponentials could be fitted to their plasma response curves, despite having a reasonable number of measurements within the 4-hour period.

To check on this, the author has generated a data sequence by sampling the averaged plasma response curve of Berk *et al.* (1969):

$$x_1(t) = 0.66\,\mathrm{e}^{-0.044t} + 0.31\,\mathrm{e}^{-0.010t} + 0.03\,\mathrm{e}^{-0.0012t} \tag{8.34}$$

(time in minutes) at the sample times used by Cobelli, Frezza and Tiribelli for the normal subject shown in their Fig. 3 ($t = 5, 10, 15, 20, 30, 45, 60, 90, 120, 180, 210$ and 240 min). Equation (8.34) is normalized so that $x_1(0^+) = 1$, while AUC $= 15 + 31 + 25 = 71$; note that the third term, while contributing only 3 % of $x_1(0^+)$, contributes over 35 % of AUC. Taking the (noise-free) data sequence correct to four decimal places, the following two-exponential response was fitted using a nonlinear curve-fitting algorithm of the Numerical Algorithms Group (NAG):

$$x_1(t) = 0.706\,\mathrm{e}^{-0.041t} + 0.291\,\mathrm{e}^{-0.00744t} \tag{8.35}$$

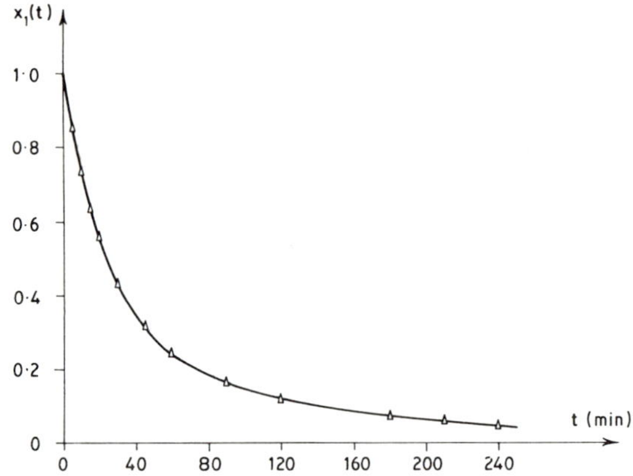

Figure 8.15 Data sequence generated from 3-exponential curve (8.34) and 2-exponential fit (8.35) up to $t = 240$ min.

for which $x_1(0^+) = 0.997$ and AUC = 56.3. Attempts to fit a third exponential resulted in the same two exponentials plus a very small amplitude, rapidly decaying exponential making no difference, to four significant figures, to the sum of squared error. The third exponential was not a slowly-decaying one, as might have been expected. The agreement between eqn (8.35) and the data sequence is excellent (see Fig. 8.15).

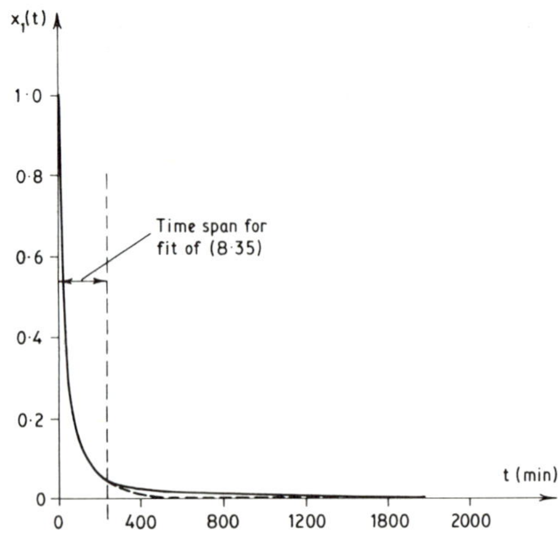

Figure 8.16 Unit impulse responses of eqns (8.34) and (8.35) plotted up to $t = 2000$ min. (———) 3-exponential curve (8.34); (– – – –) 2-exponential fit (8.35).

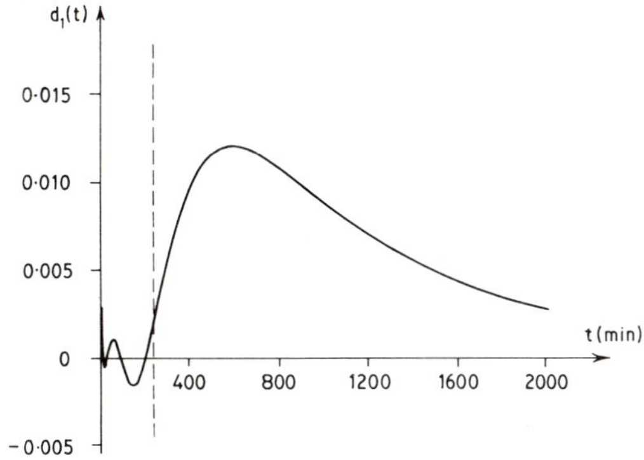

Figure 8.17 Difference $d_1(t)$, eqn (8.36), between 3-exponential curve (8.34) and 2-exponential curve (8.35).

The reason for the discrepancy in AUC becomes clear when comparing the responses eqns (8.34) and (8.35) in Fig. 8.16 and examining, in Fig. 8.17, the difference

$$d_1(t) = 0.66\,e^{-0.044t} + 0.31\,e^{-0.010t} + 0.03\,e^{-0.0012t} - 0.706\,e^{-0.041t} - 0.291\,e^{-0.00744t}. \qquad (8.36)$$

Both figures are extended well beyond the time span of the data sequence and it may be seen that, while there is good agreement within the four-hour period, there is a small, but slowly decaying difference beyond it.

Berk *et al.* proposed a three-compartment mammillary model to describe their experiment. The central (perturbed and observed) compartment (no. 1) contains the plasma and one of the peripheral compartments (no. 2) represents the liver. Elimination, assumed to be by conjugation, is from compartment 2 only. Rate constants for such a model with response given by eqn (8.34) are shown in Fig. 8.18(a). (The rate constants are slightly different from those given by Berk *et al.* who quoted averages for each rate constant for models corresponding to the 19 individual responses in their study, rather than the rate constants given here for a model of the averaged response.) Cobelli, Frezza and Tiribelli proposed a similar structure without the second peripheral compartment (no. 3), and rate constants for such a model with response given by eqn (8.35) are shown in Fig. 8.18(b). Despite the similarity of the responses within the data range, there are noticeable differences in the rate constants k_{12}, k_{21} and k_{02} between the two models.

It is also interesting to note the differences between the two models of the

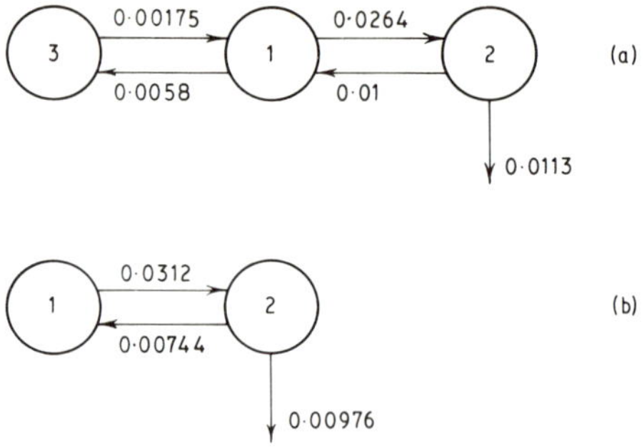

Figure 8.18 Compartmental models corresponding to eqns (8.34) and (8.35). (a) Mammillary three-compartment model, 3-exponential curve (8.34); (b) Two-compartment model, 2-exponential fit (8.35).

material flows. Taking elimination of bilirubin ($k_{02}x_2(t)$) as an example, the model of Fig. 8.18(a) gives:

$$\text{Elimination rate} = k_{02}x_2(t)$$
$$= 0.0113(-0.767\,e^{-0.044t} + 0.728\,e^{-0.010t} + 0.039\,e^{-0.0012t}) \tag{8.37}$$

while the model of Fig. 8.18(b) gives:

$$\text{Elimination rate} = k_{02}x_2(t)$$
$$= 0.00976[0.93(-e^{-0.041t} + e^{-0.00744t})]. \tag{8.38}$$

These are shown in Fig. 8.19, from which it may be seen that there are noticeable differences both within and beyond the time span of the original data. (In both cases, the total quantity eliminated, $\int_0^\infty k_{02}x_2(t)\,dt$, is equal to 1.)

The example has shown that, while Cobelli, Frezza and Tiribelli are correct in their conclusion that with data limited to a four-hour period after injection a plasma bilirubin clearance curve can be fitted by only two exponentials, limiting the data in this way leads to small but noticeable differences in material flows and to significant underestimation of AUC. We have already noted in Section 8.2.1 that unconjugated bilirubin production rate (u.b.p.) is inversely proportional to AUC, so that, using the fit of eqn (8.35), u.b.p. would be overestimated by 26% if we assume that u.b.p. estimated from eqn (8.34) is correct.

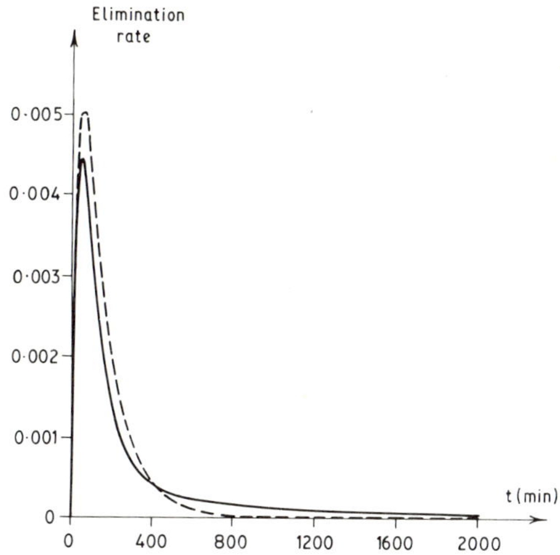

Figure 8.19 Elimination rates $k_{02}x_2(t)$ of models of Fig. 8.18. (———) 3-compartment model; (– – – –) 2-compartment model.

8.2.5 *The effects of poorly spaced samples*

Even when the overall time span of the data covers the complete response, the spacing of observations within the sample may not be adequate to allow all parameters to be estimated with reasonable accuracy. A particular example where this frequently occurs is in pharmacokinetics trials reporting plasma drug concentration following extravascular administration where, all too often, insufficient observations are made on the rising part of the plasma response curve to allow accurate estimation of dead time and the most rapid time constant.

This will now be illustrated with two examples from pharmacokinetics studies with which the author has been associated. In each case, a typical response has been selected at random; the data are listed in Table 8.1. For both responses, it was required to fit the simplest two-exponential response compatible with the data:

$$x(t) = 0, \qquad 0 \leqslant t < \tau \qquad (8.39a)$$

$$= A[\exp(-\alpha(t-\tau)) - \exp(-\beta(t-\tau))], \qquad t > \tau. \qquad (8.39b)$$

Prior information was available that the slower decay (α) was associated with drug elimination (so resolving the local identifiability problem). The faster decay and the dead time τ then model the "absorption phase" which, in both cases, includes tablet dissolution, passage of the drug to the gastro-intestinal

TABLE 8.1 Data sets discussed in Section 8.2.5

Drug A		Drug B	
t (h)	C (μg ml^{-1})	t (h)	C (μg ml^{-1})
0	0	0	0
0.5	0.25	0.167	0
1.0	0.85	0.333	1.5
2.0	1.03	0.5	6.1
3.0	0.57	0.75	6.1
4.0	0.41	1.0	11
6.0	0.25	2.0	14
9.0	0.09	3.0	16
12.0	0.06	4.0	15
15.0	0.03	6.0	12
18.0	0.00	8.0	10
		10.0	8.4
		12.0	7.9
		15.0	5.9
		24.0	2.8
		30.0	1.4
		36.0	0.9
		48.0	0.4

tract, absorption through the gut wall and transport via the liver to the systemic plasma measuring site. We will now consider the findings for each drug in turn.

Drug A. There are only nine non-zero values, only two of which are on the rise to the peak. In the curve-fitting, values of A, α and β were computed for fixed values of τ supplied by the user, the program minimizing

$$\mathrm{SSQ} = \sum_i (y(t_i) - x_1(t_i))^2 \tag{8.40}$$

(where $y(t_i)$ is the data point at $t = t_i$ and $x_1(t_i)$ is the value of $x_1(t)$ at $t = t_i$, computed from eqns (8.39)) with respect to A, α and β. Graphs of SSQ, A and α and β versus τ are shown in Figs 8.20, 8.21 and 8.22. Minimum SSQ of 0.02975 occurred at $\tau = 24$ min (0.4 h), with $A = 2.054$, $\alpha = 0.4218$ h^{-1} and $\beta = 1.771$ h^{-1}.

The first feature to note is that, near the minimum, changing τ has little effect on α, the smaller decay constant (larger time constant) which is associated with drug elimination and whose value affects the shape of the curve as t becomes large. However, τ has a significant effect on both β, the larger decay constant, associated with the "absorption phase" and A. In turn

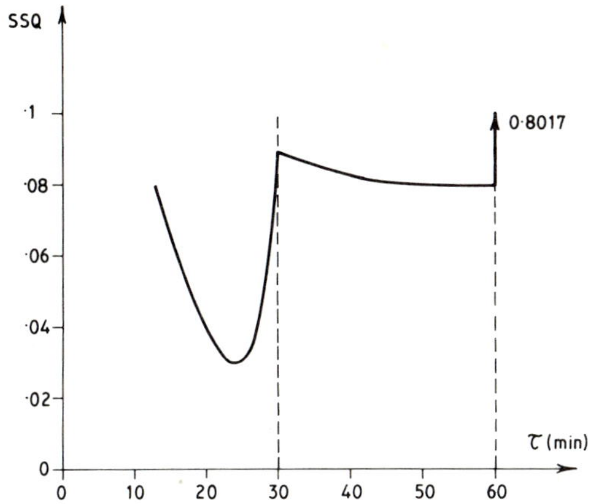

Figure 8.20 Sum of squared error SSQ, eqn (8.40), v. time delay τ for Drug A data.

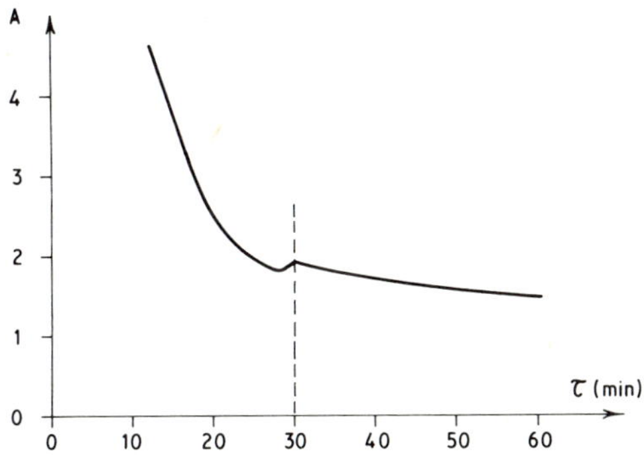

Figure 8.21 Amplitude A, eqn (8.39b), v. τ for Drug A data.

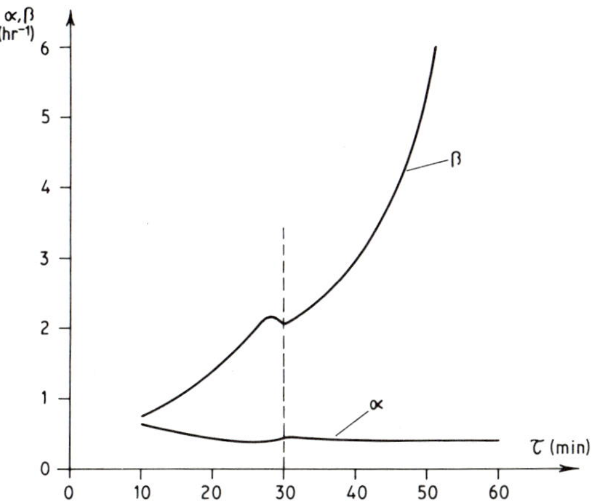

Figure 8.22 Decay constants α and β, eqn (8.39b), v. τ for Drug A data.

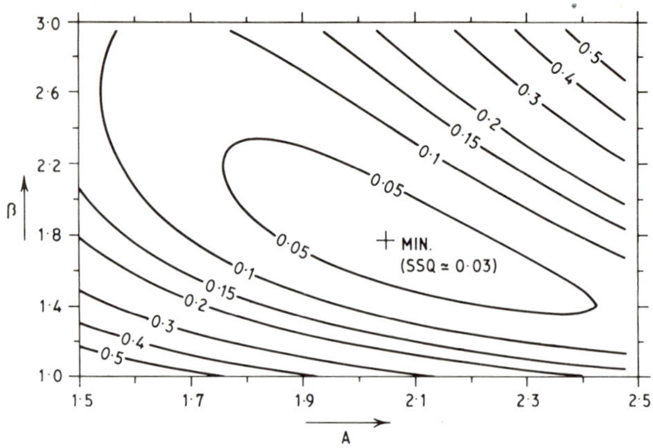

Figure 8.23 Contours of SSQ for varying β and A for Drug A data. ($\tau = 0.4$ h, $\alpha = 0.4218$ h^{-1}).

with τ fixed, changing α has considerable effect on SSQ, but changing β and A has less effect. This is illustrated in Fig. 8.23 which shows contours of SSQ for varying β and A with τ and α fixed at their optimum values of 0.4 h and 0.4218 h^{-1} respectively. The optimum response is compared with the data in Fig. 8.24 from which it may be seen that a function of the form of eqns (8.39) does not fit this data set particularly well.

The second feature of interest is the non-monotonic variation of the parameters as τ is varied, which is brought about by the paucity of the data. When τ is increased beyond 30 min (0.5 h), $x_1(0.5) = 0$ so that the data point $y(0.5) = 0.25$ contributes 0.0625 to SSQ, while when τ is increased beyond 60 min, the data point $y(1.0)$ contributes a further $(0.85)^2 = 0.7225$ to SSQ. As τ approaches 60 min, the curve fitted becomes more nearly first-order with β becoming very large. The fit for $\tau = 59$ min is shown in Fig. 8.25.

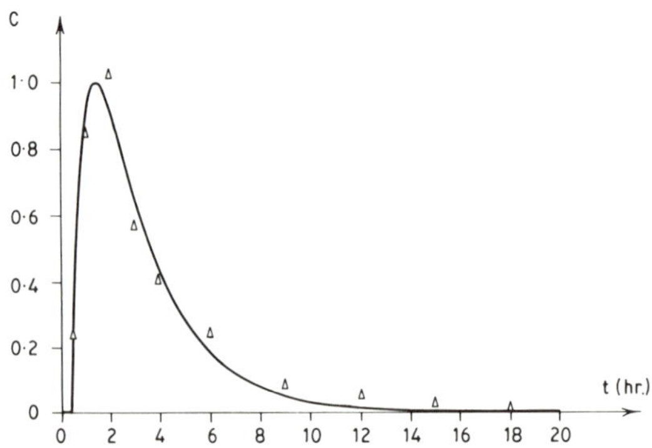

Figure 8.24 Optimum curve fit of form of eqns (8.39) to Drug A data.

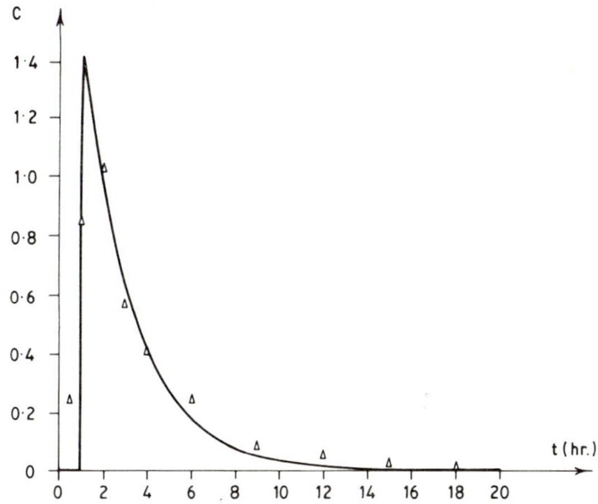

Figure 8.25 Curve fit of form of eqns (8.39) to Drug A data, with $\tau = 59$ min.

Drug B. The second data set has sixteen non-zero values, including five on the rise to the peak and a further zero value at $t = 10$ min which helps with the estimation of the larger decay constant. Minimum SSQ occurs at $\tau =$ 14 min (0.2333 h), at which $A = 20.89$, $\alpha = 0.0883\ \text{h}^{-1}$ and $\beta =$ $1.051\ \text{h}^{-1}$. Graphs of SSQ, A and α and β against τ are shown in Figs 8.26, 8.27 and 8.28. The response of the best fit to the data is shown in Fig. 8.29, from which it may be seen that the fit is reasonably good. It is interesting to note that even with this data set (comparatively large in terms of pharmacokinetic experiments), if τ had been taken as 10 min, SSQ would have increased

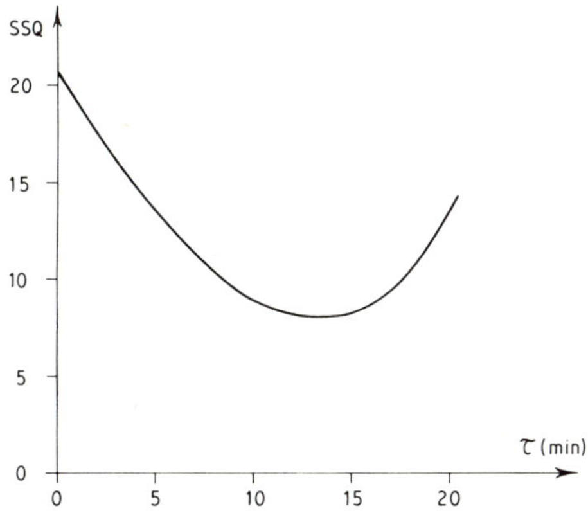

Figure 8.26 Sum of squared error SSQ, eqn (8.40), v. τ for Drug B data.

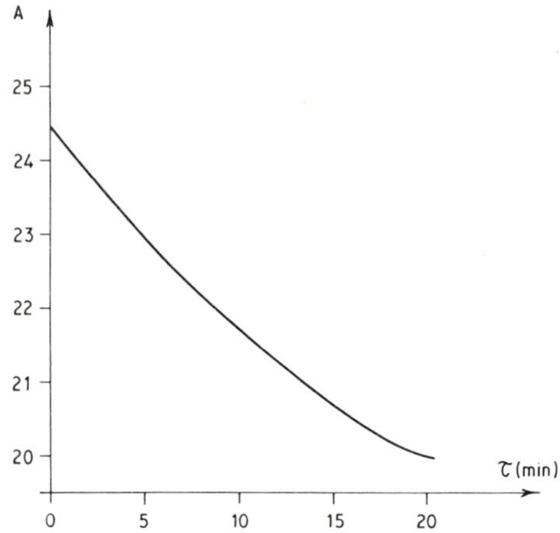

Figure 8.27 Amplitude A, eqn (8.39b), v. τ for Drug B data.

only a little, while β would have been reduced from 1.051 to 0.912 and A increased from 20.89 to 21.71.

8.2.6. *Theoretical evaluation of the effects of small samples*

Most demonstrations of the problems of numerical identifiability have been empirical in nature, considering, as in the preceding examples in this chapter, specific data sets. In two recent papers, though, Van den Bos (1980,

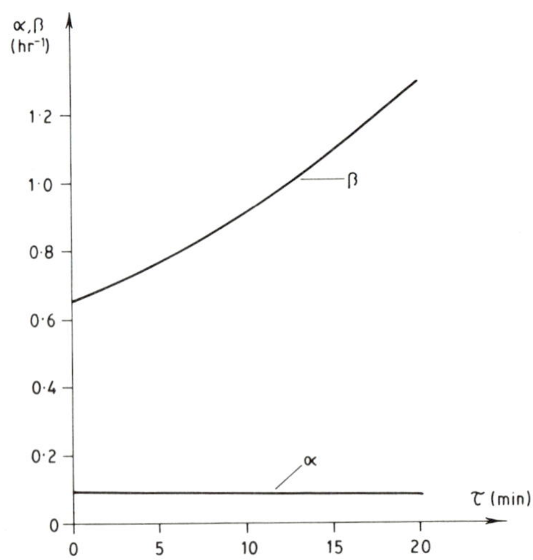

Figure 8.28 Decay constants α and β, eqn (8.39b), v. τ for Drug B data.

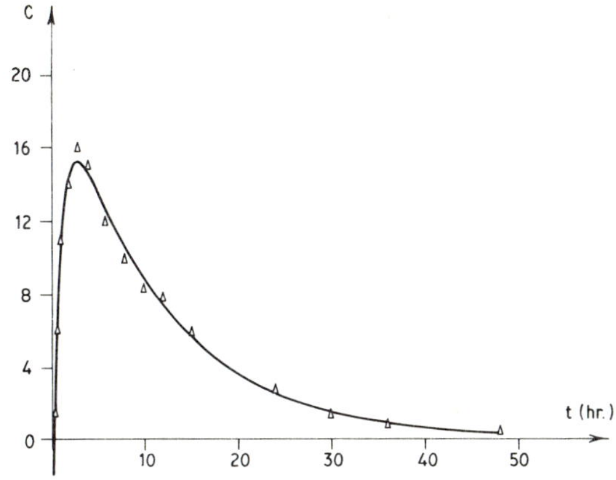

Figure 8.29 Optimum curve fit of form of eqns (8.39) to Drug B data.

1981) has developed analytical expressions for the effects of the smallness of the number of samples on the estimated values of amplitudes and exponents in multi-exponential curve fits. He shows that small-sample nonlinear least-squares can produce results having a structure different from that underlying the observations. A theoretical explanation is given for these structural differences.

Several numerical examples are given. In the earlier paper (Van den Bos, 1980), ideal data were generated by sampling

$$x_1(t) = 0.7\,\mathrm{e}^{-0.4t} + 0.3\,\mathrm{e}^{-0.32t} \tag{8.41}$$

at $t = 1, 2, \ldots, 10$. For the fit to the data, the amplitudes were assumed known exactly and only the exponents α, β in

$$x_1(t) = 0.7\,\mathrm{e}^{-\alpha t} + 0.3\,\mathrm{e}^{-\beta t} \tag{8.42}$$

were estimated. The contours of the least-squares criterion

$$\mathrm{SSQ}(\alpha, \beta) = \sum_{t=1}^{10} (y(t) - 0.7\,\mathrm{e}^{-\alpha t} - 0.3\,\mathrm{e}^{-\beta t})^2 \tag{8.43}$$

were plotted and found to have several minima, shown in the paper. The problem was made much worse in subsequent fits to the data when noise was added and was further complicated when it was required to estimate the amplitudes as well as the exponents.

From the foregoing in this chapter, the problems encountered in this example are not particularly surprising in view of the closeness of the decay constants (0.4 and 0.32) in the model generating the data sequence. The Laplace transform of eqn (8.41) is

$$X_1(s) = \frac{s + 0.344}{(s + 0.4)(s + 0.32)} \tag{8.44}$$

so that there is a close pole-zero cancellation.

In the later paper (Van den Bos, 1981), the data were generated by sampling

$$x_1(t) = 0.5\,\mathrm{e}^{-0.4t} + 0.3\,\mathrm{e}^{-0.12t} + 0.2\,\mathrm{e}^{-0.16t} \tag{8.45}$$

at $t = 1, 2, \ldots 20$. Even with amplitudes known exactly, and only exponents to be estimated, a lower-order model often resulted when noise was added to the data sequence. As in the previous example, this is not surprising because the last two decay constants in eqn (8.45) are close together; the Laplace transform is

$$X_1(s) = \frac{(s + 0.27)(s + 0.14)}{(s + 0.4)(s + 0.16)(s + 0.12)} \tag{8.46}$$

so that there is a very close pole-zero cancellation. Van den Bos notes that the situation would be improved if more observations were used and if the difference between the decay constants were increased.

The theoretical expressions developed by Van den Bos are relatively complicated, but it is not the intention that experimenters should use them every time they do a multi-exponential fit; rather that the expressions should add weight to the need (illustrated elsewhere with empirical examples) for great caution when fitting exponentials to limited data samples. Local minima which may well be far away from the global minimum must be avoided (best done by using a large number of initial estimates for the parameters) and it is necessary to recognize that the curve fitted may be of lower order than that from which the data are generated.

8.2.7. *How many exponentials?*

Let us now return to the question of the number of exponentials which can realistically be fitted to a single impulse response. In theory, if noise-free, very … records were available over a very long time, many expo… be fitted to a single response, but as we have seen in Sections … the number is limited in practice by amplitude quantization … spacing of samples in time and in some instances by limited time … available data.

Even with well-designed sampling schedules (short intervals between samples when the response is changing rapidly), it is rarely possible to fit more than three exponentials to a single response. Often an estimation will fail to converge if it is attempted to fit more, or if it does converge two decay constants will be very close together and/or an amplitude will be very small. Occasionally it is possible to fit four exponentials, and we will now examine such an example from the literature to see what qualities were needed in the data to enable this many exponentials to be fitted.

Foster *et al.* (1979) fitted the following curve to the plasma response to a rapidly administered intravenous injection of ^{69m}Zn:

$$x_1(t) = 0.790\,e^{-176t} + 0.175\,e^{-73.4t} + 0.022\,e^{-5.87t} + 0.013\,e^{-0.053t} \tag{8.47}$$

where the amplitude has been normalized so that $x_1(0^+) = 1$ and time (t) is in days. In the response, relatively noise-free samples were available from a few minutes after the injection up to five days after. Inter-sample intervals were a few minutes soon after the injection, but increased to a day between four and five days after.

The spread of time constants is large, in the ratios 1:2.4:30:3321. The contributions of the four terms to AUC are 0.0045, 0.0024, 0.0037 and 0.2453 (a total of 0.2559) so that, while the fourth term contributes only 1.3% of $x_1(0^+)$, it accounts for almost 96% of AUC. Although the amplitudes of the third and fourth terms are small, there is, with the large range of time constants, no particularly close pole-zero cancellation in the Laplace transform of $x_1(t)$:

$$X_1(s) = \frac{(s + 92.66)(s + 9.475)(s + 1.171)}{(s + 176)(s + 73.4)(s + 5.87)(s + 0.053)}. \tag{8.48}$$

The only problem with the data set is that, by the end of the measurement period ($t = 5$), the final term in $x_1(t)$ has decayed to only $e^{-0.265} = 0.767$ of its initial value, so that, even with this long response record, good initial estimates would be required to permit its accurate estimation by a nonlinear least-squares algorithm.

Returning to the general problem of number of exponentials, there are no absolute criteria for choosing the number, but there are generally two aims, firstly that the sum of squared deviations should be minimal and secondly that the scatter of the observed data around the fitted curve should be random (i.e. there should be no systematic error). Some guidance on model order can be obtained by using a statistical test based on the F-ratio (see, for example, Mandel, 1964, p. 164) which is appropriate when different orders of one type of model (such as a sum of exponentials) are being tested. The

F-ratio is then given by:

$$F = \frac{SSQ_j - SSQ_k}{df_j - df_k} \times \frac{df_k}{SSQ_k} \tag{8.49}$$

where SSQ_j and SSQ_k are the residual sums of squares of the models of order j and k ($k > j$); and df_j and df_k are the corresponding numbers of degrees of freedom (note that $df_j > df_k$). If the value of F is not significant at the chosen level of significance, the variation explained by the model of order k is not significantly greater than that explained by the model of order j.

An underlying assumption of the F-test is that the residuals are normally distributed and this may not be a valid assumption in many cases. For example, in the curve fit illustrated in Fig. 8.24, the errors appear systematic rather than random. A particular value of SSQ can, of course, arise either through random variability or through a systematic error in the fit, and this makes it essential to examine the original data and the model fit at each stage of the fitting procedure rather than relying exclusively on statistical tests such as the F-ratio. Such examination is most conveniently done using an interactive program.

8.2.8 *Summary*

In Sections 8.2.2 and 8.2.6, several pitfalls of compartmental modelling have been presented, and have been illustrated through the use of numerical examples. The pitfalls have been due to measurement accuracy and noise, the use of a limited time range of data and the effects of sampling. At this point, it would be as well to summarize the lessons to be learnt.

We have seen that, to obtain good estimates of the amplitudes and exponents of a multi-exponential curve, measurements must begin early enough for a rapid transient and/or an initial time delay to be observed and must continue long enough for any slow transient to be identified. If, as is almost always the case, the output is sampled, the samples must be gathered at short intervals when the output is changing rapidly but can be spaced at wider intervals when the output is varying slowly. It is necessary to provide good initial estimates of both amplitudes and decay constants when using a nonlinear least-squares algorithm and to start the algorithm from a number of different initial estimates. (Rapid convergence to the same values from a wide range of initial estimates is an indication that the model is well defined.) Even when the above precautions are taken, problems can still be expected when decay constants are fairly close together and, as shown theoretically by Van den Bos (1980, 1981) it is possible, if small samples are used, for the wrong model order to be indicated, even when the data are noise-free.

With the exception of Van den Bos' work, the material presented has been illustrative rather than theoretical. Most of the material previously appearing in the literature has been directed at the accuracy with which amplitudes and decay constants of a multi-exponential curve may be estimated. In this chapter, rate constants for linear, time-invariant compartmental models have been calculated for several of the examples and it has been shown that by the time three compartments are involved, there can be wide differences in some (but not necessarily all) rate constants between two models having quite similar impulse responses.

8.3 The information matrix approach to experiment design

There are some theoretical tools which allow progress to be made in the optimization of experiment design in a parameter estimation exercise. These have been developed with particular reference to compartmental models by DiStefano and his co-workers (Mori and DiStefano, 1979; DiStefano, 1981). Similar development, with emphasis on pharmacokinetic experiments, is described by D'Argenio (1981).

The mathematically tractable results are based on the sensitivity function matrix $\mathbf{S}$ (Reid, 1977) which gives the sensitivity of the output $y_i(\mathbf{p}, t)$, $i = 1,2, \ldots, m$ to the parameters p_j, $j = 1, 2, \ldots, n$:

$$\mathbf{S}(\mathbf{p}, t) = \begin{bmatrix} \dfrac{\partial y_1}{\partial p_1} & \cdots & \dfrac{\partial y_1}{\partial p_n} \\ \vdots & & \\ \dfrac{\partial y_m}{\partial p_1} & \cdots & \dfrac{\partial y_m}{\partial p_n} \end{bmatrix}. \tag{8.50}$$

A small perturbation in the parameters about an initial value $\mathbf{p}^0$, described by

$$\Delta\mathbf{p} = \mathbf{p} - \mathbf{p}^0 \tag{8.51}$$

gives rise to a small perturbation in the output:

$$\Delta\mathbf{y} = \mathbf{y}(\mathbf{p}, t) - \mathbf{y}(\mathbf{p}^0, t) = \mathbf{S}\,\Delta\mathbf{p}. \tag{8.52}$$

Consider the measurement equations for observations $\mathbf{z}(t_k)$, $k = 1, 2, \ldots, N$, given by

$$\mathbf{z}(t_k) = \mathbf{y}(\mathbf{p}, t_k) + \mathbf{e}(t_k) \tag{8.53}$$

where $\mathbf{e}(t_k)$ is the observation noise. The Fisher information matrix $\mathbf{F}$, a measure of the information concerning the unknown parameters available in the noisy data, is given by

$$\mathbf{F}(\mathbf{p}) = E\left\{\left[\frac{\partial \log f(\mathbf{Z}|\mathbf{p})}{\partial \mathbf{p}}\right]^{\mathrm{T}}\left[\frac{\partial \log f(\mathbf{Z}|\mathbf{p})}{\partial \mathbf{p}}\right]\right\} \tag{8.54}$$

where $\mathbf{Z} = [\mathbf{z}(t_1)^{\mathrm{T}}\mathbf{z}(t_2)^{\mathrm{T}} \ldots \mathbf{z}(t_N)^{\mathrm{T}}]^{\mathrm{T}}$; $f(\mathbf{Z}\,|\,\mathbf{p})$ is the conditional probability density function of $\mathbf{Z}$ given $\mathbf{p}$; and E is the expectation (i.e. averaging) operator, taken with respect to the above density function. (The gradient operation has been taken as a row vector.) This is an uninviting expression because f may be a complicated function of $\mathbf{p}$ and hence difficult to determine. It simplifies though if the noise has zero mean, unit variance and identical normal distribution at each t_k and if the values $\mathbf{e}(t_k)$ are uncorrelated. Then

$$\mathbf{F}(\mathbf{p}) = \sum_{k=1}^{N}\left[\frac{\partial \mathbf{y}(\mathbf{p}, t_k)}{\partial p_i}\right]\left[\frac{\partial \mathbf{y}(\mathbf{p}, t_k)}{\partial p_j}\right]^{\mathrm{T}} = \mathbf{S}^{\mathrm{T}}\mathbf{S}. \tag{8.55}$$

Use is then made of the Cramér–Rao theorem (Cramér, 1946) which states that the covariance matrix of parameter estimates,

$$\operatorname{cov}(\mathbf{p}) \equiv E[(\mathbf{p} - \bar{\mathbf{p}})(\mathbf{p} - \bar{\mathbf{p}})^{\mathrm{T}}], \quad \text{where} \quad \bar{\mathbf{p}} = E(\mathbf{p})$$

has a lower bound of $\mathbf{F}^{-1}$ (see, for example, Eykhoff (1974), Section 11.2.1).

The design optimization objective is to maximize $\mathbf{F}$ (or more usually to minimize $\mathbf{F}^{-1}$) with respect to the design variables, which are the inputs $\mathbf{u}(t)$, the observation interval T, the number of samples N, the sampling schedule SS and the sample accuracy. In practice, the complexity of $\mathbf{F}$ as a function of these variables requires an iterative solution, with $\mathbf{F}$ being evaluated repeatedly to find the best choice of design variables. The numerical complexity of the problem renders it highly impractical to perform this optimization with respect to more than one or two variables at a time. Such an optimization with respect to SS and $\mathbf{u}(t)$ has been reported by Mori and DiStefano (1979) who applied the procedure to a compartmental model of triiodothyronine kinetics (DiStefano *et al.*, 1975), the model being the upper half of Fig. 4.22.

For any model described by the equations:

$$\dot{\mathbf{x}}(\mathbf{p}, t) = \mathbf{A}(\mathbf{p})\mathbf{x}(\mathbf{p}, t) + \mathbf{B}\mathbf{u}(\mathbf{p}) \tag{8.56a}$$

$$\mathbf{y}(\mathbf{p}, t) = \mathbf{C}(\mathbf{p})\mathbf{x}(\mathbf{p}, t) \tag{8.56b}$$

the sensitivity values are evaluated from:

$$\frac{\partial \mathbf{y}(\mathbf{p}, t_k)}{\partial p_i} = \mathbf{C}(\mathbf{p})\frac{\partial \mathbf{x}(\mathbf{p}, t_k)}{\partial p_i} + \frac{\partial \mathbf{C}(\mathbf{p})}{\partial p_i}\mathbf{x}(\mathbf{p}, t_k) \tag{8.57}$$

where $\frac{\mathrm{d}\mathbf{x}}{\partial p_i}$ is calculated at each t_k from

$$\frac{\partial \dot{\mathbf{x}}}{\partial \mathbf{p}} = \mathbf{A}\frac{\partial \mathbf{x}}{\partial \mathbf{p}} + \frac{\partial \mathbf{A}}{\partial \mathbf{p}}\mathbf{x} + \frac{\partial}{\partial \mathbf{p}}(\mathbf{Bu}). \tag{8.58}$$

Equation (8.58) requires some prior estimates of the parameters $\mathbf{p}$, and generally, the better these initial estimates, the more effective is the design optimization at reducing parameter variability.

Unconstrained optimization usually produces the result for the sampling schedule that if there are n parameters to be estimated and there are N ($\geqslant n$) samples, then it is best to determine n points on the impulse response as closely as possible by sampling more than once at each point to make up the N samples. Practical considerations often make it necessary to impose some constraint on the resulting sampling schedule. For example, Mori and DiStefano (1979) report an optimization with the constraint that sampling intervals should not be less than 10 min with only one sample per measurement point, a constraint representative of those which may occur in physiological modelling. They still found that the optimized schedule resulted in more accurate parameter estimation than a schedule designed on an intuitive basis.

The main limitation of the technique is that it is essential to know the correct model (in particular, the correct model order); as emphasized by D'Argenio (1981), it is possible to produce less reliable parameter estimates than those from conventional sampling schemes if optimal sampling times are calculated using the wrong model. Another limitation is that the degree of improvement through the optimization depends on how closely the initial guess for the parameter values agrees with the true values. In practice, particularly in modelling in physiology, intersubject variability can make the improvement smaller than expected. It should also be remembered that the simplified expression for $\mathbf{F}$ given in eqn (8.55) assumes uncorrelated, normally-distributed noise and that $\mathbf{F}^{-1}$ is only a lower bound on the covariance and is not the covariance itself.

The merit of the technique is that it is possible to try several *ad hoc* choices of design variables before conducting an experiment.

8.4 Selection from models which are input–output equivalent

In Section 8.2, we have examined how the quality of data can affect the fitting of responses, which, in the case of linear, time-invariant compartmental models consist of sums of exponentials if the inputs are steps or impulses. In this Section, we turn to a different problem, that of choosing between two or more compartmental models which seem equally plausible to describe the experimental situation and which, for the possible perturbations and possible observations, are equivalent. This problem has nothing to do with the quality of the experimental data although, as we will see, the choice could often be resolved by observing more responses.

Consider firstly an experiment in which a unit impulse has been administered to a system, and a response

$$y(t) = 0.9\,e^{-t} + 0.1\,e^{-0.1t} \tag{8.59}$$

has been fitted to the response data. It is known that the input fraction and the observation gain are both equal to 1. The form of $y(t)$ suggests an open two-compartment model with input to, and observation of, the same compartment (no. 1, say). Excretion could be from compartment 1 only, or from compartment 2 only, or from both. Excluding the third possibility on the grounds that the rate constants cannot be individually identified (although bounds can be found, as in Chapter 7), the first two possibilities are, in the absence of any other information, equally plausible.

The values of the rate constants for model A (excretion from compartment 1 only) are:

$$k_{01} = 0.526; \quad k_{12} = 0.19; \quad k_{21} = 0.384$$

while those for model B (excretion from compartment 2 only) are:

$$k_{02} = 0.11; \quad k_{12} = 0.08; \quad k_{21} = 0.91.$$

Note the substantial differences between the models in k_{21} and k_{12}. The two models could be distinguished if it were possible to observe compartment 2 also, with a known observation gain. From eqn (3.27b),

$$x_2(t) = \frac{k_{21}}{0.9}(e^{-0.1t} - e^{-t})$$

giving $$x_2(t) = 0.427(e^{-0.1t} - e^{-t}) \quad \text{for model A} \tag{8.60a}$$

and $$x_2(t) = 1.011(e^{-0.1t} - e^{-t}) \quad \text{for model B} \tag{8.60b}$$

which are compared in Fig. 8.30.

The elimination rates to the environment are also different. For model A,

$$\text{Elimination rate} = k_{01}x_1(t) = 0.474\,e^{-t} + 0.0526\,e^{-0.1t} \tag{8.61a}$$

while for model B,

$$\text{Elimination rate} = k_{02}x_2(t) = 0.1112(e^{-0.1t} - e^{-t}). \tag{8.61b}$$

Although the elimination rates are different, the total amount eliminated to

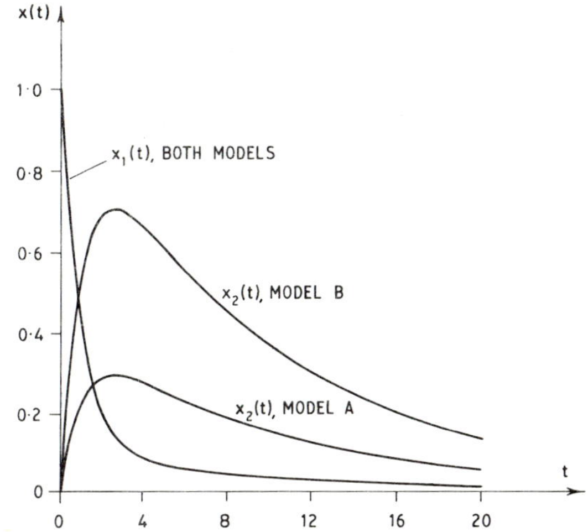

Figure 8.30 Responses $x_1(t)$ and $x_2(t)$ for models A and B, with $x_1(t)$ given by eqn (8.59) for both models and $x_2(t)$ given by eqn (8.60a) for model A and eqn (8.60b) for model B.

the environment, $\int_0^\infty k_{01}x_1(t)\,dt$ for model A and $\int_0^\infty k_{02}x_2(t)\,dt$ for model B, is the same (and equal to 1) in both cases.

In some cases, it would be necessary to know the observation gain of the second observation fairly accurately to distinguish between models. To illustrate this, consider a second experiment in which the response of a system to a unit-impulse perturbation is

$$y_1(t) = 0.7\,e^{-5t} + 0.2\,e^{-t} + 0.1\,e^{-0.1t}, \tag{8.62}$$

the input fraction and observation gain again being 1. The form of $y_1(t)$ indicates an open three-compartment model with the same compartment perturbed and observed. Suppose that there are grounds for assuming a catenary structure, with an extreme compartment (no. 1, say) perturbed and observed but with no excretion from that compartment. It is not known whether excretion is from compartment 2, or from compartment 3, or from both. If the rate constants are to be identifiable, the last of these possibilities must be excluded.

Rate constants for model C, with excretion from compartment 2 only and for model D, with excretion from compartment 3 only, are shown in Fig. 8.31(a) and (b). From the analysis of Chapter 4, the responses $x_2(t)$ (as well as $x_1(t)$) are the same for the two models, but $x_3(t)$ for model C is only $\frac{0.558}{0.825}\,x_3(t)$ for model D. Responses for the two models are shown in Fig.

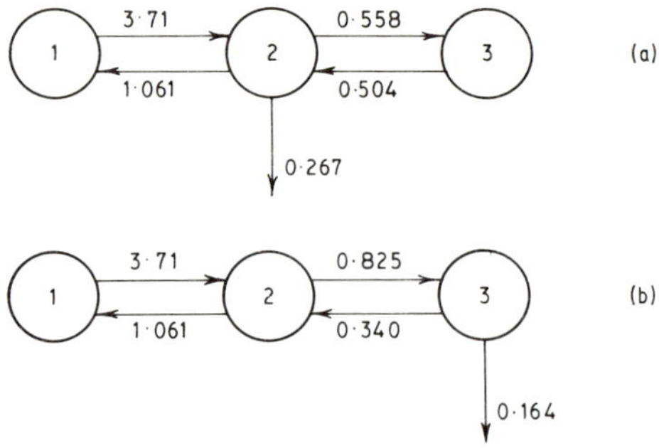

Figure 8.31 Three-compartment models both with $x_1(t)$ given by eqn (8.62). (a) Model C; (b) Model D.

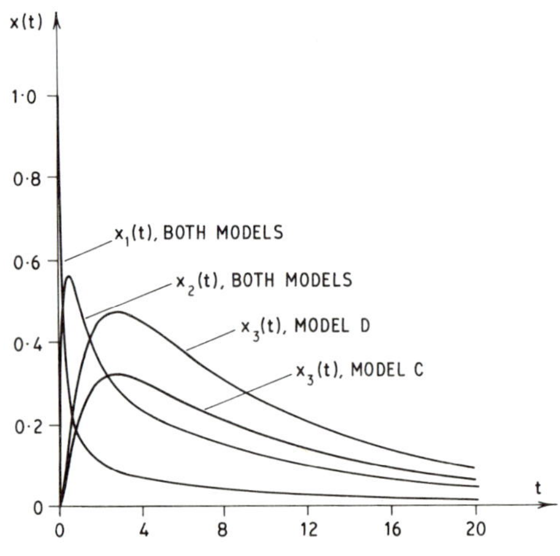

Figure 8.32 Responses for models C and D.

8.32. Thus the two models could be distinguished by observing compartment 3, but with only the small difference between the responses $\left(\dfrac{0.558}{0.825} = 0.676\right)$ the observation gain c_3 would have to be known quite accurately.

For "black-box" modelling, the differences between models A and B and between models C and D would be immaterial because the input–output behaviour is the same. For models where the parameters have physical or physiological significance, differences in k_{12} and k_{21} between models A and B and in k_{23} and k_{32} between models C and D could be important.

9 Nonlinear Compartmental Models

9.1 The form of the mass balance equations

In Chapters 2 to 8, we have been concerned with linear, time-invariant models, for which a substantial body of theory exists concerning both analysis and parameter identification. In this chapter and the next, we will consider some departures from linearity and time-invariance, discussing nonlinear compartmental models in this chapter and time-varying compartmental models and stochastic compartmental models in Chapter 10.

We have already seen in Chapter 1 that the most general form of compartmental equations for a system with p compartments is:

$$\frac{\mathrm{d}x_i}{\mathrm{d}t} = f_{i0} + \sum_{\substack{j=1 \\ j \neq 1}}^{p} (f_{ij} - f_{ji}) - f_{0i}, \qquad i = 1, 2, \ldots, p \tag{9.1}$$

where x_i is the amount of material in compartment i

f_{ij} is the flow rate to compartment i from compartment j

and subscript 0 denotes the environment.

This form of model is too general to be useful, and to make progress, it is necessary to relate the flows to the state variables $x_1 \ldots x_p$. The two most common ways of doing this are to assume either that the flows are donor-controlled (depending on the state of the donor compartment only) or that each rate coefficient depends on several of the states (the state-dependent model).

For the donor-controlled equations, the flows are of the form

$$f_{ij} = k_{ij} f_j(x_j), \quad i = 0, 1, \ldots, p; j = 1, 2, \ldots, p; i \neq j \tag{9.2}$$

where k_{ij} is a non-negative constant and $f_j(x_j)$ is a positive function of the state variable x_j. Thus, for compartment i,

$$\frac{\mathrm{d}x_i}{\mathrm{d}t} = f_{i0} + \sum_{\substack{j=1 \\ j \neq 1}}^{p} [k_{ij} f_j(x_j) - k_{ji} f_i(x_i)] - k_{0i} f_i(x_i). \tag{9.3}$$

The stability of such systems has been examined by Maeda, Kodama and Ohta (1978) who show that there is a stable solution (with no limit cycle) for constant input flows (including zero input). A similar result has been shown for such systems with time delays (Lewis and Anderson, 1980). An important example of a donor-controlled nonlinearity occurs when one or more of the inter-compartmental flows is capacity-limited and may be modelled by a Michaelis–Menten relationship:

$$f_{ij} = \frac{V_m}{K_m + x_j(t)} x_j(t). \tag{9.4}$$

Models with Michaelis–Menten elimination kinetics are particularly important in pharmacokinetics, and will be considered further in Section 9.4.

The state-dependent equations are similar to those for the linear, time-invariant system, except that the rate coefficients, instead of being constant, are dependent on the states $x_1 \ldots x_p$ so that

$$\frac{\mathrm{d}x_i}{\mathrm{d}t} = f_{i0} + \sum_{\substack{j=1 \\ j \neq i}}^{p} [k_{ij}(x_1 \ldots x_p)x_j - k_{ji}(x_1 \ldots x_p)x_i] - k_{0i}(x_1 \ldots x_p)x_i. \tag{9.5}$$

The stability of such systems has been considered by Ladde (1976a,b). In many instances, the rate coefficients are not a function of all the compartmental quantities; in a particular link, the flow rate may only depend on the quantities in the donor and receptor compartments. An example is Langmuir saturation (Krüger-Thiemer *et al.*, 1964) for which

$$f_{ij} = k_{ij}(1 - \sigma_i x_i(t))x_j(t) \tag{9.6}$$

where σ_i is a constant such that $(1 - \sigma_i x_i(t)) \geqslant 0$. When the quantity $x_i(t)$ in the receptor compartment is small, the transfer rate is approximately $k_{ij}x_j(t)$ (i.e. first-order (linear) kinetics) but when $x_i(t)$ approaches its saturation level $\frac{1}{\sigma_i}$, f_{ij} tends to zero.

9.2 Identifiability of nonlinear compartmental systems

The parameters to be identified are the rate constants for all linear links in the model, and for the nonlinear links, parameters describing the nonlinearity; the form of the nonlinearity will have been assumed. Numerical problems impose a severe practical limitation on what can be identified. The most basic requirement is that the observed responses with the nonlinearity present must be significantly different from those with linear transfers replacing the nonlinearities. This means that comparatively few nonlinearities in any one model can be identified, and that the relevant compartmental quantities must vary sufficiently to span a substantial part of any nonlinear characteristic.

For any nonlinear system, the principle of superposition no longer holds (doubling the input no longer doubles the responses) nor do relationships such as the step response being the integral of the impulse response. Deterministic identifiability thus presents even more of a problem than it did for linear systems. The Laplace transform method presented in Chapter 6 no longer works for nonlinear systems. Grewal and Glover (1976) derive sufficient conditions for local identifiability of nonlinear sets of dynamic equations based on linearization about an operating point, but they do not provide a workable algorithm for testing identifiability. Such an algorithm, based on the Taylor series of the observations about $t = 0^+$, has been provided by Pohjanpalo (1978).

Specifically, for an observation $y_i(t)$,

$$y_i(t) = y_i(0^+) + t\dot{y}_i(0^+) + \frac{t^2}{2!}\ddot{y}_i(0^+) + \frac{t^3}{3!}\dddot{y}_i(0^+) + \ldots . \tag{9.7}$$

The successive derivatives are, in principle, measurable, and they contain information about the parameters to be identified. Note that the Pohjanpalo test is the equivalent of expanding the Laplace transform of the observation vector as a power series in s^{-1}, since, taking Laplace transforms of eqn (9.7),

$$Y_i(s) = \frac{y_i(0^+)}{s} + \frac{\dot{y}_i(0^+)}{s^2} + \frac{\ddot{y}_i(0^+)}{s^3} + \frac{\dddot{y}_i(0^+)}{s^4} + \ldots . \tag{9.8}$$

To illustrate the basic principles of the Taylor series method, without the added complication of nonlinearities, we will consider firstly two examples of linear, time-invariant models.

Example 1. General two-compartment linear, time-invariant system. The model of Fig. 3.6 is described by the equations

$$\dot{x}_1(t) = a_{11}x_1(t) + a_{12}x_2(t) + b_1u_1(t) \tag{9.9a}$$

$$\dot{x}_2(t) = a_{21}x_1(t) + a_{22}x_2(t) + b_2u_2(t). \tag{9.9b}$$

Consider an experiment with $u_1(t) = D_1\,\delta(t)$, $u_2(t) = 0$ and compartment 1 (only) observed with observation gain c_1, for which

$$\dot{x}_1(t) = a_{11}x_1(t) + a_{12}x_2(t), \qquad t > 0; \qquad x_1(0^+) = b_1D_1 \tag{9.10a}$$

$$\dot{x}_2(t) = a_{21}x_1(t) + a_{22}x_2(t), \qquad t > 0; \qquad x_2(0^+) = 0 \tag{9.10b}$$

with observation $$y_1(0^+) = c_1b_1D_1, \tag{9.10c}$$

the latter allowing c_1b_1 to be identified, assuming D_1 is known. From eqns (9.10),

$$\dot{x}_1(0^+) = a_{11}x_1(0^+) + a_{12}x_2(0^+) = a_{11}x_1(0^+) \tag{9.11a}$$

$$\dot{x}_2(0^+) = a_{21}x_1(0^+) + a_{22}x_2(0^+) = a_{21}x_1(0^+) \tag{9.11b}$$

with
$$\dot{y}_1(0^+) = a_{11}(c_1b_1D_1) \tag{9.11c}$$

so that a_{11} can be identified.

Differentiating eqn (9.10a),

$$\ddot{x}_1(t) = a_{11}\dot{x}_1(t) + a_{12}\dot{x}_2(t), \qquad t > 0 \tag{9.12}$$

so that
$$\ddot{x}_1(0^+) = a_{11}\dot{x}_1(0^+) + a_{12}\dot{x}_2(0^+)$$
$$= a_{11}\dot{x}_1(0^+) + a_{12}a_{21}x_1(0^+)$$

and
$$\ddot{y}_1(0^+) = a_{11}\dot{y}_1(0^+) + a_{12}a_{21}y_1(0^+) \tag{9.13}$$

in which everything except $a_{12}a_{21}$ is known, so that $a_{12}a_{21}$ can be identified.

Differentiating eqn (9.10b),

$$\ddot{x}_2(t) = a_{21}\dot{x}_1(t) + a_{22}\dot{x}_2(t)$$

so that
$$\ddot{x}_2(0^+) = a_{21}\dot{x}_1(0^+) + a_{22}a_{21}x_1(0^+). \tag{9.14}$$

Differentiating eqn (9.12),

$$\dddot{x}_1(t) = a_{11}\ddot{x}_1(t) + a_{12}\ddot{x}_2(t), \qquad t > 0$$

so that

$$\dddot{x}_1(0^+) = a_{11}\ddot{x}_1(0^+) + a_{12}\ddot{x}_2(0^+).$$

Substituting from eqn (9.14),

$$\dddot{x}_1(0^+) = a_{11}\ddot{x}_1(0^+) + a_{12}a_{21}\dot{x}_1(0^+) + a_{22}(a_{12}a_{21})x_1(0^+),$$

so that

$$\dddot{y}_1(0^+) = a_{11}\ddot{y}_1(0^+) + a_{12}a_{21}\dot{y}_1(0^+) + a_{22}(a_{12}a_{21})y_1(0^+) \tag{9.15}$$

in which everything except a_{22} is known, so that a_{22} can be identified. Hence parameter combinations a_{11}, $(a_{12}a_{21})$ and a_{22} can be identified in turn, together with the product c_1b_1. The higher derivatives in this example contain only expressions in the parameter combinations already identified and do not allow individual identification of the rate constants k_{01}, k_{21}, k_{02} and k_{12}. The higher derivatives could, in theory, provide some model validation, since if the values of a_{11}, $(a_{12}a_{21})$ and a_{22} calculated from them did not agree with those calculated from the lower derivatives, it would indicate that the general two-compartment model is not an appropriate description for the experiment.

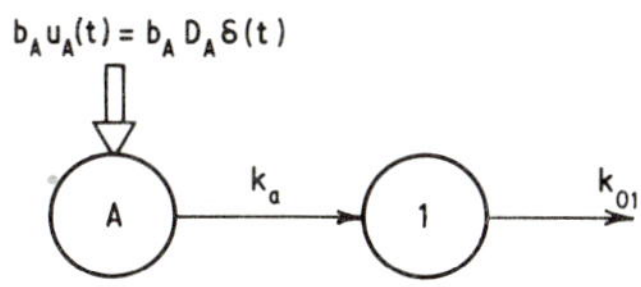

Figure 9.1 One-compartment linear model with first-order input.

Example 2. One-compartment model with first-order input. As we have seen in Chapter 3, a first-order input (modelling, for example, a dose rapidly administered via an extravascular route) can be viewed as an extra compartment, as shown in Fig. 9.1, for which the equations are

$$\dot{x}_A(t) = -k_a x_A(t), \qquad t > 0 \tag{9.15a}$$

with
$$x_A(0^+) = b_A D_A \tag{9.15b}$$

and
$$\dot{x}_1(t) = k_a x_A(t) - k_{01}x_1(t), \qquad t > 0 \tag{9.15c}$$

with
$$x_1(0^+) = 0. \tag{9.15d}$$

Assume that only compartment 1 is observed. It is slightly easier, for the Taylor series analysis, to consider only $x_1(t)$, the equation for which is, from eqns (9.15),

$$\dot{x}_1(t) = -k_{01}x_1(t) + b_A D_A k_a \,\mathrm{e}^{-k_a t}, \qquad t > 0 \tag{9.16a}$$

with
$$x_1(0^+) = 0. \tag{9.16b}$$

Then
$$y_1(0^+) = 0 \tag{9.17}$$

and
$$\dot{y}_1(0^+) = c_1\dot{x}_1(0^+) = c_1 b_A D_A k_a \tag{9.18}$$

which allows $c_1 b_A k_a$ to be identified, assuming the size of input administered (D_A) is known.

Differentiating eqn (9.16a),

$$\ddot{x}_1(t) = -k_{01}\dot{x}_1(t) - b_A D_A k_a^2 \,\mathrm{e}^{-k_a t}, \qquad t > 0 \tag{9.19}$$

so that

$$\ddot{x}_1(0^+) = -k_{01}\dot{x}_1(0^+) - b_A D_A k_a^2$$

and
$$\ddot{y}_1(0^+) = -k_{01}(c_1 b_A D_A k_a) - k_a(c_1 b_A D_A k_a) \tag{9.20}$$

from which $(k_{01} + k_a)$ can be identified.

Differentiating eqn (9.19),

$$\dddot{x}_1(t) = -k_{01}\ddot{x}_1(t) + b_A D_A k_a^3 \,\mathrm{e}^{-k_a t}, \qquad t > 0$$

so that

$$\dddot{x}_1(0^+) = -k_{01}\ddot{x}_1(0^+) + b_A D_A k_a^3$$

and
$$\dddot{y}\,(0^+) = k_{01}(k_{01} + k_a)(c_1 b_A D_A k_a) + k_a^2(c_1 b_A D_A k_a)$$
$$= [(k_{01} + k_a)^2 - k_{01}k_a]\cdot c_1 b_A D_A k_a \quad (9.21)$$

from which $k_{01}k_a$ can be identified. Thus the sum $(k_{01} + k_a)$ and the product $k_{01}k_a$ can be identified, and the model is locally identifiable with two solutions (k_{01} and k_a are indistinguishable) unless $c_1 b_A D_A$ is known, in which case k_a is uniquely identifiable from eqn (9.18) and k_{01} is uniquely identifiable from eqn (9.20).

These two linear examples have served to illustrate that, for linear, time-invariant models, there seems to be no advantage in this approach over the Laplace-transform approach. For nonlinear systems and time-varying systems, however, the Laplace-transform approach is no longer applicable, whereas the Taylor-series approach can still be used in most situations. The conditions for applicability are discussed in Proposition 3 of Pohjanpalo (1978, p. 24). For a Maclaurin series to represent a function $f(t)$ on the time interval from 0 to T, $f(t)$ and its derivatives $f^{(n)}(t) \equiv \dfrac{d^n f}{dt^n}$ must be bounded:

$$|f^{(n)}(t)| < MT^{mn}, \qquad n = 1, 2, \ldots, \infty \quad (9.22)$$

for some M and m. This applies to most cases of practical interest, although, as we will see in Chapter 10, it does exclude one simple time-varying system.

The Taylor series approach will now be applied to a nonlinear system.

Example 3. Two-compartment model with compartment 2 Langmuir-saturating. The model shown in Fig. 9.2 is described by the equations

$$\dot{x}_1(t) = -[k_{01} + k_{21}(1 - \sigma_2 x_2(t))]x_1(t) + b_1 u_1(t) \quad (9.23a)$$

$$\dot{x}_2(t) = k_{21}(1 - \sigma_2 x_2(t))x_1(t). \quad (9.23b)$$

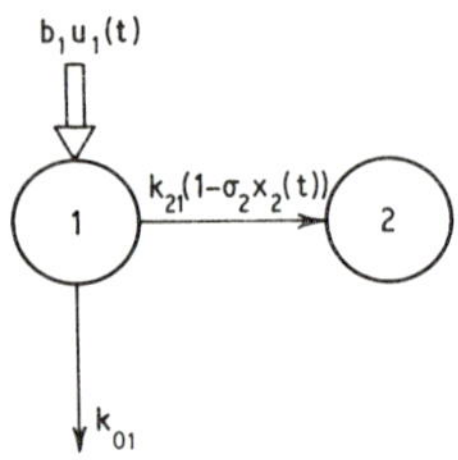

Figure 9.2 Two compartment model with one Langmuir-saturating compartment.

Let compartment 1 be impulsively perturbed ($x_1(0^+) = b_1 D_1$) and observed; compartment 2 is neither perturbed nor observed and $x_2(0^+) = 0$. The requirement is to identify k_{01}, k_{21} and σ_2. From the above,

$$y_1(0^+) = c_1 b_1 D_1 \tag{9.24}$$

which allows $c_1 b_1$ to be identified, as in Example 1. From eqn (9.23a),

$$\dot{x}_1(0^+) = -(k_{01} + k_{21})x_1(0^+)$$

so that

$$\dot{y}_1(0^+) = -(k_{01} + k_{21})y_1(0^+) \tag{9.25}$$

whence $(k_{01} + k_{21})$ can be identified. From eqn (9.23b),

$$\dot{x}_2(0^+) = k_{21}x_1(0^+). \tag{9.26}$$

Differentiating eqn (9.23a),

$$\ddot{x}_1(t) = -[k_{01} + k_{21}(1 - \sigma_2 x_2(t))]\dot{x}_1(t) + \sigma_2 k_{21}\dot{x}_2(t)x_1(t), \qquad t > 0 \tag{9.27}$$

so that, substituting from eqn (9.26),

$$\ddot{x}_1(0^+) = -(k_{01} + k_{21})\dot{x}_1(0^+) + \sigma_2 k_{21}^2 x_1^2(0^+)$$

giving

$$\ddot{y}_1(0^+) = -(k_{01} + k_{21})\dot{y}_1(0^+) + \sigma_2 k_{21}^2 y_1(0^+)x_1(0^+). \tag{9.28}$$

To identify any of the parameters from this equation, $x_1(0^+)$ or c_1 must be known. Since $x_1(0^+) = b_1 D_1 = \dfrac{y_1(0^+)}{c_1}$, this means that the input fraction b_1 or the observation gain c_1 must be known. In most cases, b_1 will already be known (usually equal to 1). In general, to complete any nonlinear system identification, it is necessary to know the state of the observed compartment and its derivatives as well as the observation and its derivatives.

To proceed further, let us assume that either b_1 or c_1 and hence $x_1(0^+)$ is known, so that $\sigma_2 k_{21}^2$ can be identified from eqn (9.28). Differentiating eqn (9.23b),

$$\ddot{x}_2(t) = k_{21}(1 - \sigma_2 x_2(t))\dot{x}_1(t) - \sigma_2 k_{21}\dot{x}_2(t)x_1(t), \qquad t > 0$$

so

$$\ddot{x}_2(0^+) = k_{21}\dot{x}_1(0^+) - \sigma_2 k_{21}^2 x_1^2(0^+). \tag{9.29}$$

Differentiating eqn (9.27),

$$\dddot{x}_1(t) = -[k_{01} + k_{21}(1 - \sigma_2 x_2(t))]\ddot{x}_1(t) + \sigma_2 k_{21}\dot{x}_2(t)\dot{x}_1(t)$$
$$+ \sigma_2 k_{21}\dot{x}_2(t)\dot{x}_1(t) + \sigma_2 k_{21}\ddot{x}_2(t)x_1(t), \qquad t > 0$$

so that
$$\begin{aligned}\dddot{x}_1(0^+) &= -(k_{01}+k_{21})\ddot{x}_1(0^+) + 2\sigma_2 k_{21}^2 x_1(0^+)\dot{x}_1(0^+)\\ &\quad + \sigma_2 k_{21} x_1(0^+)[k_{21}\dot{x}_1(0^+) - \sigma_2 k_{21}^2 x_1^2(0^+)]\\ &= -(k_{01}+k_{21})\ddot{x}_1(0^+) + 3\sigma_2 k_{21}^2 x_1(0^+)\dot{x}_1(0^+)\\ &\quad - \sigma_2 k_{21}(\sigma_2 k_{21}^2)x_1^3(0^+)\end{aligned} \tag{9.30}$$

with corresponding $\dddot{y}_1(0^+) = c_1\dddot{x}_1(0^+)$. In this equation, everything is known except $\sigma_2 k_{21}$. This completes the identification, since $\sigma_2 k_{21}^2$ and $(k_{01} + k_{21})$ have already been identified. The parameters are globally identifiable from this experiment.

This example has illustrated one of the main problems of the Taylor-series method, that the expressions in the parameters to be identified can rapidly become complicated as the differentiation proceeds. This point will be reinforced when the method is applied in Sections 9.4.2 and 9.4.3 to models with Michaelis–Menten elimination kinetics. Further applications of the Taylor-series method are described by Pohjanpalo (1978).

9.3 Perturbation equations and tracer equations

Both perturbation and tracer equations apply to the situation where a compartmental system has been operating at a stable steady state $\bar{x}_1, \ldots, \bar{x}_p$, with steady input flows $\bar{f}_{10}, \ldots, \bar{f}_{p0}$. For the perturbation equations, there is then (at $t = 0$) a small perturbation of the steady state, while tracer equations apply specifically to the kinetics of a small amount of labelled material added to the system at $t = 0$. While the perturbation quantities can be positive or negative, the tracer quantities must be non-negative.

Consider firstly the perturbation equations, with a small perturbation of the steady state at $t = 0$:

$$x_i(0) = \bar{x}_i + \tilde{x}_i(0), \qquad i = 1, 2, \ldots, p \tag{9.31a}$$

so that, for $t > 0$,

$$x_i(t) = \bar{x}_i + \tilde{x}_i(t), \qquad i = 1, 2, \ldots, p. \tag{9.31b}$$

For the donor-controlled system (eqn (9.3)), the steady state for compartment i is described by

$$0 = \bar{f}_{i0} + \sum_{\substack{j=1\\ j\neq i}}^{p} [k_{ij} f_j(\bar{x}_j) - k_{ji} f_i(\bar{x}_i)] - k_{0i} f_i(\bar{x}_i). \tag{9.32}$$

The perturbation equations are then obtained by Taylor series expansion of eqn (9.3), neglecting second-order and higher-order terms:

$$\frac{d\tilde{x}_i}{dt} = \sum_{\substack{j=1 \\ j \neq i}}^{p} [k_{ij} f'_j(x_j)\tilde{x}_j - k_{ji} f'_i(x_i)\tilde{x}_i] - k_{0i} f'_i(x_i)\tilde{x}_i, \qquad t \geqslant 0 \tag{9.33}$$

where

$$f'_j(x_j) = \left. \frac{\partial f(x_j)}{\partial x_j} \right|_{x_j = \bar{x}_j}. \tag{9.34}$$

Similar perturbation equations are readily derived for the state-dependent equations of the form of eqn (9.5).

As an example, consider the nonlinear two-compartment model shown in Fig. 9.3. Compartment 1 has a second-order excretion to the environment and transfers material to compartment 2 at a capacity-limited rate. Compartment 2 has a zero-order excretion to the environment and transfers material to compartment 1 in first-order (linear) fashion. The equations for the model are thus

$$\dot{x}_1 = -k_1 x_1^2 - \frac{V_m x_1}{K_m + x_1} + k_{12} x_2 + u_1 \tag{9.35a}$$

$$\dot{x}_2 = \frac{V_m x_1}{K_m + x_1} - k_{12} x_2 - k_0 + u_2. \tag{9.35b}$$

The steady-state equations are

$$0 = -k_1(\bar{x}_1)^2 - \frac{V_m \bar{x}_1}{K_m + \bar{x}_1} + k_{12}\bar{x}_2 + \bar{u}_1 \tag{9.36a}$$

and

$$0 = \frac{V_m \bar{x}_1}{K_m + \bar{x}_1} - k_{12}\bar{x}_2 - k_0 + \bar{u}_2. \tag{9.36b}$$

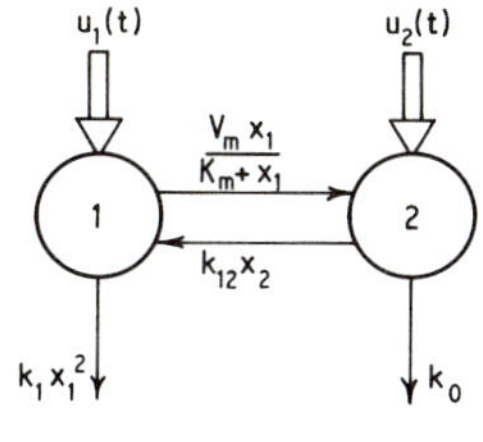

Figure 9.3 Nonlinear two-compartment model (quantities shown are flow rates).

The corresponding perturbation equations are

$$\frac{\mathrm{d}\tilde{x}_1}{\mathrm{d}t} = -\left(2k_1\bar{x}_1 + \frac{V_m K_m}{(K_m + \bar{x}_1)^2}\right)\tilde{x}_1 + k_{12}\tilde{x}_2 + \tilde{u}_1 \tag{9.37a}$$

$$\frac{\mathrm{d}\tilde{x}_2}{\mathrm{d}t} = \frac{V_m K_m}{(K_m + \bar{x}_1)^2}\tilde{x}_1 - k_{12}\tilde{x}_2 + \tilde{u}_2. \tag{9.37b}$$

Note that the parameter k_0, describing the zero-order elimination rate from compartment 2, does not appear in the perturbation equations. Further examples of perturbation equations are given by DiStefano (1976).

The deviations $\tilde{x}_i$ in the perturbation equations are deviations from a steady state; they are changes in quantities of material in a compartment and they can be negative. In a tracer experiment, on the other hand, a small amount of labelled material is added to the system, and the tracer equations describe the time course of the labelled material, so the tracer quantities must be non-negative. In deriving the tracer equations, it is assumed that the system is initially in a steady state, that there is complete mixing of the labelled and unlabelled material, that the mixing is rapid compared with the rates of transfer of material between compartments and that the amount of labelled material added to any compartment is negligibly small compared with the quantity of unlabelled material.

Following the derivation given by Sandberg (1978), let a very small amount of labelled material be injected at $t = 0$ into one or more compartments and let $\delta_i(t)$ denote the amount of labelled material in compartment i at time t, in the same units as $\bar{x}_i$. Assume that the labelled material has the same transport properties as the unlabelled material (see Jacquez, 1972, Chapter 5 for a discussion of isotope fractionation effects). For donor-controlled equations of the form of eqn (9.3), the flow rate from compartment i to compartment j is approximately $k_{ji}f_i(\bar{x}_i)$, so that, in a small time interval Δt around time t, the amount of material transferred is approximately $k_{ji}f_i(\bar{x}_i)\cdot\Delta t$, of which the fraction $\dfrac{\delta_i(t)}{\bar{x}_i}$ is labelled. Thus the flow rate of labelled material is $\dfrac{\delta_i(t)\cdot k_{ji}f_i(\bar{x}_i)}{\bar{x}_i}$ and the tracer equations are

$$\frac{\mathrm{d}\delta_i}{\mathrm{d}t} = \sum_{\substack{j=1 \\ j\neq i}}^{p}\left[k_{ij}f_j(\bar{x}_j)\frac{\delta_j}{\bar{x}_j} - k_{ji}f_i(\bar{x}_i)\frac{\delta_i}{\bar{x}_i}\right] - k_{0i}f_i(\bar{x}_i)\frac{\delta_i}{\bar{x}_i},$$

$$t > 0;\ i = 1, 2, \ldots, p. \tag{9.38}$$

Returning to the example of eqns (9.35), assume that a small quantity of tracer is rapidly administered at $t = 0$ to compartment 1. The tracer equations (for $t > 0$) are then:

$$\frac{d\delta_1}{dt} = -k_1(\bar{x}_1)^2\frac{\delta_1}{\bar{x}_1} - \frac{V_m\bar{x}_1}{K_m + \bar{x}_1}\frac{\delta_1}{\bar{x}_1} + k_{12}\bar{x}_2\frac{\delta_2}{\bar{x}_2}$$

$$= -\left(k_1\bar{x}_1 + \frac{V_m}{K_m + \bar{x}_1}\right)\delta_1 + k_{12}\delta_2 \qquad (9.39a)$$

$$\frac{d\delta_2}{dt} = \frac{V_m}{K_m + \bar{x}_1}\delta_1 - k_{12}\delta_2 - k_0 \qquad (9.39b)$$

with $\delta_1(0) = D_1$; $\delta_2(0) = 0$.

The development given above of the tracer equations has assumed that the tracee (the unlabelled material) is in a steady state. In some experiments, it is possible to measure the tracee as well as the tracer, and occasionally it is found that the tracee is not in a steady state (see, for example, Smith and Llaurado, 1974; Staub *et al.*, 1981). In this case, it is recommended that the same compartmental structure should be used to describe the kinetics of both the tracee and the tracer (Ottaway and Mowbray, 1973); the initial conditions will be markedly different and there may be a difference in the input to the model from external sources. An example of a system with tracee not in a steady state (Staub *et al.*, 1981) is discussed in Application Example 3 in Chapter 11.

9.4 Models with Michaelis–Menten elimination kinetics

We will conclude this chapter with a more detailed examination of one particular type of nonlinearity; that in which the elimination rate is capacity-limited and can be described by Michaelis–Menten kinetics (Michaelis and Menten, 1913). Such a system is of considerable practical importance, since there is strong evidence (Wagner, 1973b) that many substances are eliminated from the body with capacity-limited rates, whether through metabolism or renal excretion.

Most applications which have assumed Michaelis–Menten elimination kinetics have also assumed a one-compartment model (Fig. 9.4) so that the quantity $x_1(t)$ in the compartment is described by the equation

$$\frac{dx_1}{dt} = -\frac{V_m}{K_m + x_1}x_1 + b_1u_1(t). \qquad (9.40)$$

A plot of the elimination rate v. x_1 is shown in Fig. 9.5. In a number of applications in medicine, the initial rates of removal of a substance rapidly

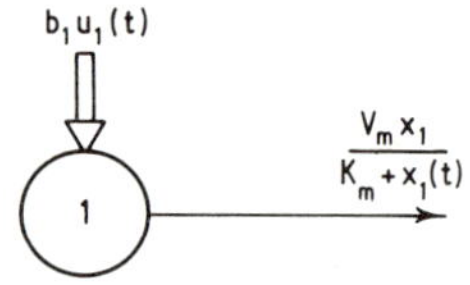

Figure 9.4 One compartment model with Michaelis-Menten elimination kinetics (quantities shown are flow rates).

administered at several different dose levels are measured. One of the best-known substances used in this manner to test liver function is Indocyanine Green (ICG)—see, for example, Paumgartner *et al.* (1970) and Martin *et al.* (1976). With plasma response values available at several different dose levels, some form of plot is made comparing the initial removal rate $(\dot{x}_1(0))$ and dose. The best-known is a plot of $(\dot{x}_1(0))^{-1}$ v. $(\text{dose})^{-1}$ (Lineweaver and Burk, 1934) from which the parameters V_m and K_m can be estimated easily. More recently, Atkins and Nimmo (1975) compared seven methods of fitting the Michaelis–Menten equation and showed that, in many instances, the Lineweaver–Burk plot is not the best way of treating the data, and that a different plot (Eisenthal and Cornish-Bowden, 1974) proves better.

The importance for drug dosage regimen design of recognizing that an elimination is capacity-limited has been graphically illustrated by Wagner (1978) in a figure reproduced here as Fig. 9.6, which shows the plasma concentration plotted against dose number for a rapidly administered dose

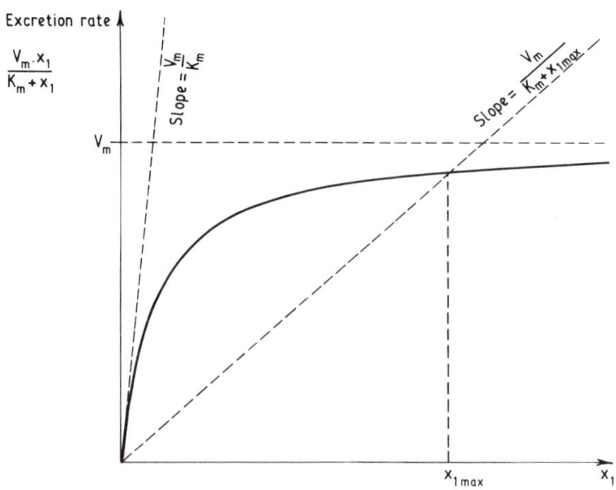

Figure 9.5 Michaelis-Menten saturation curve, showing slopes corresponding to $x_1 = 0$ and $x_1 = x_{1\,\text{MAX}}$.

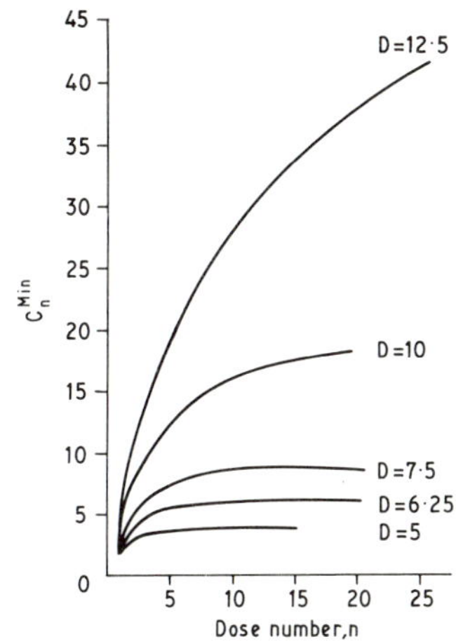

Figure 9.6 Build-up of concentration in one-compartment model with Michaelis-Menten elimination kinetics ($V_m = 15$, $K_m = 12$) with doses (shown on curves) administered at regular intervals of 1 time unit. (Reproduced from Wagner (1978), with permission.)

given at regular intervals of 1 time unit. The system is as described in eqn (9.40), with $V_m = 15$, $K_m = 12$ and the curves shown are for doses of 5, 6.25, 7.5, 10 and 12.5. Note how the concentration at any particular value of t increases considerably more than the increase of dose. (The concentrations shown are those just prior to the administration of a dose.) It would thus be all too easy to overdose using a regimen designed on the assumption of linear elimination; overdosing for this reason has been reported (see, for example, Garrettson and Jusko, 1975).

In pharmacokinetics, interest has been focused on estimating the parameters V_m and K_m from the response to a *single* dose (Wagner, 1973a,b). Although this approach has been criticized recently (Tong and Metzler, 1980), it seems valid provided the checklist given by Wagner (1973b) is followed. The two most important parts of this are that, for the particular drug, there should be evidence from animal studies at different dosages that the elimination departs from linear form and that for the particular response, there should be significant departure from the response of a linear model; the latter, as we will see, requires a substantial part of an amplitude range to be covered.

The first published work on fitting V_m and K_m to a single response was by Lundquist and Wolthers (1958), applied to alcohol elimination data. They also gave analytical expressions for concentration for the one-compartment model with inputs of impulsive and constant-infusion form. These analytical expressions will be derived in Section 9.4.1, then aspects of identifiability will be discussed in Section 9.4.2, and the chapter will conclude with a discussion of some aspects of two-compartment models with Michaelis–Menten elimination kinetics.

9.4.1 *Analysis of one-compartment model with Michaelis–Menten elimination kinetics* (Fig. 9.4)

The differential equation for $x_1(t)$ is given by eqn (9.40). We will consider the impulse and constant infusion inputs in turn.

(a) Impulsive input. For an impulsive input $D_1\,\delta(t)$ to a system previously with $x_1(t) = 0$ $(t < 0)$, eqn (9.40) becomes

$$\frac{\mathrm{d}x_1}{\mathrm{d}t} = -\frac{V_m}{K_m + x_1}x_1; \qquad x_1(0^+) = D_1. \tag{9.41}$$

The variables can be separated:

$$\left(\frac{K_m}{x_1} + 1\right)\mathrm{d}x_1 = -V_m\,\mathrm{d}t$$

so that, integrating, $[K_m \ln x_1 + x_1]_{D_1}^{x_1} = -[V_m t]_0^t$

i.e.
$$t = \frac{1}{V_m}\left(D_1 - x_1 + K_m \ln\frac{D_1}{x_1}\right). \tag{9.42}$$

Wagner (1973a) notes that Henri (1902) published this equation eleven years before Michaelis and Menten published the differential form!

The integrated form thus expresses t as a function of x_1; clearly no analytical expression for x_1 as a function of t could be found. For the linear system,

$$x_1(t) = D_1 \exp(-k_{01}t)$$

so that the expression corresponding to eqn (9.42) is

$$t = \frac{1}{k_{01}} \ln\frac{D_1}{x_1}. \tag{9.43}$$

Plots of x_1 v. t are shown in Fig. 9.7 for three values of K_m; all the graphs have $D_1 = 1$ and $V_m = 1$. In Fig. 9.7(a) $K_m = 0.1$; in Fig. 9.7(b), $K_m = 1$ and in Fig. 9.7(c), $K_m = 10$. The response of the nonlinear system lies somewhere

---

Figure 9.7 Unit impulse responses of one-compartment model with Michaelis–Menten elimination characteristics (eqn (9.42)) with $V_m = 1$; (a) $K_m = 0.1$; (b) $K_m = 1$; (c) $K_m = 10$; Also shown are bounds for linear system responses, given by eqns (9.44).

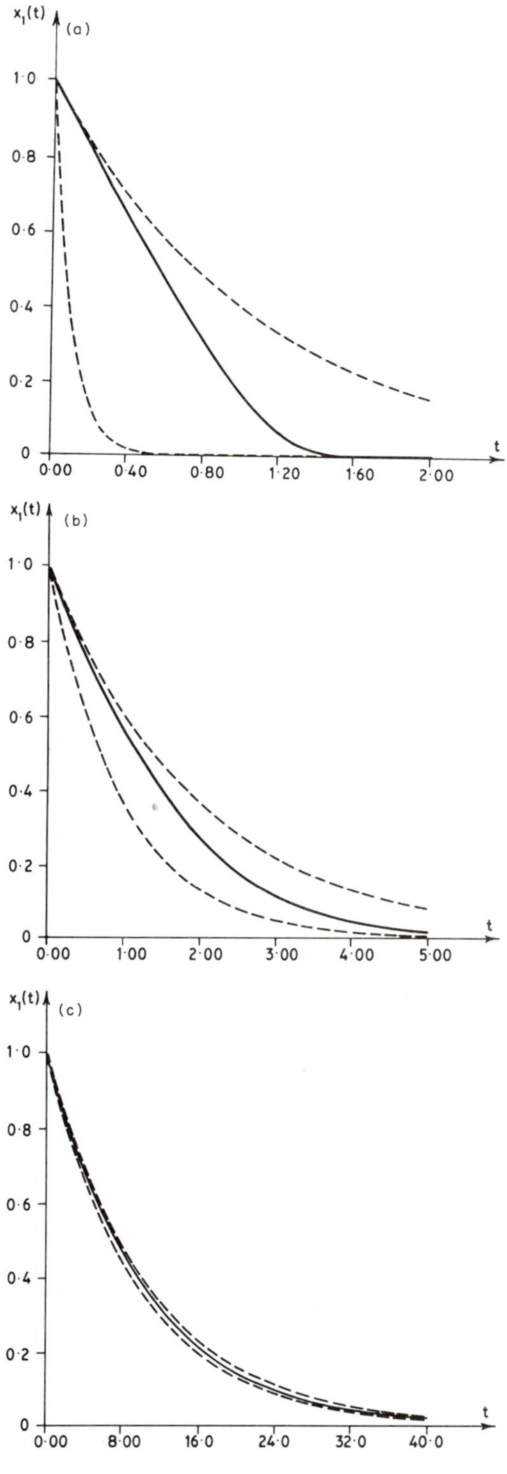
x1(t)
(a)
1·0
0·8
0·6
0·4
0·2
0
t
0·00
0·40
0·80
1·20
1·60
2·00
x1(t)
(b)
1·0
0·8
0·6
0·4
0·2
0
t
0·00
1·00
2·00
3·00
4·00
5·00
x1(t)
(c)
1·0
0·8
0·6
0·4
0·2
0
t
0·00
8·00
16·0
24·0
32·0
40·0

between two bounds which are the responses of linear systems (Tong and Metzler, 1980). The lower bound corresponds to a first-order elimination with rate constant $\frac{V_m}{K_m}$, and is the value approached when x_1 becomes small. The upper bound corresponds to a first-order elimination with rate constant $\frac{V_m}{K_m + x_{1\max}}$ (see Fig. 9.5) where $x_{1\max}$, the maximum of x_1, is equal to D_1 for an impulsive input. Thus the linear bounds shown in Fig. 9.7 are

$$\text{Lower bound} = D_1 \exp\left(-\frac{V_m}{K_m}t\right) \tag{9.44a}$$

and

$$\text{Upper bound} = D_1 \exp\left(-\frac{V_m}{K_m + D_1}t\right). \tag{9.44b}$$

The features to note in Fig. 9.7 are the initial approximately linear part of the curve, when x_1 is large in comparison with $K_m \ln \frac{D_1}{x_1}$, and the way that the upper and lower bounds get closer together as K_m increases. The latter has implications for identifiability, as we will see in Section 9.4.2. The curves of Fig. 9.7 are replotted semilogarithmically in Fig. 9.8. The departure from a straight line of the nonlinear system response is clear in Fig. 9.8(a), is less clear in Fig. 9.8(b) and is not at all clear in Fig. 9.8(c) with $K_m = 10$ (as expected with the upper and lower linear bounds so close).

(b) Constant continuous infusion. Assume that, at $t = 0$, a constant continuous infusion at rate k_i per unit time is commenced into a one-compartment model with $x(0^-) = x_0$. Equation (9.40) then becomes:

$$\frac{dx_1}{dt} = -\frac{V_m x_1}{K_m + x_1} + k_i = \frac{(k_i - V_m)x_1 + K_m k_i}{K_m + x_1}, \qquad t > 0 \tag{9.45}$$

in which t and x_1 can be separated:

$$dt = \frac{K_m\, dx_1}{(k_i - V_m)x_1 + K_m k_i} + \frac{x_1\, dx_1}{(k_i - V_m)x_1 + K_m k_i}. \tag{9.46}$$

For the special case of $k_i = V_m$ (so that the input rate equals the maximum excretion rate), the solution is

$$t = \frac{1}{k_i}(x_1(t) - x_0) + \frac{1}{2K_m k_i}(x_1^2(t) - x_0^2). \tag{9.47}$$

Figure 9.8 Responses of Fig. 9.7 replotted on log-linear scale.

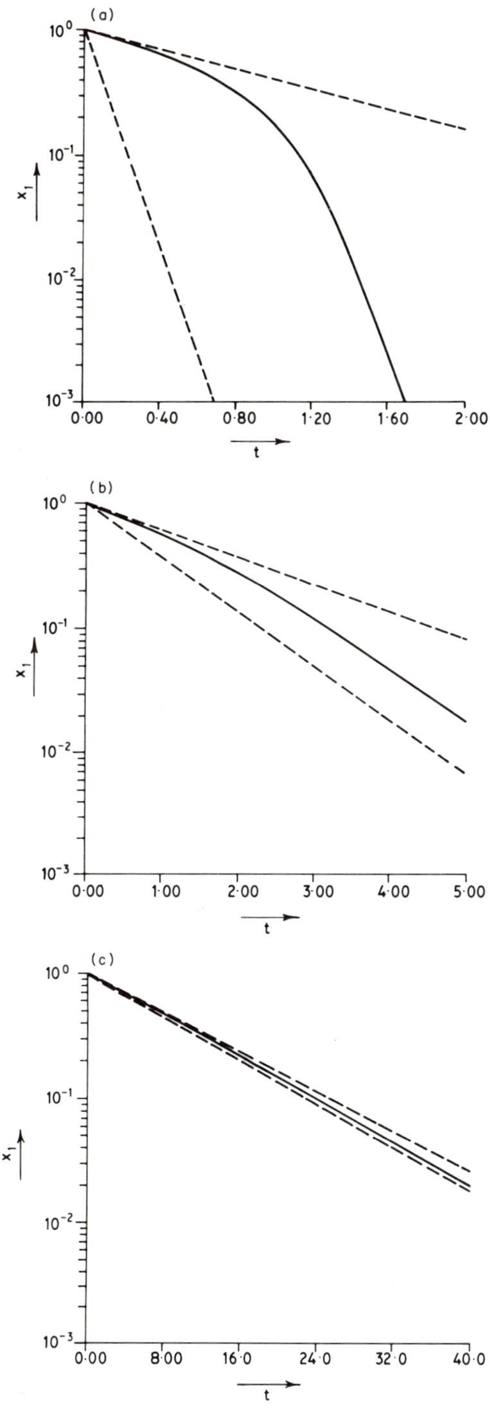

(a)
10^0
10^{-1}
10^{-2}
10^{-3}
x_1
0·00
0·40
0·80
1·20
1·60
2·00
t
(b)
10^0
10^{-1}
10^{-2}
10^{-3}
x_1
0·00
1·00
2·00
3·00
4·00
5·00
t
(c)
10^0
10^{-1}
10^{-2}
10^{-3}
x_1
0·00
8·00
16·0
24·0
32·0
40·0
t

For $k_i \neq V_m$, the integral of the first term on the right-hand side of eqn (9.46) is

$$\frac{K_m}{k_i - V_m} \ln\left[(k_i - V_m)x_1 + K_m k_i\right]$$

while the integral of the second term is

$$-\frac{K_m k_i}{(k_i - V_m)^2} \ln\left[(k_i - V_m)x_1 + K_m k_i\right] + \frac{x_1}{k_i - V_m}.$$

Adding and inserting limits,

$$t = \frac{x_1(t) - x_0}{k_i - V_m} - \frac{V_m K_m}{(k_i - V_m)^2} \ln \frac{(k_i - V_m)x_1(t) + K_m k_i}{(k_i - V_m)x_0 + K_m k_i}. \tag{9.48}$$

The quantity $x_1(t)$ only reaches a steady value if the input rate, k_i, is less than the maximum excretion rate, V_m; otherwise $x_1(t)$ increases indefinitely.

Some examples of responses for various values of k_i are given in Fig. 9.9. Upper and lower bounds for linear systems are calculated as before, the lower bound being the solution of

$$\dot{x}_1(t) = -\frac{V_m}{K_m} x_1(t) + k_i$$

i.e.

$$x_1(t) = \frac{k_i K_m}{V_m}\left(1 - e^{-(V_m/K_m)t}\right) \tag{9.49}$$

and the upper bound being the solution of

$$\dot{x}_1(t) = -\frac{V_m}{K_m + x_{1\max}} x_1(t) + k_i$$

i.e.

$$x_1(t) = \frac{k_i(K_m + x_{1\max})}{V_m}\left(1 - e^{-(V_m/(K_m + x_{1\max}))t}\right). \tag{9.50}$$

An obvious problem with a step input is how to assign a maximum value for x_1 when $k_i \geqslant V_m$; the approach adopted here is to consider an infusion of finite length and simply to take the value of the nonlinear system response at the end of the infusion as the value to be used in eqn (9.50).

In all the examples in Fig. 9.9, $x_0 = 0$ and $V_m = K_m = 1$. In Fig. 9.9(a), $k_i = 0.5$, so that the nonlinear system response is, from eqn (9.48),

$$t = -2x_1(t) - 4 \ln\left(1 - x_1(t)\right)$$

which is shown as the middle curve; the steady-state value of x_1 is 1. From eqn (9.49), the lower bound is

$$x_1(t) = 0.5(1 - e^{-t})$$

and substituting $x_{1\max} = 1$ in eqns (9.50), the upper bound is

$$x_1(t) = 1 - e^{-0.5t}.$$

In Fig. 9.9(b), $k_i = 0.9$, so that the nonlinear response is

$$t = -10x_1(t) - 100 \ln\left(1 - \frac{1}{9}x_1(t)\right)$$

for which the steady-state value is 9. The lower and upper bounds are therefore:

$$\text{Lower bound} = 0.9(1 - e^{-t})$$

$$\text{Upper bound} = 9(1 - e^{-0.1t}).$$

In Figs 9.9(c) and (d), $k_i = 1$ so that the nonlinear system response is obtained from eqn (9.47):

$$t = x_1 + \frac{x_1^2}{2}.$$

From eqn (9.49), the lower bound $= 1 - e^{-t}$. In Fig. 9.9(c), the infusion is (arbitrarily) assumed to cease at $t = 3$, for which $x_{1\max} = 1.646$, so that

$$\text{Upper bound} = 2.646(1 - e^{-0.378t}).$$

In Fig. 9.9(d), the infusion is (arbitrarily) assumed to cease at $t = 60$, for which $x_{1\max} = 10$, so that

$$\text{Upper bound} = 11(1 - e^{-0.091t}).$$

Finally, in Fig. 9.9(e), $k_i = 1.5$, so that the nonlinear system response is again obtained from eqn (9.48) and is

$$t = 2x_1 - 4 \ln(1 + 0.333x_1(t)).$$

From eqn (9.49), the lower bound is $1.5(1 - e^{-t})$. The infusion is (arbitrarily) assumed to cease at $t = 2$, for which $x_{1\max} = 2.036$, and from eqn (9.50),

$$\text{Upper bound} = (3.036)(1.5)(1 - e^{-0.329t}) = 4.554(1 - e^{-0.329t}).$$

From inspection of the five parts of Fig. 9.9, it is evident that the nonlinear responses do not differ greatly from the upper bounds if K_m and $x_{1\max}$ are similar (as in Fig. 9.9(a), (c) and (e)) and only when $x_{1\max}$ is considerably

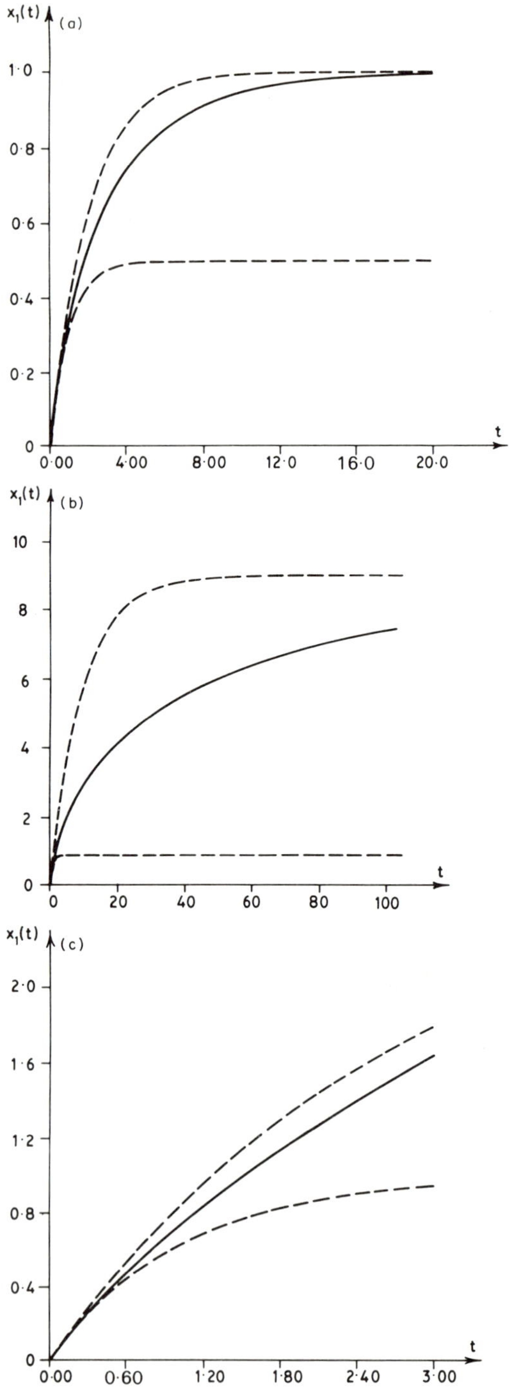

(a)
$x_1(t)$
1·0
0·8
0·6
0·4
0·2
0
t
0·00
4·00
8·00
12·0
16·0
20·0
(b)
$x_1(t)$
10
8
6
4
2
0
t
0
20
40
60
80
100
(c)
$x_1(t)$
2·0
1·6
1·2
0·8
0·4
0
t
0·00
0·60
1·20
1·80
2·40
3·00

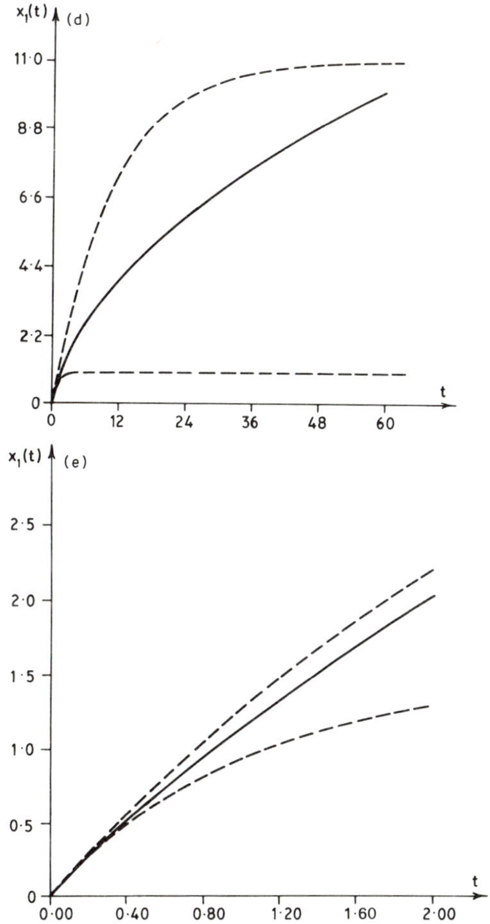

Figure 9.9 Step response of one-compartment model with Michaelis–Menten elimination characteristics with $V_m = K_m = 1$. Nonlinear responses, eqn (9.48); lower bounds, eqn (9.49); upper bounds, eqn (9.50). (a) $k_i = 0.5$; (b) $k_i = 0.9$; (c) $k_i = 1.0$, assuming infusion finishes at $t = 3$; (d) $k_i = 1.0$, assuming infusion finishes at $t = 60$; (e) $k_i = 1.5$, assuming infusion finishes at $t = 2$.

greater than K_m (as in Fig. 9.9(b) and (d)) does the nonlinear curve differ sufficiently from the upper bound to give some hope that the parameters V_m and K_m describing the nonlinearity could be identified with reasonable accuracy.

(c) Other forms of input. Lundquist and Wolthers (1958) present an analytical solution for $u_1(t) = Kx_1(t)$, where K is a constant, for which a solution by separation of variables $x_1(t)$ and t is again possible. This case corresponds to an input rate proportional to the quantity in the compartment. This solution does not appear to be of any great practical significance, and will not, therefore, be set out here.

Apart from this, and the impulse and step inputs considered above, it is not usually possible to find an analytical solution for x_1 in terms of t or vice versa. No analytical solution exists, for example, for the case of first-order input rate (important in biomedicine for modelling an input via an extravascular route) for which the differential equation is

$$\dot{x}_1 = -\frac{V_m}{K_m + x_1} x_1 + k_a b_A D_A \, e^{-k_a t} \tag{9.51}$$

where D_A is the dose administered, b_A is the input fraction, and k_a is the absorption rate constant.

A solution can still be computed numerically, and in Fig. 9.10(a) and (b) are shown solutions, for both of which $k_a = b_A = D_A = 1$, so that the nonlinear response is the solution of

$$\dot{x}_1 = -\frac{V_m}{K_m + x_1} x_1 + e^{-t}. \tag{9.52}$$

The upper and lower linear bounds are found as before. For the lower bound, the differential equation is

$$\dot{x}_1 = -\frac{V_m}{K_m} x_1 + k_a b_A D_A \, e^{-k_a t} \tag{9.53}$$

while for the upper bound, it is

$$\dot{x}_1 = -\frac{V_m}{K_m + x_{1\max}} x_1 + k_a b_A D_A \, e^{-k_a t}. \tag{9.54}$$

In Fig. 9.10(a), $V_m = 1$ and $K_m = 0.1$. Solving eqn (9.53), the lower bound is

$$x_1(t) = \tfrac{1}{9}(e^{-t} - e^{-10t}).$$

From the solution of eqn (9.52), $x_{1\max} = 0.148$, so that the differential equation for the upper bound is

$$\dot{x}_1 = -4.03x_1 + e^{-t},$$

with solution $$x_1(t) = 0.33(e^{-t} - e^{-4.03t}).$$

In Fig. 9.10(b), $V_m = K_m = 1$, so that the lower bound is

$$x_1(t) = t\, e^{-t}.$$

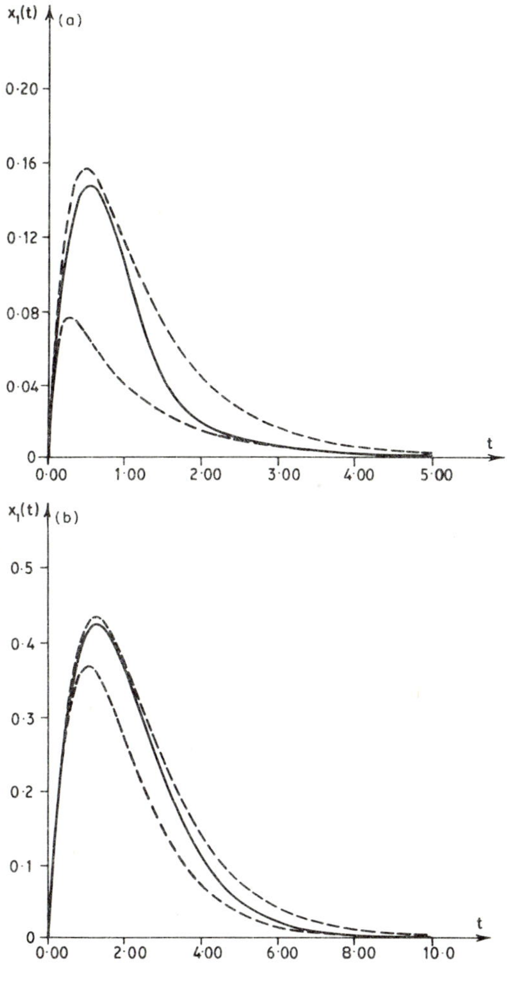

Figure 9.10 Responses of one-compartment model with Michaelis–Menten elimination kinetics to a first-order input, $b_1u_1(t) = e^{-t}$. Nonlinear responses obtained by numerical solution of eqn (9.52). Lower bounds: solution of eqn (9.53); Upper bounds: solution of eqn (9.54). (a) $V_m = 1$, $K_m = 0.1$; (b) $V_m = 1$, $K_m = 1$.

This time, the maximum value of the nonlinear response is 0.426, so that the differential equation for the upper bound is

$$\dot{x}_1 = -0.701x_1 + e^{-t},$$

with solution

$$x_1(t) = 3.347(e^{-0.701t} - e^{-t}).$$

The bounds would of course be more widely separated if there were a greater difference between K_m and $K_m + x_{1\max}$.

An example of fitting Michaelis–Menten parameters to drug response curves is considered in Application Example 6 in Chapter 11.

9.4.2 *Identifiability of a one-compartment model with Michaelis–Menten elimination kinetics* (Fig. 9.4)

Aspects of numerical identifiability have become apparent as we have proceeded through the examples of Section 9.4.1. In order to be able to determine V_m and K_m with reasonable accuracy, the nonlinear response must depart significantly from the form resulting from linear dynamics, and this requires that K_m and $K_m + x_{1\max}$ must be significantly different.

Turning now to deterministic identifiability, we will consider three forms of input—impulsive, constant ($t > 0$) and first-order.

(a) Impulsive input. We have already seen in Example 3 of Section 9.2, that, to be able to identify a nonlinear system, it is usually necessary to know, in principle, the state $x_1(0^+)$ and its derivatives as well as the observation $y_1(0^+)$ and its derivatives. Let us reinforce that conclusion by considering an impulsive input applied to the model of Fig. 9.4, so that

$$x_1(0^+) = b_1 D_1 \tag{9.55}$$

where b_1 is the input fraction, and

$$y_1(0^+) = c_1 b_1 D_1 \tag{9.56}$$

where c_1 is the observation gain. Since D_1 is known, the product $c_1 b_1$ can be identified from eqn (9.56). From eqn (9.40),

$$\dot{x}_1(t) = -\frac{V_m}{K_m + x_1(t)} x_1(t), \qquad t > 0 \tag{9.57}$$

so that

$$\dot{y}_1(0^+) = -\frac{V_m}{K_m + x_1(0^+)} y_1(0^+) \tag{9.58}$$

from which $\dfrac{V_m}{K_m + x_1(0^+)}$ can be identified.

Differentiating eqn (9.57),

$$\ddot{x}_1(t) = \frac{-V_m(K_m + x_1(t))\dot{x}_1(t) + V_m x_1(t)\dot{x}_1(t)}{(K_m + x_1(t))^2}, \qquad t > 0$$

$$= -\frac{V_m K_m}{(K_m + x_1(t))^2}\dot{x}_1(t) \tag{9.59}$$

so that
$$\ddot{y}_1(0^+) = -\frac{V_m K_m}{(K_m + x_1(0^+))^2}\,\dot{y}_1(0^+) \tag{9.60}$$

from which $\frac{V_m K_m}{(K_m + x_1(0^+))^2}$ can be identified.

If $x_1(0^+)$ were known, this would complete the identification, since V_m and K_m could be identified from eqns (9.58) and (9.60), but let us investigate whether the third derivative would have given V_m and K_m when $x_1(0^+)$ was unknown. Differentiating eqn (9.59),

$$\dddot{x}_1(t) = \frac{-V_m K_m (K_m + x_1(t))^2 \ddot{x}_1(t) + V_m K_m \dot{x}_1(t) 2(K_m + x_1(t))\dot{x}_1(t)}{(K_m + x_1(t))^4}, \qquad t > 0$$

$$= -\frac{V_m K_m}{(K_m + x_1(t))^2}\,\ddot{x}_1(t) + \frac{2V_m K_m}{(K_m + x_1(t))^3}\,(\dot{x}_1(t))^2 \tag{9.61}$$

so that
$$\dddot{y}_1(0^+) = -\frac{V_m K_m}{(K_m + x_1(0^+))^2}\,\ddot{y}_1(0^+) + \frac{2V_m K_m}{(K_m + x_1(0^+))^3}\,\dot{x}_1(0^+)\dot{y}_1(0^+) \tag{9.62}$$

so that the system is still unidentifiable, because the coefficient of $\dot{y}_1(0^+)$ contains the additional unknown $\dot{x}_1(0^+)$.

(b) Constant input, $t > 0$. The differential equation for x_1 is eqn (9.45), with $x_1(0^+) = 0$. Assume that the observation gain c_1 is known so that $\dot{x}_1(0^+)$, $\ddot{x}_1(0^+)\ldots$ are known from the observation and its derivatives at $t = 0^+$. (Note again that deterministic identifiability applies to almost all parameter values, so that special cases such as $k_i = V_m$ are excluded.) From eqn (9.45),

$$\dot{x}_1(0^+) = k_i \text{ (which gives } k_i \text{ if it is not already known)}$$

and
$$\ddot{x}_1(t) = -\frac{V_m K_m}{(K_m + x_1(t))^2}\,\dot{x}_1(t), \qquad t > 0 \tag{9.63}$$

so that
$$\ddot{x}_1(0^+) = -\frac{V_m}{K_m}\,k_i, \text{ which gives } \frac{V_m}{K_m}.$$

Differentiating eqn (9.63)

$$\dddot{x}_1(t) = -\frac{V_m K_m}{(K_m + x_1(t))^2}\,\ddot{x}_1(t) + \frac{2V_m K_m}{(K_m + x_1(t))^3}\,(\dot{x}_1(t))^2, \qquad t > 0 \tag{9.64}$$

so that
$$\dddot{x}_1(0^+) = -\frac{V_m}{K_m}\ddot{x}_1(0^+) + \frac{2V_m}{K_m^2}(\dot{x}_1(0^+))^2 \tag{9.65}$$

which gives $\frac{V_m}{K_m^2}$ and completes the identification; V_m, K_m (and k_i) are globally identifiable.

(c) First-order input. The differential equation for x_1 is eqn (9.51) with $x_1(0^+) = 0$. Assume that the observation gain c_1 is known (so that $\dot{x}_1(0^+)$, $\ddot{x}_1(0^+)$ ... are known from the observations). The parameters to be identified are V_m, K_m and k_a. From eqn (9.51),

$$\dot{x}_1(0^+) = k_a b_A D_A \tag{9.66}$$

which gives k_a provided the input fraction b_A and the dose D_A are known. It is less reasonable to assume that b_A is known than it was in (a) since, in biomedical experiments, a first-order input usually models a dose administered via an extravascular route for which the input fraction is also the bioavailability fraction, often unknown. Nevertheless, b_A must be known for the remaining parameters to be identified, as we will see in the remainder of the analysis.

Differentiating eqn (9.51),

$$\ddot{x}_1(t) = -\frac{V_m K_m}{(K_m + x_1(t))^2}\dot{x}_1(t) - k_a^2 b_A D_A\, e^{-k_a t} \qquad t > 0 \tag{9.67}$$

so that
$$\ddot{x}_1(0^+) = -\frac{V_m}{K_m}\dot{x}_1(0^+) - k_a^2 b_A D_A \tag{9.68}$$

whence $\frac{V_m}{K_m}$ can be identified.

Differentiating eqn (9.67),

$$\dddot{x}_1(t) = -\frac{V_m K_m}{(K_m + x_1(t))^2}\ddot{x}_1(t) + \frac{2V_m K_m}{(K_m + x_1(t))^3}(\dot{x}_1(t))^2 + k_a^3 b_A D_A\, e^{-k_a t}, \qquad t > 0 \tag{9.69}$$

so that
$$\dddot{x}_1(0^+) = -\frac{V_m}{K_m}\ddot{x}_1(0^+) + \frac{2V_m}{K_m^2}(\dot{x}_1(0^+))^2 + k_a^3 b_A D_A \tag{9.70}$$

whence $\frac{V_m}{K_m^2}$ can be identified. This completes the identification; the parameters V_m, K_m and k_a are all globally identifiable from this experiment provided the input fraction is known.

9.4.3 *Two-compartment model with Michaelis–Menten elimination kinetics*

The constants V_m and K_m have traditionally been estimated assuming a one-compartment model, but the pharmacokinetics of some drugs are almost certainly better described by a two-compartment model with Michaelis–Menten elimination kinetics. Sedman and Wagner (1974) present some simulation results which show that treatment of two-compartment data by one-compartment analysis can result in erroneous values for V_m and K_m. They suggest carrying out experiments at several different dose levels, to establish the structure of the model and then to estimate V_m and K_m.

Numerical problems would certainly make the estimation of V_m and K_m from a single response quite difficult for a two-compartment model, as we will now see. Two such models are considered, and these are shown in Figs 9.11(a) and (b). For Model A, elimination is from compartment 1 only ($k_{02} = 0$), and the differential equations are

$$\dot{x}_1(t) = -\left(k_{21} + \frac{V_m}{K_m + x_1(t)}\right)x_1(t) + k_{12}x_2(t) + b_1u_1(t) \quad (9.71a)$$

$$\dot{x}_2(t) = k_{21}x_1(t) - k_{12}x_2(t) + b_2u_2(t) \quad (9.71b)$$

while for Model B, elimination is from compartment 2 only, so that

$$\dot{x}_1(t) = -k_{21}x_1(t) + k_{12}x_2(t) + b_1u_1(t) \quad (9.72a)$$

$$\dot{x}_2(t) = k_{21}x_1(t) - \left(k_{12} + \frac{V_m}{K_m + x_2(t)}\right)x_2(t) + b_2u_2(t). \quad (9.72b)$$

There seems to be no analytical solution for either of these pairs of equations, and a numerical solution has to be computed.

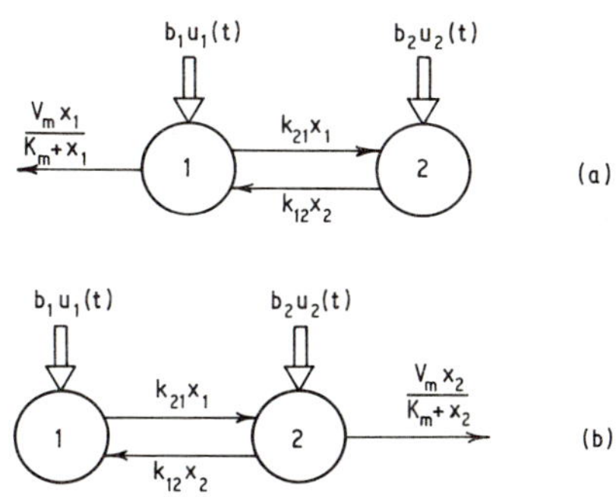

Figure 9.11 Two-compartment models with Michaelis–Menten elimination kinetics (quantities shown are flow rates); (a) Model A; (b) Model B.

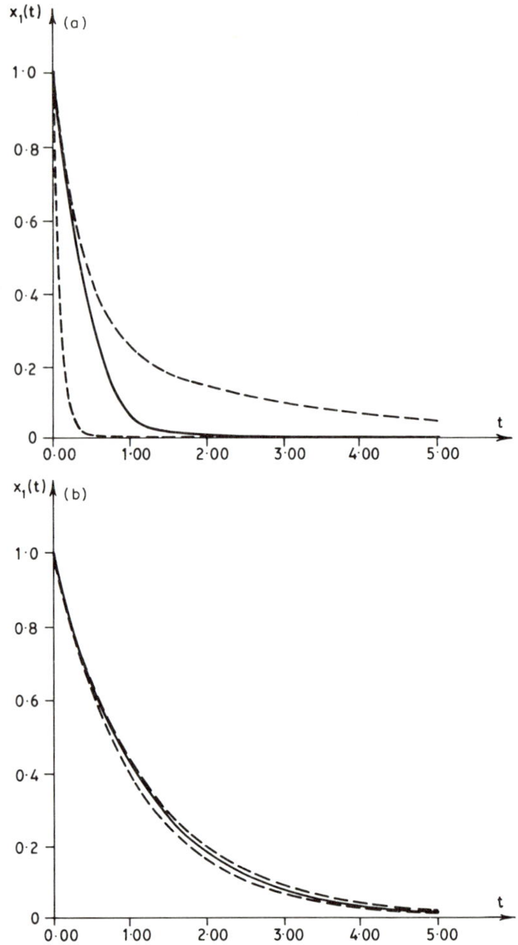

Figure 9.12 Responses $x_1(t)$ and linear bounds of the models of Fig. 9.11 to a unit impulse perturbation of compartment 1. $V_m = 1$, $K_m = 0.1$, $k_{12} = k_{21} = 1$. (a) Model A; (b) Model B.

Responses of $x_1(t)$ following unit impulsive perturbation of compartment 1 ($b_1u_1(t) = \delta(t)$; $b_2u_2(t) = 0$) are shown in Figs 9.12 and 9.13, together with linear system bounds. In Fig. 9.12, the parameter values were $k_{21} = k_{12} = V_m = 1$, $K_m = 0.1$. The nonlinear response $x_1(t)$ and the linear system bounds are shown in Fig. 9.12(a) for Model A and in Fig. 9.12(b) for Model B. In Fig. 9.12(a), $x_{1\max}$ was, of course 1, while in Fig. 9.12(b), $x_{2\max}$ was 0.1228.

In Fig. 9.13, similar responses are shown for $k_{21} = k_{12} = V_m = K_m = 1$.

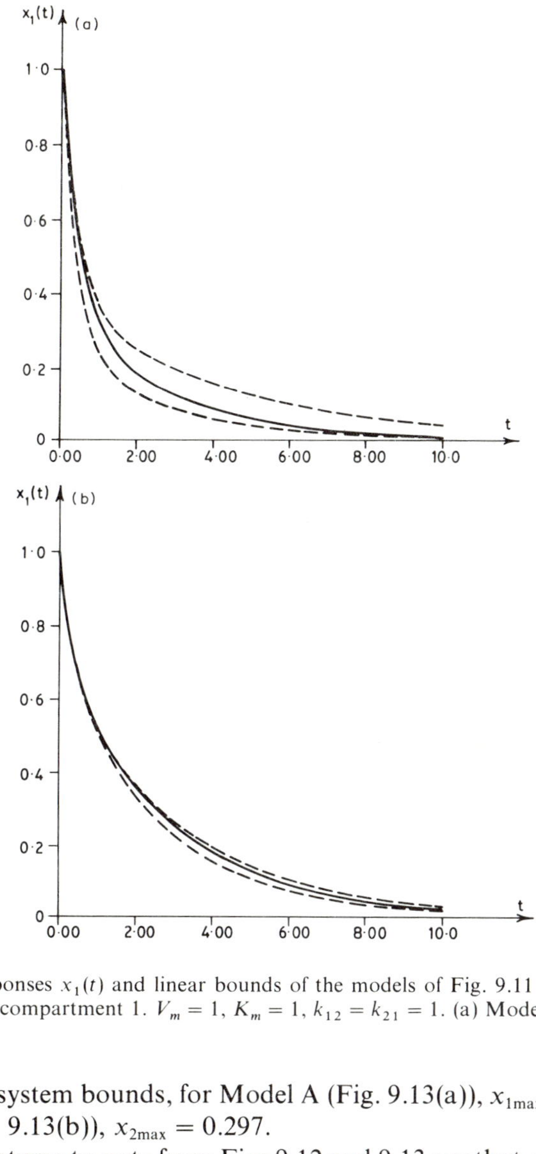

Figure 9.13 Responses $x_1(t)$ and linear bounds of the models of Fig. 9.11 to a unit impulse perturbation of compartment 1. $V_m = 1$, $K_m = 1$, $k_{12} = k_{21} = 1$. (a) Model A; (b) Model B.

For the linear system bounds, for Model A (Fig. 9.13(a)), $x_{1\max} = 1$, while for Model B (Fig. 9.13(b)), $x_{2\max} = 0.297$.

The main features to note from Figs 9.12 and 9.13 are that, for both values of K_m, the linear system bounds are very close together for Model B, so that it would be very difficult (numerically) to estimate V_m and K_m for excretion from one compartment using a response from perturbation and observation of another compartment. In Fig. 9.13(a), with $K_m = 1$, the bounds are getting rather close together even for Model A; they would be further apart for a larger perturbation. The bounds are also dependent on the values of the rate constants k_{12} and k_{21}.

The two models also bring out a very interesting point concerning deterministic identifiability analysis using the Taylor-series approach. One would expect the analysis for Model A to require about the same amount of effort as that for Model B, but that does not prove to be the case. Consider identifiability from an experiment in which $b_1u_1(t) = D_1\,\delta(t)$ (D_1 known), $b_2u_2(t) = 0$, and with compartment 1 (only) observed, with known observation gain. Consider Model B first for which

$$x_1(0^+) = D_1 \quad \text{and} \quad x_2(0^+) = 0.$$

From, eqn (9.72a),

$$\dot{x}_1(0^+) = -k_{21}x_1(0^+), \text{ which gives } k_{21}.$$

From eqn (9.72b),

$$\dot{x}_2(0^+) = k_{21}x_1(0^+). \tag{9.73}$$

Differentiating eqn (9.72a),

$$\ddot{x}_1(t) = -k_{21}\dot{x}_1(t) + k_{12}\dot{x}_2(t), \qquad t > 0 \tag{9.74}$$

so that, substituting from eqn (9.73),

$$\ddot{x}_1(0^+) = -k_{21}\dot{x}_1(0^+) + k_{12}k_{21}x_1(0^+) \tag{9.75}$$

in which everything except k_{12} is known, so that k_{12} can be identified.

Differentiating eqn (9.72b),

$$\ddot{x}_2(t) = k_{21}\dot{x}_1(t) - k_{12}\dot{x}_2(t) - \frac{V_mK_m}{(K_m + x_2(t))^2}\,\dot{x}_2(t) \tag{9.76}$$

so that
$$\ddot{x}_2(0^+) = k_{21}\dot{x}_1(0^+) - \left(k_{12} + \frac{V_m}{K_m}\right)\dot{x}_2(0^+)$$

$$= k_{21}\dot{x}_1(0^+) - \left(k_{12} + \frac{V_m}{K_m}\right)k_{21}x_1(0^+). \tag{9.77}$$

Differentiating eqn (9.74),

$$\dddot{x}_1(t) = -k_{21}\ddot{x}_1(t) + k_{12}\ddot{x}_2(t), \qquad t > 0 \tag{9.78}$$

so that, substituting from eqn (9.77)

$$\dddot{x}_1(0^+) = -k_{21}\ddot{x}_1(0^+) + k_{12}k_{21}\dot{x}_1(0^+) - k_{12}k_{21}\left(k_{12} + \frac{V_m}{K_m}\right)x_1(0^+) \tag{9.79}$$

whence $\dfrac{V_m}{K_m}$ can be identified.

Differentiating eqn (9.76),

$$\ddot{x}_2(t) = k_{21}\ddot{x}_1(t) - k_{12}\ddot{x}_2(t) - \frac{V_m K_m}{(K_m + x_2(t))^2}\ddot{x}_2(t) + \frac{2V_m K_m}{(K_m + x_2(t))^3}(\dot{x}_2(t))^2 \tag{9.80}$$

so that

$$\begin{aligned}\ddot{x}_2(0^+) &= k_{21}\ddot{x}_1(0^+) - \left(k_{12} + \frac{V_m}{K_m}\right)\ddot{x}_2(0^+) + \frac{2V_m}{K_m^2}(\dot{x}_2(0^+))^2 \\ &= k_{21}\ddot{x}_1(0^+) - \left(k_{12} + \frac{V_m}{K_m}\right)\left[k_{21}\dot{x}_1(0^+) - \left(k_{12} + \frac{V_m}{K_m}\right)k_{21}x_1(0^+)\right] \\ &\quad + \frac{2V_m}{K_m^2}k_{21}^2(x_1(0^+))^2.\end{aligned} \tag{9.81}$$

Differentiating eqn (9.78),

$$\dddot{x}_1(t) = -k_{21}\ddot{x}_1(t) + k_{12}\ddot{x}_2(t), \qquad t > 0 \tag{9.82}$$

so that

$$\dddot{x}_1(0^+) = -k_{21}\ddot{x}_1(0^+) + k_{12}\ddot{x}_2(0^+). \tag{9.83}$$

From the form of $\ddot{x}_2(0^+)$ given in eqn (9.81), it may be seen that this equation gives $\dfrac{V_m}{K_m^2}$ and this completes the identification: k_{12}, k_{21}, V_m and K_m are globally identifiable from this experiment.

Consider now Model A. For the same experiment,

$$x_1(0^+) = D_1 \quad \text{and} \quad x_2(0^+) = 0.$$

From eqn (9.71a),

$$\dot{x}_1(0^+) = -\left(k_{21} + \frac{V_m}{K_m + x_1(0^+)}\right)x_1(0^+). \tag{9.84}$$

From eqn (9.71b),

$$\dot{x}_2(0^+) = k_{21}x_1(0^+). \tag{9.85}$$

Differentiating eqn (9.71a),

$$\ddot{x}_1(t) = -\left(k_{21} + \frac{V_m K_m}{(K_m + x_1(t))^2}\right)\dot{x}_1(t) + k_{12}\dot{x}_2(t), \qquad t > 0 \tag{9.86}$$

so that, substituting from eqn (9.85),

$$\ddot{x}_1(0^+) = -\left(k_{21} + \frac{V_m K_m}{(K_m + x_1(0^+))^2}\right)\dot{x}_1(0^+) + k_{12}k_{21}x_1(0^+). \tag{9.87}$$

Differentiating eqn (9.71b),

$$\ddot{x}_2(t) = k_{21}\dot{x}_1(t) - k_{12}\dot{x}_2(t) \tag{9.88}$$

so that

$$\ddot{x}_2(0^+) = k_{21}\dot{x}_1(0^+) - k_{12}k_{21}x_1(0^+). \tag{9.89}$$

Differentiating eqn (9.86)

$$\ddot{x}_1(t) = -\left(k_{21} + \frac{V_m K_m}{(K_m + x_1(t))^2}\right)\dot{x}_1(t) + \frac{2V_m K_m}{(K_m + x_1(t))^3}(\dot{x}_1(t))^2 + k_{12}\dot{x}_2(t), \qquad t > 0 \tag{9.90}$$

so that, substituting from eqn (9.89),

$$\ddot{x}_1(0^+) = -\left(k_{21} + \frac{V_m K_m}{(K_m + x_1(0^+))^2}\right)\dot{x}_1(0^+) + \frac{2V_m K_m}{(K_m + x_1(0^+))^3}(\dot{x}_1(0^+))^2 + k_{12}k_{21}\dot{x}_1(0^+) - k_{12}^2 k_{21} x_1(0^+). \tag{9.91}$$

Differentiating eqn (9.88),

$$\dddot{x}_2(t) = k_{21}\ddot{x}_1(t) - k_{12}\ddot{x}_2(t)$$

so that

$$\ddot{x}_2(0^+) = k_{21}\ddot{x}_1(0^+) - k_{12}k_{21}\dot{x}_1(0^+) + k_{12}^2 k_{21} x_1(0^+). \tag{9.92}$$

Differentiating eqn (9.90),

$$\begin{aligned}\dddot{x}_1(t) = &-k_{21}\ddot{x}_1(t) \\ &+ \frac{-V_m K_m (K_m + x_1(t))^2 \ddot{x}_1(t) + V_m K_m \dot{x}_1(t) 2(K_m + x_1(t))\dot{x}_1(t)}{(K_m + x_1(t))^4} \\ &+ \frac{\{2V_m K_m (K_m + x_1(t))^3 2\dot{x}_1(t)\ddot{x}_1(t) - 2V_m K_m (\dot{x}_1(t))^2 3(K_m + x_1(t))^2 \dot{x}_1(t)\}}{(K_m + x_1(t))^6} \\ &+ k_{12}\ddot{x}_2(t), \qquad t > 0 \\ = &-\left(k_{21} + \frac{V_m K_m}{(K_m + x_1(t))^2}\right)\ddot{x}_1(t) + \frac{6V_m K_m \ddot{x}_1(t)}{(K_m + x_1(t))^3}\dot{x}_1(t) \\ &- \frac{6V_m K_m (\dot{x}_1(t))^3}{(K_m + x_1(t))^4} + k_{12}\ddot{x}_2(t), \qquad t > 0.\end{aligned} \tag{9.93}$$

Substituting for $x_2(0^+)$ from eqn (9.92),

$$\dddot{x}_1(0^+) = -\left(k_{21} + \frac{V_m K_m}{(K_m + x_1(0^+))^2}\right)\ddot{x}_1(0^+) + \frac{6V_m K_m \ddot{x}_1(0^+)\dot{x}_1(0^+)}{(K_m + x_1(0^+))^3}$$
$$- \frac{6V_m K_m (\dot{x}_1(0^+))^3}{(K_m + x_1(0^+))^4} + k_{12}k_{21}\ddot{x}_1(0^+) - k_{12}^2 k_{21}\dot{x}_1(0^+)$$
$$+ k_{12}^3 k_{21} x_1(0^+). \tag{9.94}$$

Thus, the four equations relating the unknown parameters to the first four derivatives of $x_1(t)$ at $t = 0^+$ are eqns (9.84), (9.87), (9.91) and (9.94).

Although there are four equations in the four unknowns, checking for redundancy and multiple solutions proves exceedingly difficult and has so far defeated the attempts of the author and others.

Nevertheless, the point has been made that in some cases, as in Model B, the deterministic identifiability analysis proves reasonably straightforward, while in others superficially very similar, as in Model A, it proves very difficult indeed. In the analysis for Model B, unknown equantities k_{21}, k_{12}, $\frac{V_m}{K_m}$ and $\frac{V_m}{K_m^2}$ entered the equations one at a time, whereas in the analysis for Model A, three of the four parameters to be identified already appeared in the first derivative (eqn (9.84)). By the time $x_1(t)$ had been differentiated four times (eqn (9.94)) parameter combinations of degree four had appeared, with none of the parameters known from the earlier derivatives.

10 Time-varying Compartmental Models and Stochastic Compartmental Models

10.1 Time-varying compartmental models

10.1.1 *Some general comments*

As we have seen in Section 1.1, a linear but time-varying compartmental model has rate coefficients which are functions of time:

$$k_{ij} = k_{ij}(t) \tag{10.1}$$

but which are not functions of the states. Engineers regard such a model as linear, but in some other applications areas, it might be described as nonlinear. The time dependence arises from external influences such as change in ambient temperature, intake of food or variation of light. The mass balance for compartment i is thus:

$$\frac{dx_i}{dt} = \sum_{\substack{j=1 \\ j\neq 1}}^{p} (k_{ij}(t)x_j - k_{ji}(t)x_i) - k_{0i}(t)x_i + u_i(t). \tag{10.2}$$

Mazanov (1976) showed that this model is stable, subject to the limitation that the total instantaneous flow rate from any compartment is greater than or equal to the sum of the flow rates from that compartment to the other compartments, i.e.

$$-a_{ii}(t)x_i(t) \geqslant \sum_{\substack{j=1 \\ j\neq 1}}^{p} k_{ji}(t)x_i(t). \tag{10.3}$$

In theory, the time-varying rate coefficients can take any value, but in practice they will be non-negative. Mazanov also showed that the model remains stable when there are time delays in the intercompartmental transfers.

A very important form of time variation is that in which the rate constants vary with time with period T:

$$k_{ij}(t) = k_{ij}(t + T). \tag{10.4}$$

Many bodily functions vary with a period of approximately 24 h (circadian rhythms) and a similar periodicity can be expected in many ecosystems due to variation in light and temperature. There is also strong evidence that some bodily functions vary periodically with periods shorter than 24 h. Mulholland and Keener (1974) considered compartmental systems with periodically-varying rate coefficients, in the particular context of ecosystem modelling, and showed that if there is no input flow, the compartmental quantities in an open system decay asymptotically to zero, while if there are periodic inputs, the quantities are periodic with the same period once the effects of initial conditions have died away.

Another form of time variation which seems to occur often is that where some (or all) of the rate coefficients suddenly change from one steady value to another. This is a particular feature of many pharmacokinetics trials, and the change may be caused by exercise by a subject previously at rest or the intake of food by a subject who had previously been fasting. In such instances, a change in the smooth response of drug in plasma to an administration occurs at the same time for each subject in a multi-subject trial. As an illustration, consider the one-compartment model with first-order input shown in Fig. 10.1. Two responses of $x_A(t)$ and $x_1(t)$ following a unit impulsive input to compartment 1 are shown in Figs 10.2(a) and (b). For both curves, $k_a = 1$ and $k_{01} = 2$ for $t < 1.5$; for the lower curves in each figure, k_a and k_{01} remain at these values for $t > 1.5$ while for the upper curves, they are halved. Although such a large change may seem extreme, the upper response in Fig. 10.2(b) is by no means unrepresentative of an actual plasma response in a pharmacokinetics trial.

This form of time variation, with rate coefficients suddenly changing from one (steady) value to another, is easy to analyse, using the linear methods in Chapters 3 and 4. For the time variation in the example above,

$$\dot{x}_A(t) = -x_A(t); \qquad x_A(0) = 1, \qquad 0 < t < 1.5 \tag{10.5a}$$

$$\dot{x}_1(t) = x_A(t) - 2x_1(t); \qquad x_2(0) = 0, \qquad 0 < t < 1.5 \tag{10.5b}$$

with solution

$$x_A(t) = e^{-t}, \qquad 0 < t < 1.5 \tag{10.6a}$$

$$x_1(t) = e^{-t} - e^{-2t}, \qquad 0 < t < 1.5. \tag{10.6b}$$

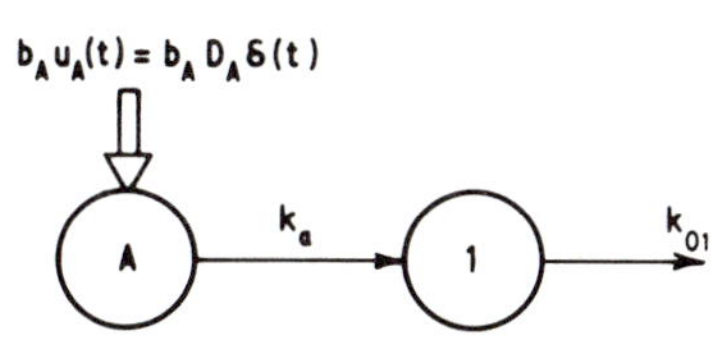

Figure 10.1 One-compartment model with first-order input.

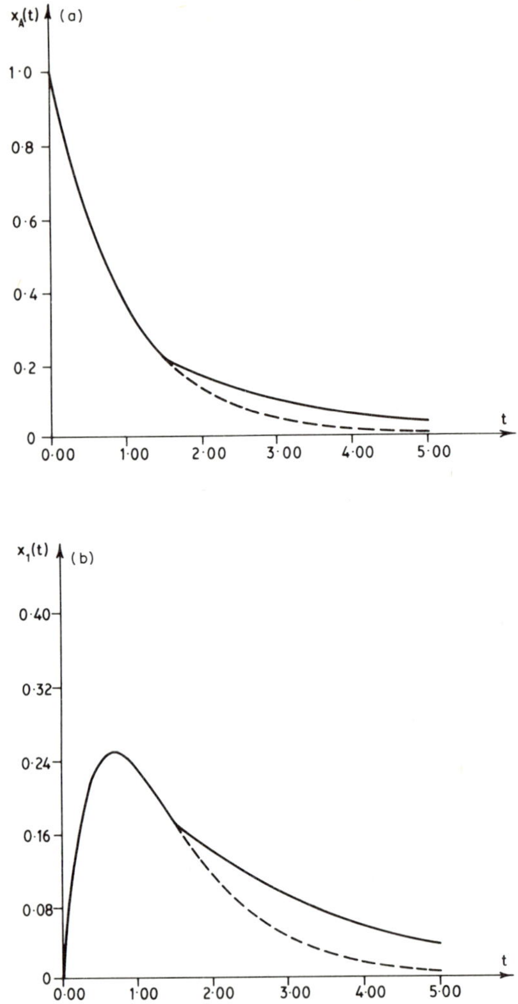

Figure 10.2 Responses of model of Fig. 10.1 to unit impulsive input to compartment 1; (a) $x_A(t)$, (b) $x_1(t)$. (———) $k_a = 1$, $k_{01} = 2$, $t < 1.5$, $k_a = 0.5$, $k_{01} = 1$, $t > 1.5$; (----) $k_a = 1$, $k_{01} = 2$ throughout.

For $t \geqslant 1.5$, defining a variable $t' = t - 1.5$,

$$\dot{x}_A(t') = -0.5x_A(t'); \qquad x_A(0) = e^{-1.5} = 0.223 \tag{10.7a}$$

$$\dot{x}_1(t') = 0.5x_A(t') - x_1(t'); \qquad x_1(0) = e^{-1.5} - e^{-3} = 0.173 \tag{10.7b}$$

which have solution

$$x_A(t') = 0.223\ \mathrm{e}^{-0.5t'} \tag{10.8a}$$

$$x_1(t') = 0.223\ \mathrm{e}^{-0.5t'} - 0.050\ \mathrm{e}^{-t'}. \tag{10.8b}$$

More general forms of time variation are much more difficult to analyse, as we will now see.

10.1.2 *Analysis of time-varying compartmental systems*

In Section 2.3, we saw, that, for a linear time-*invariant* system described by

$$\dot{\mathbf{x}} = \mathbf{Ax} + \mathbf{Bu} \tag{10.9}$$

the time solution is given by

$$\mathbf{x}(t) = \mathbf{\Phi}(t)\mathbf{x}(0^-) + \int_{0^-}^{t} \mathbf{\Phi}(t - \lambda)\mathbf{Bu}(\lambda)\,\mathrm{d}\lambda \tag{10.10}$$

where the transition matrix,

$$\mathbf{\Phi}(t) = \mathrm{e}^{\mathbf{A}t} \tag{10.11}$$

can be found by a number of methods including Laplace transformation of the transfer function matrix. For the linear time-varying system,

$$\dot{\mathbf{x}}(t) = \mathbf{A}(t)\mathbf{x} + \mathbf{B}(t)\mathbf{u} \tag{10.12}$$

the time solution can still be written in terms of a transition matrix $\mathbf{\Phi}(t, t_0)$

$$\mathbf{x}(t) = \mathbf{\Phi}(t, t_0)\mathbf{x}(t_0) + \int_{t_0}^{t} \mathbf{\Phi}(t, t_0)\mathbf{B}(\lambda)\mathbf{u}(\lambda)\,\mathrm{d}\lambda \tag{10.13}$$

but the relationship between $\mathbf{\Phi}(t, t_0)$ and $\mathbf{A}(t)$ is now much more complicated (Richards, 1979, p. 142):

$$\mathbf{\Phi}(t, t_0) = \mathbf{I} + \int_{t_0}^{t} \mathbf{A}(\tau)\,\mathrm{d}\tau + \int_{t_0}^{t} \mathbf{A}(\tau)\left(\int_{0}^{\tau_1} A(\tau_2)\,\mathrm{d}\tau_2\right)\mathrm{d}\tau_1 + \ldots \tag{10.14}$$

In most cases, this is a difficult expression to evaluate and it is necessary to resort to a computer to obtain a numerical solution for a time-varying system.

Some progress can be made towards an analytical solution if $\mathbf{A}(t)$ and the matrix $\mathbf{V}$ defined by

$$\mathbf{V} = \int_{t_0}^{t} \mathbf{A}(\lambda)\,\mathrm{d}\lambda \tag{10.15}$$

commute (that is, $\mathbf{AV} = \mathbf{VA}$). Jacquez (1972, p. 149) shows that the solution for the homogeneous equation ($\mathbf{u} = 0$ in eqn (10.12)) is then

$$\mathbf{x}(t) = \mathrm{e}^{\mathbf{V}}\, \mathbf{x}(t_0), \tag{10.16}$$

where $t_0 = 0^-$ for a washout experiment or $t_0 = 0^+$ following an impulsive perturbation. Unfortunately, $\mathbf{A}$ and $\mathbf{V}$ commute in only very few cases, but one example where they do is if

$$\mathbf{A}(t) = \mathbf{K}\psi(t) \tag{10.17}$$

where $\mathbf{K}$ is a constant matrix and $\psi(t)$ is a (constant) function of time. The transition matrix is then

$$\mathbf{\Phi}(t) = \exp\left[\mathbf{K} \int_{t_0}^{t} \psi(\lambda)\, \mathrm{d}\lambda\right] \tag{10.18}$$

and the solution for $\mathbf{x}(t)$ is the same as that for the time-invariant system ($\psi(t) = 1$), but with t replaced by $\int_{t_0}^{t} \psi(\lambda)\, \mathrm{d}\lambda$.

In compartmental models, for eqn (10.17) to hold, the form and timing of time variation of all the rate coefficients must be the same. An example of practical interest where this may well occur is if all the rate coefficients are sinusoidally varying:

$$a_{ij}(t) = \alpha_{ij}(1 + \gamma \sin(\omega t + \phi)), \qquad i = 1 \ldots p; \qquad j = 1 \ldots p \tag{10.19}$$

where α_{ij} is constant and ω (the radian frequency), ϕ (the phase angle) and γ are constants. Note that all the rate coefficients are sinusoidally varying with the same frequency and phase and have the same proportion γ of cyclic variation. ($\gamma \leqslant 1$ to ensure non-negativity of the rate coefficients.) It is important to stress that $\mathbf{A}$ and $\mathbf{V}$ do not commute with a more general form of sinusoidal variation; in particular they do not when γ is not the same for all rate coefficients.

We will now consider two examples of the variation described by eqn (10.19).

Example 1. One-compartment model (Fig. 10.3). Suppose that the excretion rate coefficient is sinusoidally varying:

$$k_{01}(t) = k_{01}(1 + \gamma \sin(\omega t + \phi)) \tag{10.20}$$

so that

$$\psi(t) = 1 + \gamma \sin(\omega t + \phi) \tag{10.21a}$$

and

$$\int_0^t \psi(\lambda)\, \mathrm{d}\lambda = t + \frac{\gamma}{\omega}(\cos\phi - \cos(\omega t + \phi)). \tag{10.21b}$$

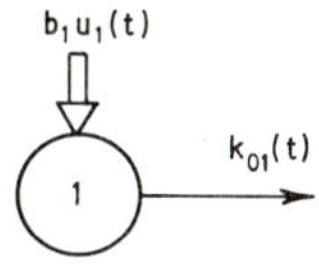

Figure 10.3 One-compartment model with time-varying excretion rate coefficient.

If $x_1(0^-) = 0$, $u_1(t) = D_1\,\delta(t)$ and $b_1 = 1$, the solution for the time-invariant system is, from Section 3.1.1,

$$x_1(t) = D_1 \exp(-k_{01}t) \tag{10.22}$$

so that the corresponding solution for the time-varying system is

$$x_1(t) = D_1 \exp(-k_{01}t) \exp\left[-\frac{k_{01}\gamma}{\omega}(\cos\phi - \cos(\omega t + \phi))\right]. \tag{10.23}$$

(For the one-compartment model, this solution can, of course, also be obtained by direct solution of the differential equation for $x_1(t)$.)

Some features of this solution are illustrated in Figs 10.4 and 10.5, in both of which $k_{01} = \gamma = 1$. In these figures, the response for the time-invariant system ($\gamma = 0$; $\psi(t) = 1$) is also shown for comparison. In Figs 10.4(a) to (c), the effects of injecting at different times during the cycle is shown; for these figures, $\omega = 2\pi$ (i.e. cyclic frequency = 1).

In Fig. 10.4(a), $\phi = 0$, so that the perturbation is applied as $k_{01}(t)$ is rising through its mean value of 1.0 and $k_{01}(t) = 1 + \sin 2\pi t$. From eqn (10.23),

$$x_1(t) = D_1 \exp(-t) \exp\left(-\frac{(1 - \cos 2\pi t)}{2\pi}\right).$$

Now, $1 - \cos 2\pi t$ varies between 0 (at $t = 0, 1, 2, \ldots$) and 2 (at $t = 0.5, 1.5, 2.5, \ldots$) and the second exponential thus varies between 1 and $\exp\left(-\frac{1}{\pi}\right)$. The decay thus has the time-invariant response $D_1 \exp(-t)$ as its *upper* bound (reached at $t = 0, 1, 2, \ldots$), and is less than the time-invariant response by a factor of $\exp\left(-\frac{1}{\pi}\right) = 0.727$ at $t = 0.5, 1.5, 2.5, \ldots$,. The area under the curve (AUC) is smaller for the time-varying system.

In Fig. 10.4(b), $\phi = \frac{\pi}{2}$, so that the perturbation is applied when $k_{01}(t)$ is at its maximum value of 2.0 and $k_{01}(t) = 1 + \sin\left(2\pi t + \frac{\pi}{2}\right) = 1 + \cos 2\pi t$.

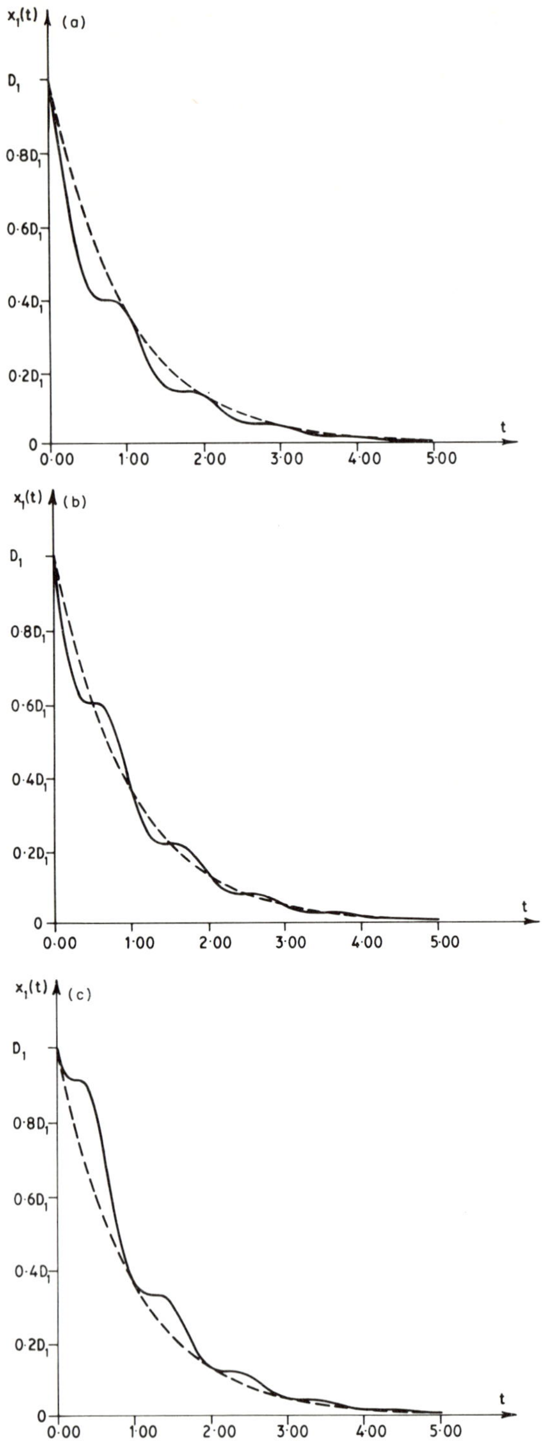

x1(t)
(a)
D1
0·8D1
0·6D1
0·4D1
0·2D1
0
t
0·00
1·00
2·00
3·00
4·00
5·00
x1(t)
(b)
D1
0·8D1
0·6D1
0·4D1
0·2D1
0
t
0·00
1·00
2·00
3·00
4·00
5·00
x1(t)
(c)
D1
0·8D1
0·6D1
0·4D1
0·2D1
0
t
0·00
1·00
2·00
3·00
4·00
5·00

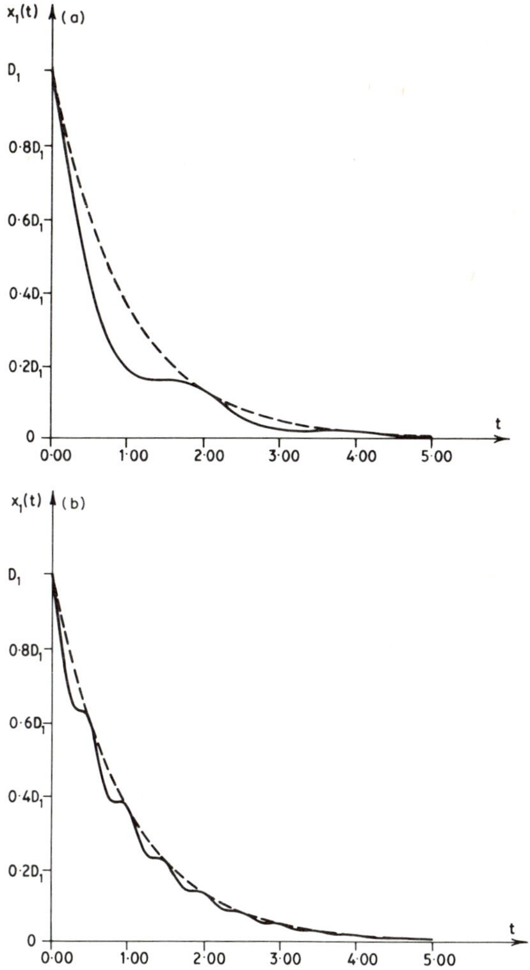

Figure 10.5 Impulse response of eqns (10.22) and (10.23) for two values of ω; (a) $\omega = \pi$; (b) $\omega = 4\pi$. For (a) and (b), $k_{01} = \gamma = 1$ and $\phi = 0$. (———) response of time-varying system; (----) response of time-invariant system.

Figure 10.4 (*opposite*) Impulse responses of eqns (10.22) and (10.23) for three values of ϕ; (a) $\phi = 0$, (b) $\phi = \frac{\pi}{2}$, (c) $\phi = \pi$. For (a), (b) and (c), $k_{01} = \gamma = 1$ and $\omega = 2\pi$. (———) response of time-varying system; (----) response of time-invariant system.

With $\phi = \frac{\pi}{2}$, $\cos\phi = 0$, so that, from eqn (10.23),

$$x_1(t) = D_1 \exp(-t) \exp\left(-\frac{-\cos\left(2\pi t + \frac{\pi}{2}\right)}{2\pi}\right)$$

$$= D_1 \exp(-t) \exp\left(-\frac{\sin 2\pi t}{2\pi}\right).$$

The second exponential varies between $\exp\left(-\frac{1}{2\pi}\right)$ at $t = 0.25, 1.25, \ldots$ and $\exp\left(\frac{1}{2\pi}\right)$ at $t = 0.75, 1.75, \ldots$, being $\exp(0) = 1$ at $t = 0, 0.5, 1.0, 1.5, \ldots$. The decay with the periodic component present thus varies on either side of the decay of the time-invariant system.

In Fig. 10.4(c), $\phi = \pi$, so that the perturbation is applied as $k_{01}(t)$ is decreasing through its mean value of 1.0 and $k_{01}(t) = 1 - \sin 2\pi t$. By analogy with the case for $\phi = 0$, the decay now has $D_1 \exp(-t)$ as its *lower* bound. The area under the curve (AUC) is larger for the time-varying system.

The effect of different frequencies with fixed phase is illustrated in Figs 10.5(a), 10.4(a) and 10.5(b), for all of which $k_{01} = \gamma = 1$ and $\phi = 0$, so that, from eqn (10.23),

$$x_1(t) = D_1 \exp(-t) \exp\left(-\frac{(1 - \cos\omega t)}{\omega}\right).$$

The (radian) frequency ω is π in Fig. 10.5(a), 2π in Fig. 10.4(a) and 4π in Fig. 10.5(b), corresponding to cyclic frequencies of 0.5, 1 and 2 Hz.

Example 2. Two-compartment model (Fig. 10.6). Consider a time-varying two-compartment system with

$$\mathbf{A}(t) = \begin{bmatrix} \alpha_{11} & \alpha_{12} \\ \alpha_{21} & \alpha_{22} \end{bmatrix} (1 + \gamma \sin(\omega t + \phi)). \qquad (10.24)$$

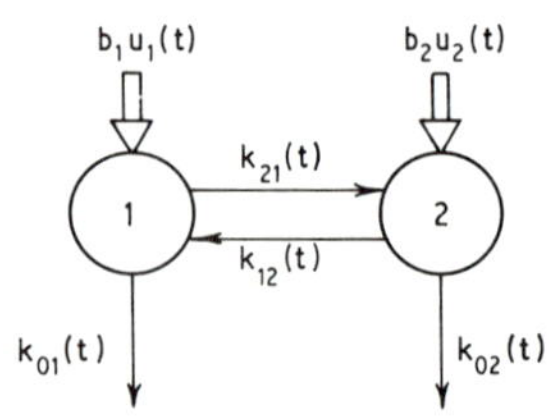

Figure 10.6. Two-compartment model with time-varying rate coefficients.

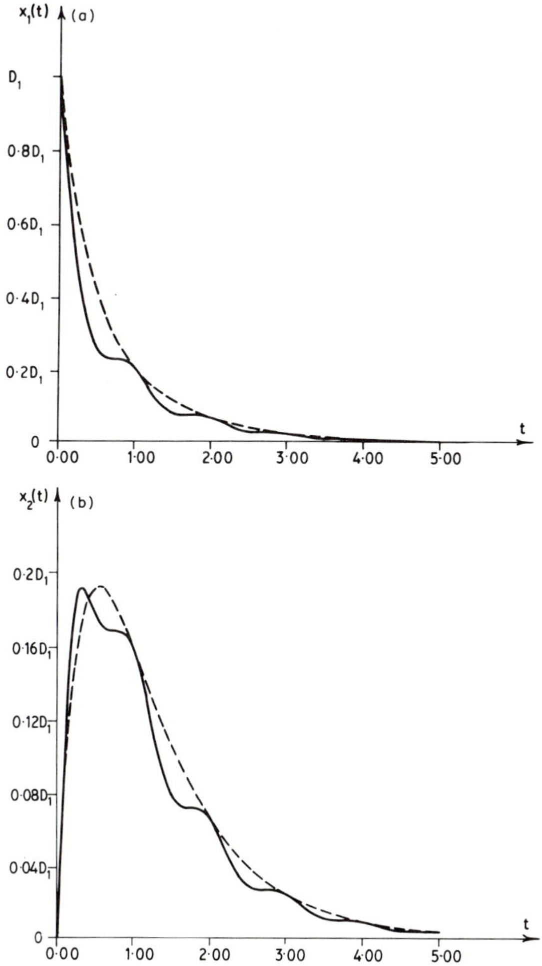

Figure 10.7 Impulse responses of eqns (10.26) and (10.28); (a) $x_1(t)$, (b) $x_2(t)$; (———) response of time-varying system; (– – – –) response of time-invariant system.

Then, as noted above, the solutions for $x_1(t)$ and $x_2(t)$ can be found from those for the corresponding time-invariant system (Section 3.2.1) with t replaced by $\left(t + \frac{\gamma}{\omega}(\cos\phi - \cos(\omega t + \phi))\right)$.

In Figs 10.7(a) and (b), the responses $x_1(t)$ and $x_2(t)$ to an impulse perturbation of compartment 1 (with $b_1 = 1$, and $u_2(t) = 0$) are shown for

$k_{01}(t) = k_{21}(t) = k_{12}(t) = k_{02}(t) = 1 + \sin 2\pi t$, for which

$$\mathbf{A}(t) = \begin{bmatrix} -2 & 1 \\ 1 & -2 \end{bmatrix}(1 + \sin 2\pi t). \tag{10.25}$$

For the time-invariant system, we know from Example 3.1 that

$$x_1(t) = \tfrac{1}{2}D_1[\exp(-t) + \exp(-3t)] \tag{10.26a}$$

and

$$x_2(t) = \tfrac{1}{2}D_1[\exp(-t) - \exp(-3t)] \tag{10.26b}$$

(which are shown dotted in Figs 10.7(a) and (b)), so for the time-varying system, with

$$\int_0^t \psi(\lambda)\, d\lambda = t + \frac{1}{2\pi}(1 - \cos 2\pi t) \tag{10.27}$$

the solutions are:

$$x_1(t) = \frac{D_1}{2}\left[\exp(-t)\exp\left(-\frac{1}{2\pi}(1 - \cos 2\pi t)\right)\right] + \frac{D_1}{2}\left[\exp(-3t)\exp\left(-\frac{3}{2\pi}(1 - \cos 2\pi t)\right)\right] \tag{10.28a}$$

and

$$x_2(t) = \frac{D_1}{2}\left[\exp(-t)\exp\left(-\frac{1}{2\pi}(1 - \cos 2\pi t)\right)\right] - \frac{D_1}{2}\left[\exp(-3t)\exp\left(-\frac{3}{2\pi}(1 - \cos 2\pi t)\right)\right]. \tag{10.28b}$$

From Fig. 10.7, it is evident that the faster of the two exponential decays would be difficult to estimate with any accuracy with the time variation present.

10.1.3 *Deterministic identifiability*

Very few deterministic identifiability analyses of time-varying systems have appeared in the literature. For systems with $\mathbf{A}(t)$ of the form of eqn (10.17), Cobelli, Lepschy and Romanin-Jacur (1978a) have shown that, subject to certain conditions, the parameters of the constant matrix $\mathbf{K}$ and of the time function $\psi(t)$ can be identified provided the corresponding linear, time-invariant model ($\psi(t) = 1$) is identifiable. The conditions are that the functional form of $\psi(t)$ should be known and that the elements of the input and observation matrices $\mathbf{B}$ and $\mathbf{C}$ should be known and should be constant.

For more general systems, the Laplace transform identifiability technique is not applicable, but the Taylor-series approach (Pohjanpalo, 1978) is applicable provided the function and its derivatives are bounded as in eqn (9.22). One rather innocent-looking system excluded by this requirement is

$$\dot{x}(t) = \frac{2}{t^3}\,x(t); \qquad x(0) = 0 \tag{10.29}$$

for which the solutions are

$$x(t) = \alpha \exp\left(-1/t^2\right) \tag{10.30}$$

(where α is a constant) which does not satisfy (9.22) except when $\alpha = 0$; $x(t)$ and its derivatives are all zero at $t = 0$. For such functions, it may still be possible to use the Taylor-series approach, expanding about some time other than zero. Two examples of the application of the Taylor-series approach will now be given.

Example A. One-compartment system with sinusoidally-varying rate coefficient. Consider the one-compartment model of Fig. 10.3 with

$$k_{01}(t) = k_{01}(1 + \gamma \sin(\omega t + \phi))$$

so that

$$\dot{x}_1(t) = -k_{01}(1 + \gamma \sin(\omega t + \phi))x_1(t) + b_1 u_1(t). \tag{10.31}$$

Let us check the finding of Cobelli, Lepschy and Romanin-Jacur (1978a) for this system, by assuming that the input fraction b_1 and the observation gain c_1 are known and see whether k_{01}, γ, ω and ϕ can be identified from the response to an impulsive perturbation at $t = 0$. For this perturbation,

$$x_1(0^+) = b_1 D_1$$

is known, and

$$\dot{x}_1(t) = -k_{01}(1 + \gamma \sin(\omega t + \phi))x_1(t), \qquad t > 0 \tag{10.32}$$

so that

$$\dot{x}_1(0^+) = -k_{01}(1 + \gamma \sin \phi)x_1(0^+)$$

whence $k_{01}(1 + \gamma \sin \phi)$ can be identified.

Differentiating eqn (10.32),

$$\ddot{x}_1(t) = -k_{01}(1 + \gamma \sin(\omega t + \phi))\dot{x}_1(t)$$
$$- k_{01}\gamma\omega \cos(\omega t + \phi)x_1(t), \qquad t > 0 \tag{10.33}$$

so that

$$\ddot{x}_1(0^+) = -k_{01}(1 + \gamma \sin \phi)\dot{x}_1(0^+) - k_{01}\gamma\omega \cos \phi \; x_1(0^+)$$

in which everything except $k_{01}\gamma\omega \cos \phi$ is known.

Differentiating eqn (10.33),

$$\ddot{x}_1(t) = -k_{01}(1+\gamma\sin(\omega t+\phi))\ddot{x}_1(t) - 2k_{01}\gamma\omega\cos(\omega t+\phi)\dot{x}_1(t) + k_{01}\gamma\omega^2\sin(\omega t+\phi)x_1(t), \qquad t>0 \tag{10.34}$$

so that $$\ddot{x}_1(0^+) = -k_{01}(1+\gamma\sin\phi)\ddot{x}_1(0^+) - 2k_{01}\gamma\omega\cos\phi\,\dot{x}_1(0^+) + k_{01}\gamma\omega^2\sin\phi\,x_1(0^+)$$

in which everything except $k_{01}\gamma\omega^2\sin\phi$ is known.

Differentiating eqn (10.34),

$$\dddot{x}_1(t) = -k_{01}(1+\gamma\sin(\omega t+\phi))\ddot{x}_1(t) - 3k_{01}\gamma\omega\cos(\omega t+\phi)\ddot{x}_1(t) + 3k_{01}\gamma\omega^2\sin(\omega t+\phi)\dot{x}_1(t) + k_{01}\gamma\omega^3\cos(\omega t+\phi)x_1(t), \qquad t>0$$

so that $$\dddot{x}_1(0^+) = -k_{01}(1+\gamma\sin\phi)\ddot{x}_1(0^+) - 3k_{01}\gamma\omega\cos\phi\,\ddot{x}_1(0^+) + 3k_{01}\gamma\omega^2\sin\phi\,\dot{x}_1(0^+) + k_{01}\gamma\omega^3\cos\phi\,x_1(0^+)$$

in which everything except $k_{01}\gamma\omega^3\cos\phi$ is known.

Thus, in the successive stages of differentiation, parameter combinations $k_{01}(1+\gamma\sin\phi)$, $k_{01}\gamma\omega\cos\phi$, $k_{01}\gamma\omega^2\sin\phi$ and $k_{01}\gamma\omega^3\cos\phi$ can be identified, from which the individual parameters k_{01}, γ, ω and ϕ can be uniquely identified. This confirms the result of Cobelli, Lepschy and Romanin-Jacur (1978a) for this system.

Example B. Two-compartment system with sinusoidally-varying excretion rate coefficient. Consider now the two-compartment model shown in Fig. 10.8, in which $k_{01}=0$, k_{12} and k_{21} are constants and the excretion rate coefficient $k_{02}(t)$ is varying sinusoidally. Note that $\mathbf{A}(t)$ for this system is *not* of the form of eqn (10.17), so that the simple method for obtaining an analytical solution is not applicable. Let compartment 1 be perturbed impulsively ($u_1(t)=D_1\,\delta(t)$) with known input fraction and assume that compartment 1 (only) is observed, with known observation gain. Even with these assumptions, identifiability analysis using the Taylor-series approach

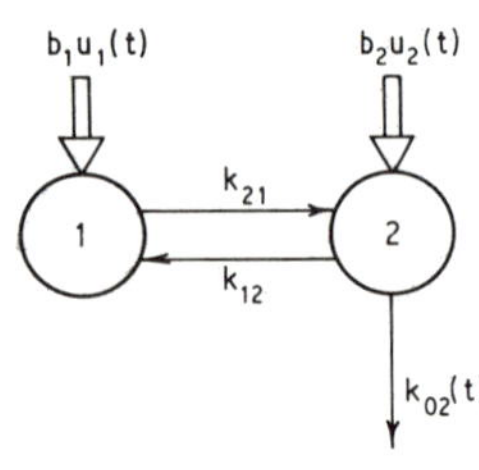

Figure 10.8 Two-compartment model with time-varying excretion rate coefficient.

becomes extremely lengthy if all six parameters ($k_{02}, k_{12}, k_{21}, \gamma, \omega$ and ϕ) are to be identified, so let us reduce the number by assuming that ω and ϕ are known. The assumption that ω is known is reasonable, since it will be the only oscillation in the observation (the eigenvalue of the time-invariant model being real). The assumption that ϕ is known is rather less reasonable, since it is equivalent to knowing at what point in the cyclic variation of $k_{02}(t)$ the input is administered; the assumption is made only to make the identifiability analysis tractable, and to this end, it will be further assumed that $\phi = 0$, so that $k_{02}(t) = k_{02}(1 + \gamma \sin \omega t)$. The parameters to be identified are thus k_{02}, k_{12}, k_{21} and γ. (The corresponding time-invariant system is identifiable from this experiment.)

The system and experiment are described by the equations:

$$\dot{x}_1(t) = -k_{21}x_1(t) + k_{12}x_2(t), \qquad t > 0 \tag{10.35a}$$

$$x_1(0^+) = b_1 D_1 \qquad \text{(which is known)} \tag{10.35b}$$

$$\dot{x}_2(t) = k_{21}x_1(t) - (k_{12} + k_{02}(1 + \gamma \sin \omega t))x_2(t) \tag{10.35c}$$

$$x_2(0^+) = 0. \tag{10.35d}$$

From eqn (10.35a), $\dot{x}_1(0^+) = -k_{21}x_1(0^+)$, which gives k_{21} since the observation $\dot{y}_1(0^+) = c_1 x_1(0^+)$ and c_1 and $x_1(0^+)$ are known. From eqn (10.35c), $\dot{x}_2(0^+) = k_{21}x_1(0^+)$.

Differentiating eqn (10.35a),

$$\ddot{x}_1(t) = -k_{21}\dot{x}_1(t) + k_{12}\dot{x}_2(t), \qquad t > 0 \tag{10.36}$$

so that
$$\begin{aligned}\ddot{x}_1(0^+) &= -k_{21}\dot{x}_1(0^+) + k_{12}\dot{x}_2(0^+) \\ &= -k_{21}\dot{x}_1(0^+) + k_{12}k_{21}x_1(0^+)\end{aligned}$$

in which everything except k_{12} is known, so that k_{12} can be identified.

Differentiating eqn (10.35c)

$$\begin{aligned}\ddot{x}_2(t) = k_{21}\dot{x}_1(t) &- (k_{12} + k_{02}(1 + \gamma \sin \omega t))\dot{x}_2(t) \\ &- k_{02}\gamma\omega(\cos \omega t)x_2(t)\end{aligned} \tag{10.37}$$

so that
$$\ddot{x}_2(0^+) = k_{21}\dot{x}_1(0^+) - (k_{12} + k_{02})\dot{x}_2(0^+). \tag{10.38}$$

Differentiating eqn (10.36),

$$\dddot{x}_1(t) = -k_{21}\ddot{x}_1(t) + k_{12}\ddot{x}_2(t), \qquad t > 0 \tag{10.39}$$

so that
$$\dddot{x}_1(0^+) = -k_{21}\ddot{x}_1(0^+) + k_{12}\ddot{x}_2(0^+).$$

Substituting for $\ddot{x}_2(0^+)$ from eqn (10.38),

$$\dddot{x}_1(0^+) + k_{21}\ddot{x}_1(0^+) = k_{12}k_{21}\dot{x}_1(0^+) - (k_{12} + k_{02})k_{21}k_{12}x_1(0^+)$$

which allows k_{02} to be identified.

Differentiating eqn (10.37),

$$\ddot{x}_2(t) = k_{21}\ddot{x}_1(t) - (k_{12} + k_{02}(1 + \gamma \sin \omega t))\,\ddot{x}_2(t) - 2k_{02}\gamma\omega(\cos \omega t)\dot{x}_2(t) + k_{02}\gamma\omega^2(\sin \omega t)x_2(t)$$

so that

$$\begin{aligned}\ddot{x}_2(0^+) &= k_{21}\ddot{x}_1(0^+) - (k_{12} + k_{02})\ddot{x}_2(0^+) - 2k_{02}\gamma\omega\dot{x}_2(0^+) \\ &= k_{21}\ddot{x}_1(0^+) - (k_{12} + k_{02})(k_{21}\dot{x}_1(0^+) - (k_{12} + k_{02})k_{21}x_1(0^+)) \\ &\quad - 2k_{02}\gamma\omega k_{21}x_1(0^+).\end{aligned} \tag{10.40}$$

Differentiating eqn (10.39),

$$\dddot{x}_1(t) = -k_{12}\ddot{x}_1(t) + k_{12}\ddot{x}_2(t), \qquad t > 0$$

so that

$$\dddot{x}_1(0^+) = -k_{21}\ddot{x}_1(0^+) + k_{12}\ddot{x}_2(0^+) \tag{10.41}$$

and it is clear from eqns (10.40) and (10.41) that $\gamma\omega$ (hence γ) can be identified.

The unknown parameters have entered one at a time at each stage of the differentiation, so that the identifiability analysis has been straightforward. This does not occur for the system with $k_{02} = 0$ and $k_{01}(t) = k_{01}(1 + \gamma \sin \omega t)$, for which identifiability analysis is very difficult, as it was for the system with Michaelis–Menten elimination from compartment 1 considered in Section 9.4.3.

10.2 Stochastic compartmental models

10.2.1 *Introduction*

In some applications areas, there is motivation for incorporating stochastic variability into a compartmental model. In biological experiments, for example, such variations could be caused by the presence of drugs other than the one administered, changes in cell membrane volume/area ratios and changing environmental and measurement conditions. In pharmacokinetics trials, normal healthy subjects are not expected to lie still for the duration of the trial and the very process of getting up and walking around may cause rate coefficients to vary stochastically. For some systems, particularly in ecology, variability can arise due to the relatively small number of particles of interest within each compartment.

Work on stochastic compartmental models has proceeded along two lines, the first with the stochasticity incorporated into the transfer rate coefficients and the second with it incorporated into the state variables, with the assumption of deterministic rate coefficients. For systems with stochastic rate coefficients, a considerable body of theory has been developed to deal with the variability of responses, usually inter-subject, but occasionally intra-subject, to a given perturbation of drug or tracer and this will be reviewed in Section 10.2.2. By contrast, comparatively little has appeared on the analysis of compartmental models in which the rate coefficients vary randomly in the course of a response; this case will be discussed in Section 10.2.3. The second line of work, with the stochasticity incorporated into the state variables, will be discussed in Section 10.2.4.

10.2.2 *Stochastic variability of responses*

Consider a typical pharmacokinetic trial or set of tracer experiments, with a panel of N human subjects. There are N responses to a rapidly administered drug or tracer, and curve-fitting each response gives a sum of exponentials:

$$y(t) = \sum_{i=1}^{p} A_1 \, e^{\lambda_i t}, \qquad t \geqslant 0. \tag{10.42}$$

The N values of the amplitudes A_i ($i = 1, 2, \ldots, p$) and the eigenvalues λ_i ($i = 1, 2, \ldots, p$) may be treated as sample values of random variables, and if N is large, the sample statistics can be regarded as a reasonable approximation to the population statistics. In principle, this allows the statistics of the corresponding linear, time-invariant compartmental model parameters (the rate constants) to be calculated, so that any measure of uncertainty regarding these parameters can be found. The literature describing this approach has been reviewed by Matis and Wehrly (1979).

The approach will now be illustrated using an example presented by Campello and Cobelli (1978), considering the modelling of bilirubin metabolism in eight normal subjects, eight hemolytic subjects and fifteen subjects with Gilbert's syndrome. A two-compartment model was used, with $x_1(t)$ and $x_2(t)$ representing, respectively, the amounts of labelled unconjugated bilirubin in the plasma and liver. The structure of the model was assumed known, with k_{01} zero. It was also assumed that the times of the observations were deterministic and that the observations were mutually independent, so that successive measurement errors are uncorrelated.

With a two-compartment, linear, time-invariant open model with unit-impulse perturbation of compartment 1 (input fraction = 1) and observation of compartment 1 (observation gain = 1), any particular response is given by

$$y(t) = x_1(t) = A_1\,e^{\lambda_1 t} + A_2\,e^{\lambda_2 t}, \qquad t \geqslant 0 \tag{10.43}$$

with $(A_1 + A_2) = 1$. To proceed further, a probability distribution has to be assumed for the eigenvalues (λ_1 and λ_2), but not necessarily for the amplitudes (A_1 and A_2). Campello and Cobelli assumed that the distributions associated with $\{A_i\}$ and $\{\lambda_i\}$ do not vary with time, that the random variables $\{\lambda_i\}$ are independent of each of the $\{A_i\}$ and that the random variables $\{\lambda_1\}$ and $\{\lambda_2\}$ are independent of each other. Clearly the amplitudes are not uncorrelated, since $A_1 + A_2 = 1$, but the last of the above assumptions is difficult to justify since λ_1 and λ_2 are functions of the same rate coefficients. The authors note that some possible methods for circumventing this difficulty are given by Soong and Dowdee (1974).

To implement the method, it is necessary to have more than one observation at any given time point. Campello and Cobelli then show how the problem of estimating stochastic parameters can be formulated as a deterministic estimation problem. Their minimization criterion was a weighted sum of squares of differences between actual mean values of $y(t)$ at each observed time point and mean values estimated from the model.

The obvious major problem with applying the method concerns the choice of probability distribution for the eigenvalues, in the absence of a very large number of responses. The only guidance is that the eigenvalues are non-positive, so the distribution for $(-\lambda_i)$ should be non-negative. The authors chose a positive gamma distribution, with probability density function given by

$$f(X) = \frac{1}{\beta^{\alpha+1}\Gamma(\alpha+1)}\,X^{\alpha}\,e^{-X/\beta}, \qquad 0 \leqslant X < \infty \tag{10.44a}$$

$$= 0, \qquad X < 0 \tag{10.44b}$$

where Γ is the gamma function. Thus $f(X)$ is described by two parameters, $\alpha(> -1)$ and $\beta(>0)$. The distribution has the properties that the mean of X is $\beta(\alpha + 1)$ and the variance of X is $\beta^2(\alpha + 1)$. The mean and variance of the eigenvalues for the three groups of subjects were estimated, and then the mean and variance of the transfer rate coefficients, by relating them to A_1, A_2, λ_1 and λ_2 through the usual equations (Section 3.2) and using standard expressions for the mean and variance of functions of random variables.

Campello and Cobelli (1978) were able to compare the results of this study with an earlier one (Cobelli, Frezza and Tiribelli, 1975) in which the rate constants were calculated individually for each response. They found good agreement between values from the two studies for all three groups of subjects. They concluded that the main weakness of the stochastic approach is the need to assume a probability distribution for the model parameters (usually with no good reason for choosing one rather than another); a further weakness is that distinctive information contained in the responses for each individual trial may be lost. The merit of the approach is that estimates can be found for transfer rate coefficients without having to identify values for every response, which may prove useful if a very large number of responses are available for any one condition.

10.2.3 *Stochastic rate coefficients in individual responses*

Comparatively little work seems to have appeared on the analysis of compartmental models whose rate coefficients are varying stochastically since the brief discussion by Jacquez (1972, Chapter 12) who laid the groundwork for theoretical developments. Again, the obvious problem in practice is the choice of a probability distribution for the variation.

Some idea of the effect of stochastic variation of a rate coefficient on a compartmental response can be gained by examining Figs 10.9, 10.10 and 10.11. All of these figures show the unit-impulse response of $x_1(t)$ for a one-compartment model with elimination rate coefficient given by

$$k_{01} = 1 + \varepsilon_r \tag{10.45}$$

where ε_r is a sequence of computer-generated random numbers with Gaussian distribution, zero mean and standard deviation of 0.4. The compartmental quantity is given by

$$\dot{x}_1(t) = -(1 + \varepsilon_r)x_1(t); \qquad x_1(0) = 1. \tag{10.46}$$

The quantity $-(1 + \varepsilon_r)$ is also shown on the figures with linear interpolation between successive values for clarity of presentation; in the simulation, the value is held constant during the interval between successive numbers. For each figure, sections of the random signal were chosen such that at no interval was k_{01} less than zero.

Figures 10.9(a), (b) and (c) show three typical responses for inter-sample interval, $\Delta t = 0.2$. In Fig. 10.9(a), the response with the stochastic component present is above that for k_{01} constant ($=1$) because $-(1 + \varepsilon_r)$ is above its average value, at least during the early part of the response. In Fig. 10.9(b), the response is similarly mainly below the response with $k_{01} = 1$, because $-(1 + \varepsilon_r)$ is below its average value during the early part of the response. In Fig. 10.9(c), the response is sometimes above and sometimes below that for $k_{01} = 1$. The integration process acts as a low-pass filter, and

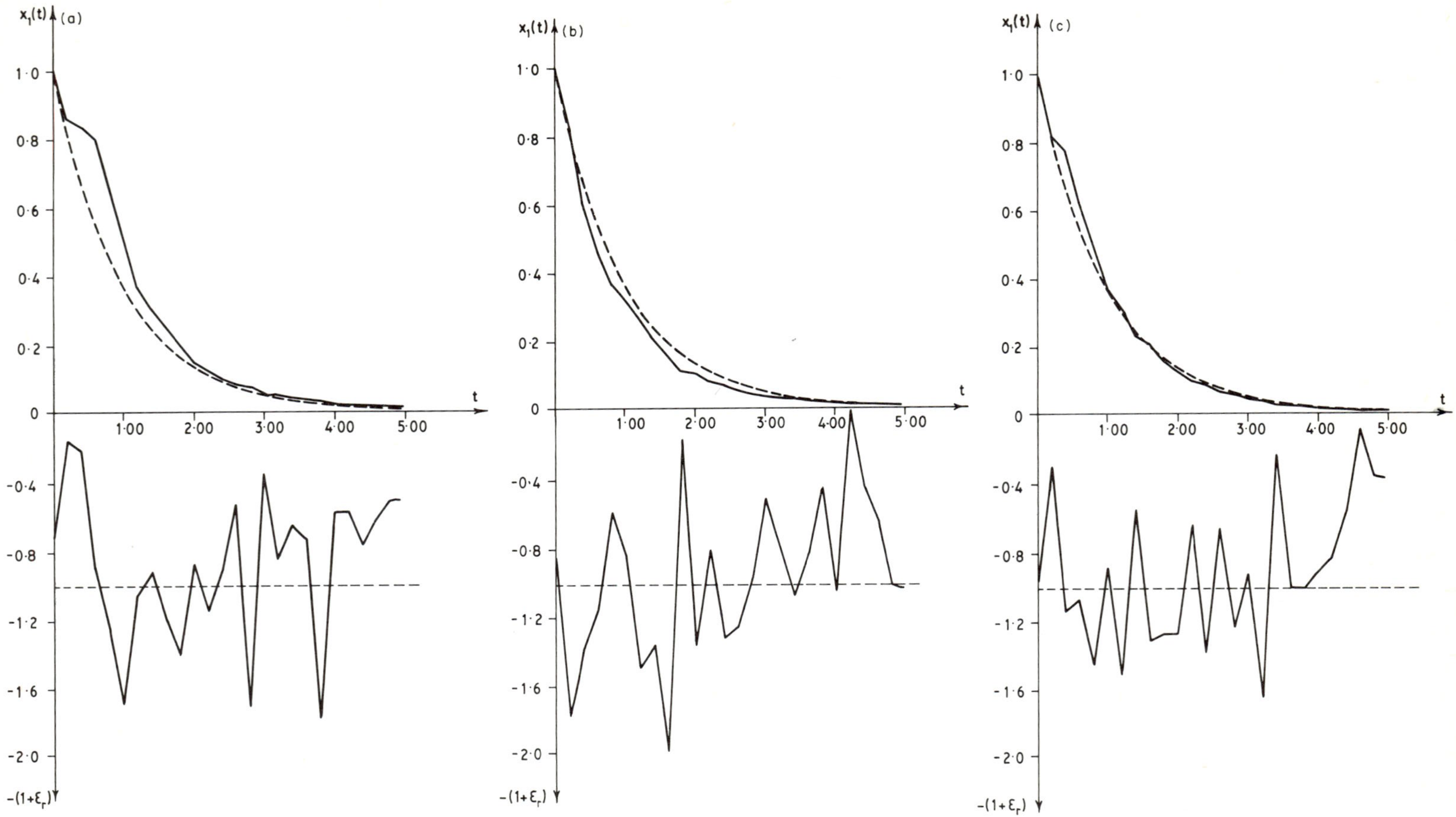

Figure 10.9 Unit impulse responses of one-compartment model with stochastically-varying rate coefficient; inter-sample interval = 0.2. ((----) $\varepsilon_r = 0$). ((a), (b) and (c) are for three different sections of ε_r).

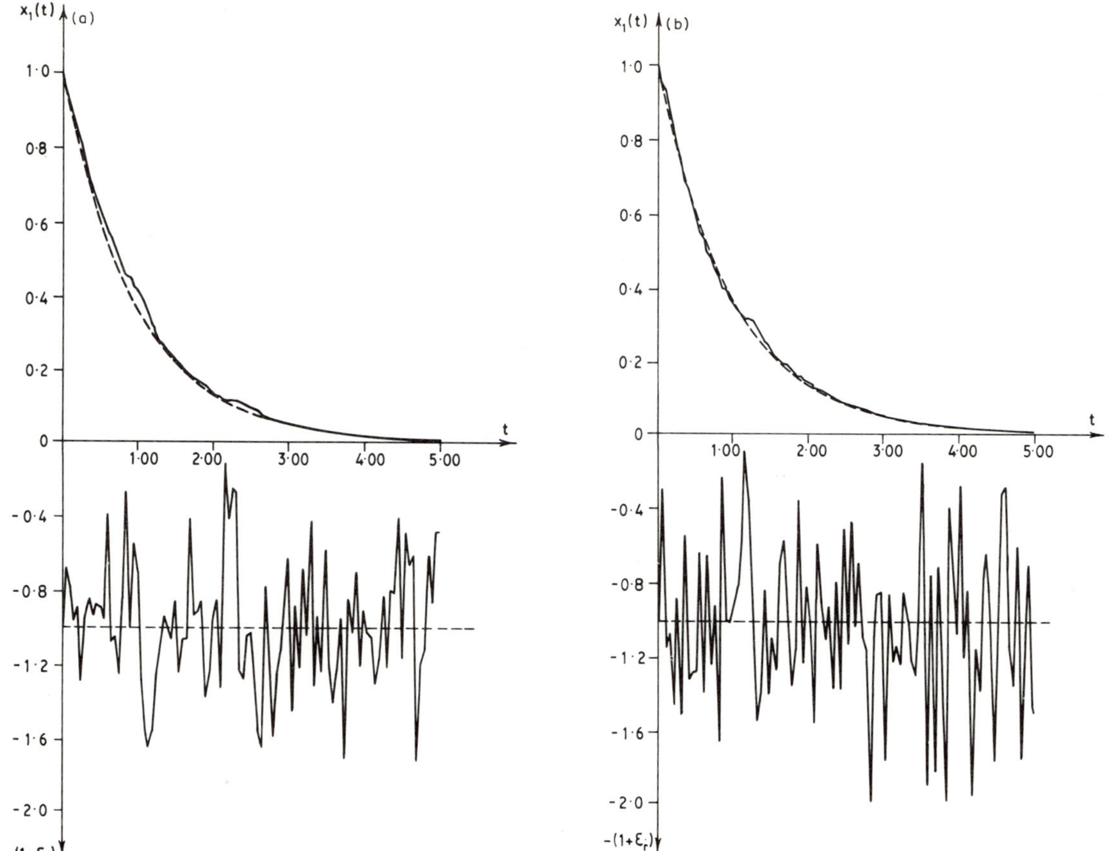

Figure 10.10 Unit impulse responses of one-compartment model with stochastically-varying rate coefficient; inter-sample interval = 0.05. ((----) $\varepsilon_r = 0$). ((a) and (b) are for two different sections of ε_r).

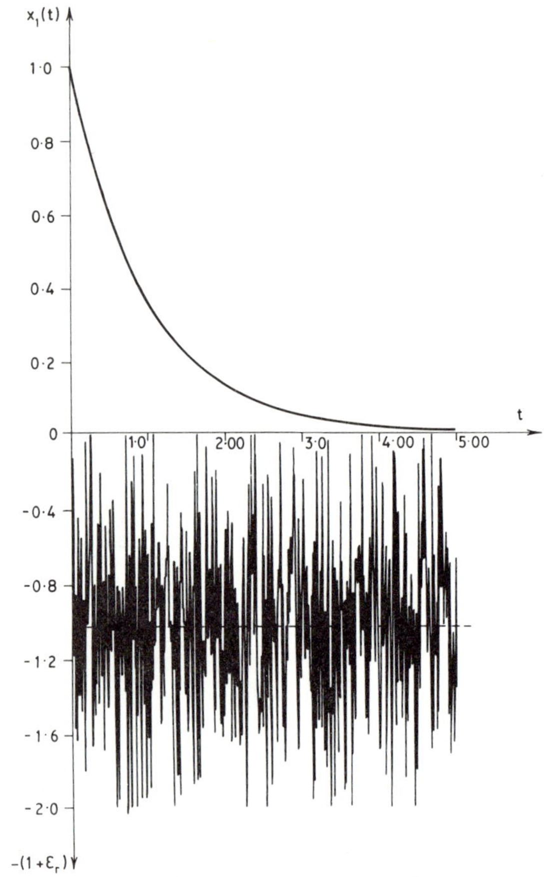

Figure 10.11 Unit impulse response of one-compartment model with stochastically-varying rate coefficient; inter-sample interval = 0.01.

this can be seen in Figs 10.10(a) and (b) for two lengths of record with $\Delta t = 0.05$ and even more noticeably in Fig. 10.11 for a length of record with $\Delta t = 0.01$. In this case, despite the substantial stochastic component, the smoothing action of the integration is such that the responses with and without the stochastic component are almost indistinguishable.

Cobelli and Morato (1978) have presented a procedure for the estimation of stochastic rate parameters, formulating the problem in terms of nonlinear filtering. The theory is developed for the case in which the rate coefficients are independent of each other, although the authors recognize that this assumption may not be valid in many practical situations.

10.2.4 *Stochastic state variables*

In the second main area of work on stochastic compartmental models, the stochasticity is incorporated via a probabilistic transfer mechanism with either constant or time-varying deterministic rate coefficients. Such models are particularly appropriate in situations where the number of particles of interest in a compartment becomes small. From the definition of a compartmental system, perfect mixing is assumed to occur in any one compartment, so that it is reasonable to assume that particles move independently of each other. The process underlying transitions of particles from one compartment to another is most often assumed to be Markovian (i.e. future transitions are independent of past states) and it is then possible to derive analytical expressions for the probability distributions of numbers of particles in the compartments. As the number of particles becomes large, the stochastic models tend in the limit towards the deterministic linear compartmental models, described by eqn (1.2) for constant rate coefficients and by eqn (10.2) for time-dependent rate coefficients.

The considerable body of theory associated with this approach has been reviewed by Purdue (1979). Typical applications are the modelling of the gastrointestinal tract of sheep (Matis and Hartley, 1971) where the particles of interest are indigestible plastic beads used as roughage substitute, the physiological modelling of carcinogenic trace pollutants in human beings (Matis, Kodell and Cardenas, 1976) where onset of carcinogenesis may correspond to a threshold number of particles in a peripheral compartment of a mammillary model and the population modelling of stomach tumour onset and growth (Tolley *et al.*, 1978) where the numbers of people in the compartments is small.

Two recent developments in stochastic compartmental modelling are of particular interest. Matis and Tolley (1979) have combined a number of possible sources of stochasticity to give a general framework for the stochastic modelling of compartmental metabolic systems. In the second development, the Markov interaction process (in which residence times in each compartment are exponential and the transition probability matrix does not depend on the elapsed residence time) has been generalized to a semi-Markov process (Weiner and Purdue, 1977a,b; Marcus, 1979). Such a generalization can be formulated either with a constant transition probability matrix and arbitrary residence time distributions or by having the transition probability matrix dependent on the residence time in the compartment (Marcus, 1979). With this additional flexibility, responses are no longer necessarily of exponential form, and a variety of responses can be accommodated (Purdue, 1979).

11 Applications of Compartmental Models

11.1 Introduction

There have been numerous examples of the applications of compartmental models in the fields of biomedicine, pharmacokinetics and ecology, and recently there has been an increasing awareness among engineers that substantial parts of many engineering systems models are of compartmental form. Early applications were usually of linear, time-invariant models, but more recently, nonlinearities have featured increasingly, particularly in pharmacokinetics.

The number of applications published runs into hundreds and it is clearly out of the question to give a comprehensive survey of these in one chapter of a book. Instead, some guidance will be given to the reader as to where such references may be found. In addition, nine Application Examples will be considered in more detail. These examples, from metabolic system modelling, pharmacokinetics, ecology and engineering, have been selected to illustrate a number of aspects of modelling technique.

11.2 Modelling of biomedical systems (excluding pharmacokinetics)

11.2.1 *Metabolic systems*

The majority of metabolic system applications are concerned with the modelling of tracer experiments, where the model fitted is linear and time-invariant. Several applications are described in two conference proceedings (Hart, 1963; Bergner and Lushbaugh, 1967) and others by Atkins (1969, Chapters 3 and 4) and Jacquez (1972, Chapter 8). In their recent review of the identification of metabolic systems, Carson, Cobelli and Finkelstein (1981) have emphasized the importance of compartmental modelling in the metabolic area. They also note that more complex models are often partly compartmental.

Most metabolic system applications adopt the procedure of deciding on the structure of the model and then designing experiments to identify the parameters; the black-box approach of designing a model to fit experimental data is comparatively rare. Many early applications considered only the responses of single compartments, representing the response of plasma concentration to a tracer dose administered intravenously or orally so that the number of parameters that could be estimated was strictly limited by numerical considerations (see Chapter 8). An example of such an application is considered in Application Example No. 1. More recently, there has been a trend towards using additional measurements, particularly as techniques of external scanning of radioactivity have become more widely used. An example of how to proceed with modelling when observations of several compartments are available is given in Application Example No. 2. As experience has been gained, there has also been an increasing trend towards incorporating prior knowledge of some rate constants into models making it possible to use models with more compartments.

Because most experiments on metabolic systems are made with labelled tracer quantities of material, so that a linear model is appropriate, applications of nonlinear models in this area have been comparatively few. As we have seen in Chapter 9, tracer equations only apply when the tracee is in a steady state, and recently, an interesting compartmental model with one nonlinear link has been used to describe calcium metabolism in calcium-deficient rats. In this example, discussed in Application Example No. 3, the tracee shows a marked circadian rhythm.

Application example 1. A classical linear, time-invariant model application (Berk *et al.*, 1969; 1970). There are many classical applications of compartmental modelling reported in the literature, in which differences in the parameters of a linear, time-invariant compartmental model are sought between normal and diseased subjects. A particularly well-documented study is that of Berk *et al.* (1969; 1970), in which a three-compartment mammillary model was used and differences found in some rate constants between normal subjects and subjects suffering from Gilbert's syndrome. The input was a rapidly-administered intravenous tracer dose of ^{14}C-labelled bilirubin.

The normal subjects (Berk *et al.*, 1969) were thirteen healthy adults aged between 21 and 24 years. A substantial number of plasma samples were taken at intervals from soon after the administration up to between 24 and 30 hours after the administration. For six of the subjects, two clearances were measured (well separated in time) so that nineteen clearances were recorded in all. It was found that three exponentials were both necessary and sufficient to describe the responses, and the values of the amplitudes (sum normalized to 1) and exponents, all (mean $\pm$ standard deviation), were

$$y_1(t) = (0.66 \pm 0.17) \exp(-0.044 \pm 0.013)t + (0.31 \pm 0.16) \exp(-0.010 \pm 0.003)t + (0.03 \pm 0.02) \exp(-0.0012 \pm 0.0002)t \qquad (11.1)$$

where the exponents are in $(\text{min})^{-1}$ and time in min. Several model-independent quantities were calculated, based on the (unnormalized) zero-time intercept and the area under the curve and, in some cases, separate checks were possible to see whether the values obtained were reasonable.

With three exponentials, an open three-compartment model was necessary and a mammillary structure was postulated, one peripheral compartment being a hepatic pool of unconjugated bilirubin and the other an extravascular pool. As we have seen in Chapter 6, if the experiment consists, as in this case, of perturbation and observation of the central compartment only, there must be prior knowledge of two rate constants to allow the other five to be identified. The prior knowledge assumed by the authors was that elimination from the plasma and extravascular pools was negligible and that elimination, presumably by conjugation, took place only from the hepatic pool. The three-compartment model considered is thus as shown in Fig. 11.1, and the local identifiability problem associated with this model (that the two peripheral compartments are indistinguishable from perturbation and observation of the central compartment only) was solved by assuming that the rate constant for transfer to the hepatic pool (no. 2) was greater than that for transfer to the extravascular pool (no. 3).

Similar experiments were conducted on eleven subjects with Gilbert's syndrome, a constitutional hepatic dysfunction characterized by the presence of chronic, low-grade unconjugated hyperbilirubinemia in otherwise healthy persons (Berk *et al.*, 1970). Again, three exponentials were necessary and sufficient to characterize the plasma clearance curves, the values of the amplitudes (sum normalized to 1) and exponents (mean $\pm$ standard deviation) being:

$$y_1(t) = (0.47 \pm 0.15) \exp(-0.024 \pm 0.008)t + (0.41 \pm 0.08) \exp(-0.0040 \pm 0.0008)t + (0.12 \pm 0.09) \exp(-0.00080 \pm 0.00020)t. \qquad (11.2)$$

Comparing the rate constants for the compartmental models, it was found that k_{21} and k_{02} were significantly reduced for the subjects with Gilbert's syndrome, the rate of both hepatic uptake (k_{21}) and conjugation (k_{02}) being only some 40% to 50% of normal. The rate constants k_{12} and k_{13} were not

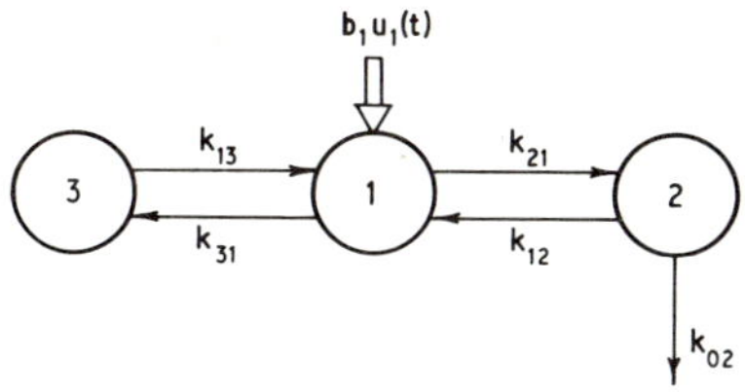

Figure 11.1 Three-compartment model of bilirubin metabolism used by Berk *et al.* (1969, 1970).

significantly different between the two sets of subjects, but the value of k_{31} (0.0016 ± 0.0006) was significantly lower in the Gilbert's subjects than for the normal subjects (0.0047 ± 0.0025). At the time of publication, Berk *et al.* (1970) could offer no explanation for this particular difference. A further 21 studies in a third group with anaemia and/or hemolysis are reported by Berk *et al.* (1970), but the rate constants for this group were indistinguishable from normal, suggesting that neither of these factors is responsible for the abnormalities noted for the subjects with Gilbert's syndrome.

The authors also calculated many quantities of interest, from either the exponential fit or the compartmental model fit. For example, since this was a tracer experiment, with compartment 1 perturbed and observed, and observation was of concentration, it was possible to estimate the initial volume of distribution V_1 from the known dose administered and the zero time intercept $y_1(0^+)$. From this, it was possible to determine the mass of bilirubin in the central compartment, which, for the normal subjects, was found to be 13 mg. It is then possible to determine the (steady-state) volumes of distribution (hence the masses) of compartments 2 and 3, using calculations similar to those in Section 6.2.6. For the normal subjects, since the mean values of the rate constants were $k_{31} = 0.0047$, $k_{13} = 0.0017$, $k_{21} = 0.023$, $k_{12} = 0.0065$, $k_{02} = 0.011$ (all in min^{-1}), the (steady-state) mass of bilirubin in compartment $3 = 13 \times \dfrac{k_{31}}{k_{13}} = 36$ mg, while that in compartment $2 = 13 \times \dfrac{k_{21}}{k_{02} + k_{12}} = 18$ mg.

The most significant point of this example is that, through the compartmental modelling, statistically significant differences have been found in two parameters of physiological interest, between normal subjects and those with Gilbert's syndrome. It is worth noting that the amplitudes and exponents of the exponential fits are markedly different between eqns (11.1) and (11.2); plotting all the responses on graph paper revealed that there was no overlap between the two groups for sixteen hours after the administration.

Application example 2. Simultaneous measurement of several quantities (Aamodt *et al.*, 1979; Foster *et al.*, 1979). An approach to model-building when several measurements are available has recently been described by Foster *et al.* (1979), fuller details of the experimental techniques being given in Aamodt *et al.* (1979). The experiments were conducted to see whether ^{69m}Zn is suitable for studying zinc metabolism in humans and to determine if the route of administration affects the kinetics. In this book, we will consider only the results for intravenous administration.

There were 17 subjects, all with a smell and/or taste dysfunction. They were given a rapid intravenous injection of ^{69m}Zn at least 14 hours after their previous meal, and measurements of zinc activity were made in whole blood, plasma, red blood cells, urine, faeces and (using external counters) over liver and thigh. Blood samples were collected frequently in the early stages and measurement was continued for five days, although by then only one blood sample per day was being collected.

The model was assumed to be mammillary in form, with the central compartment (plasma zinc activity) exchanging with other compartments in parallel. The modelling proceeded in three stages. First, the plasma activity was fitted to a sum of exponentials. Second, this curve was used as a forcing function to individual subsystems for which data were available (red blood cells, liver, and urinary and faecal extraction) and parameters of these subsystems were estimated. Third, the plasma activity curve was reconstructed using subsystems already identified and adding further ones as necessary.

In the first stage of the modelling, the plasma measurements were fitted by four exponentials:

$$x_1(t) = 0.790\,\mathrm{e}^{-176t} + 0.175\,\mathrm{e}^{-73.4t} + 0.022\,\mathrm{e}^{-5.87t} + 0.013\,\mathrm{e}^{-0.053t}. \tag{11.3}$$

The quality of the blood sample data needed to fit as many as four exponentials has already been commented upon in Section 8.2.7 of this book.

In the second stage of the modelling, this was used as a forcing function to the red cells, liver, urine and faecal subsystems. Both the urine and faecal activities could be fitted as single loss pathways, although the faecal pathway required a dead-time of 1.7 days; these are shown as compartments 2 and 4 (respectively) in Fig. 11.2. A two-compartment system was required to describe the red cell activity, and in the absence of detailed information on how ^{69m}Zn was handled by the red cells, the subsystem was modelled as compartments in series. (A similar remark also applies to the liver subsystem.) Compartment 5 exchanged rapidly with the plasma compartment, and compartment 6 exchanged rather less rapidly with compartment 5.

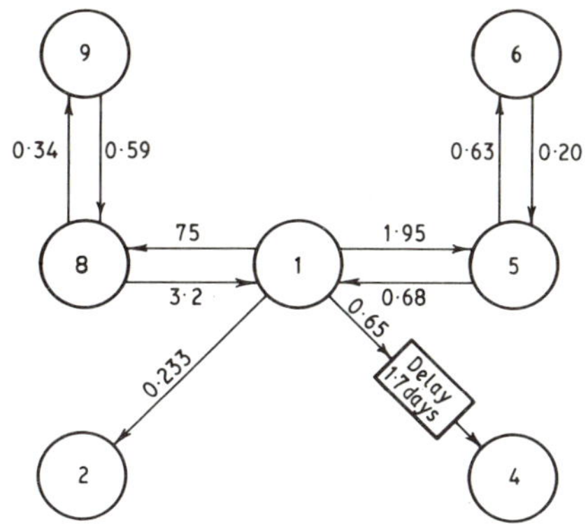

Figure 11.2 Model of zinc metabolism at the end of the second stage of the modelling procedure of Foster *et al.* (1979).

The observation is the sum of activities in compartments 5 and 6. The liver activity similarly required two compartments: no. 8 exchanging rapidly with the plasma and no. 9 exchanging less rapidly with no. 8; fractions of the plasma and red cell activities were also required to explain the liver activity. From the data available, no direct loss pathway from the liver could be resolved. The model fitted so far is thus summarized in Fig. 11.2.

The third stage of the modelling consisted of checking whether the plasma response curve could be reconstructed. It was found that it could not, and four extra compartments were added to the model, all in parallel with the central compartment and not exchanging with any compartment other than the central compartment. Three of the four extra compartments exchange rapidly with the plasma and the fourth much more slowly, thus representing zinc pools with very slow turnover, such as muscle and bone, and losses from the body other than via urine and faeces. It is not clear from the published material (Foster *et al.*, 1979) how as many as four extra compartments were fitted at this stage, but, in a private communication, Foster has explained that additional prior knowledge (for example, of the approximate total body zinc) was used along with the hitherto unused thigh activity measurement.

Two further papers have appeared recently (Aamodt *et al.*, 1982; Babcock *et al.*, 1982) describing studies of zinc metabolism in humans over a much longer period varying between 290 and 440 days. In the second paper (Babcock *et al.*, 1982), it is noted that it was possible to obtain good fits to the long-term data using a model with fewer compartments than that described above which models the shorter term dynamics.

Application example 3. A nonlinear compartmental model of calcium metabolism (Staub *et al.*, 1981). The authors describe the development of a nonlinear compartmental model of calcium metabolism in calcium-deficient rats. The observations were the plasma concentrations of calcium (tracee) and of its radionuclide ^{45}Ca (tracer) following rapid intravenous injection of ^{45}Ca, measured at various intervals up to 188 h after injection. The tracee record showed a gradual decline with large oscillations (of period 24 h) while the tracer record showed a decay with a smaller amount of oscillation also with period 24 h. With the tracee not in a steady state, the same form of nonlinear compartmental model was postulated to describe the kinetics of the tracee and the tracer.

The structure of the model chosen was catenary, and only one nonlinearity was introduced, in that part of the model representing bone growth. The nonlinearity chosen was an autocatalytic process of order 2—that is, the flow rate from compartment j to compartment i is

$$f_{ij} = k_{ij} x_i^2 x_j \tag{11.4}$$

where k_{ij} is a positive constant. The physiological reasons for choosing an autocatalytic process are discussed at some length by the authors. The reasons for the choice of order were that it has been shown that sustained oscillations can only occur if the order is greater than 1 (Sel'Kov, 1968) and 2 was selected for mathematical simplicity.

The minimum number of compartments of a catenary model with one nonlinearity which reasonably reflected tracee and tracer behaviour was 5, and the (closed) model is shown in Fig. 11.3. Compartment 1 contains the plasma, compartments 2, 3 and 4 are bone compartments and compartment 5 is a further compartment interacting with the plasma compartment. The differential equations for the compartmental quantities are:

$$\frac{\mathrm{d}x_1}{\mathrm{d}t} = -(k_{21} + k_{51})x_1 + k_{12}x_2 + k_{15}x_5 \tag{11.5a}$$

$$\frac{\mathrm{d}x_2}{\mathrm{d}t} = k_{21}x_1 - k_{12}x_2 - k_{32}x_3^2x_2 + k_{23}x_3 \tag{11.5b}$$

$$\frac{\mathrm{d}x_3}{\mathrm{d}t} = k_{32}x_3^2x_2 - (k_{23} + k_{43})x_3 + k_{34}x_4 \tag{11.5c}$$

$$\frac{\mathrm{d}x_4}{\mathrm{d}t} = k_{43}x_3 - k_{34}x_4 \tag{11.5d}$$

$$\frac{\mathrm{d}x_5}{\mathrm{d}t} = k_{51}x_1 - k_{15}x_5 \tag{11.5e}$$

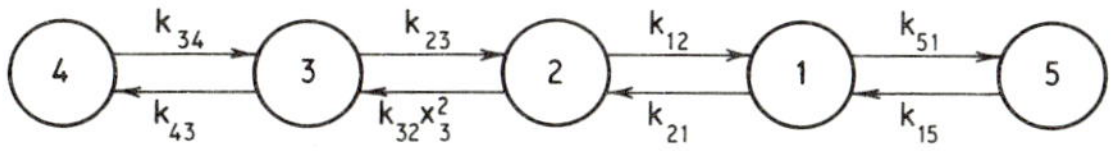

Figure 11.3 Five-compartment catenary model (with one nonlinearity) of calcium metabolism (Staub *et al.*, 1981). The flow rate from compartment 2 to compartment 3 is $k_{32}x_3^2x_2$.

The initial conditions for the tracer quantities are

$$\mathbf{x}(0^+) = [D_1 \quad 0 \quad 0 \quad 0 \quad 0]^T \tag{11.6}$$

where D_1 is the (known) dose of ^{45}Ca injected into the plasma compartment. The initial conditions for the tracee quantities, written in capital letters to distinguish them from tracer quantities, are

$$\mathbf{X}(0) = [X_1(0) \quad X_2(0) \quad X_3(0) \quad X_4(0) \quad X_5(0)]^T \tag{11.7}$$

in which only $X_1(0)$ is known.

The identification procedure gave estimates for $k_{12}, k_{21}, k_{23}, k_{32}, k_{34}, k_{43}, k_{15}, k_{51}, X_2(0), X_3(0), X_4(0)$ and $X_5(0)$. An optimization procedure was applied, with initial estimates obtained by manual search with the aim of roughly reflecting overall behaviour followed by random search around these values. The model gave a good fit to both tracee and tracer data.

The probable uniqueness of the model was studied by linearizing the model, which, for the experiment considered, gives a locally identifiable model with four solutions (Cobelli, Lepschy and Romanin-Jacur, 1979b)—see Section 6.4. The other three sets of parameters were determined, and it was found that the nonlinear models with these parameter sets did not reflect the observed behaviour. From this, the authors concluded that the solution found is probably unique. The authors also investigated the stability of the model, and found evidence for global stability of the identified model.

The example raises a number of interesting questions. For instance, could other forms of nonlinearity give the same observed responses? Could similar behaviour be found with fewer compartments if the catenary structure of the model were changed—for example, using a cyclic structure?

11.2.2 *Other biomedical applications*

The great majority of applications of compartmental models in biomedicine have been to metabolic systems, but there is now an increasing use of compartmental models in other areas of biomedicine. For example, Bajzer and Nosil (1977) have described a simple model of the lung using equations of compartmental form. The impetus for their work was the ability in recent times to measure aspects of lung function by external scanning. The subject is made to breathe air containing a radioactively-labelled gas, and the amount of radioactivity in different lung regions can be scanned using a Gamma Camera. Results are presented for a washout experiment.

A somewhat more complex compartmental model of lung function has been proposed by Bache, Gray and Murray-Smith (1981), who model the gas exchange processes and use continuous measurements of gas flow rates and gas concentration at the subject's mouth to estimate model parameters for cardiac output and other cardiopulmonary quantities. The additional complexity is brought about by having to model the solubility of the gas, in contrast to the Bajzer and Nosil model which is for insoluble radioactive gases such as ^{81m}Kr.

11.3 Pharmacokinetics

The subject of pharmacokinetics is defined by Gibaldi and Perrier (1975) as "the study of the time course of drug and metabolite levels in different fluids, tissues and excreta of the body, and of the mathematical relationships required to develop models to interpret such data". The first detailed description of pharmacokinetics in terms of a linear, time-invariant compartmental model was by Teorell (1937a,b), although he did not specifically use the term "compartment" which was only introduced some years later by Sheppard (1948).

11.3.1 *Linear pharmacokinetic models*

The approach taken in most linear pharmacokinetic modelling is black-box—that is, an experimental response is obtained, exponentials are fitted to the response, the number of compartments is matched to the number of exponentials and the compartmental model parameters are estimated. Little physiology is attached to the model, other than that the plasma is contained within the central compartment. There are a few examples of physiologically-based pharmacokinetic models; for a review, see Himmelstein and Lutz (1979).

With the black-box approach, the motivation for using linear, time-invariant compartmental models to describe drug concentration responses alone is not very clear, a fact remarked upon by Wagner (1975). Indeed, some quantities of interest, such as area under the curve (AUC) and, following extravascular administration, peak concentration and time to peak, can be obtained without even fitting exponentials to the observed responses.

The liver is the principal site of drug metabolism, i.e. biotransformation of the drug into other substances called metabolites. While linear, time-invariant compartmental models are not particularly useful for describing drug kinetics alone, they are very useful for describing the combined kinetics of a drug and its metabolite(s) (Gibaldi and Perrier, 1975). We will now consider an example of such an application.

Application example 4. A multi-species example (Boxenbaum and Riegelman, 1976). The paper describes an investigation (in man) of the pharmacokinetics of isoniazid (INH), a commonly-used antituberculosis drug, and its metabolites acetylisoniazid (AcINH), isonicotinic acid (INA) and isonicotinuric acid (INU). The four compounds were intravenously administered in turn at well-separated times to two healthy male volunteers. Human beings are (hereditary) rapid or slow acetylators of INH, and one of the volunteers was a rapid, and the other a slow, acetylator. Urinary excretion data were obtained for AcINH, INA and INU and plasma data for INH; the sum of INH and hydrazones was also measured in the urine. Here, we will discuss only the modelling for the last two compounds in the chain, INA and INU. The reader is referred to the original paper for the complete model for all four compounds (with the reminder that the order of rate-constant subscripts in pharmacokinetics is the reverse of that elsewhere, including this book).

For both INA and INU, the administration was a constant infusion for 60 min, with urine measurements being made every 15 min during, and for 60 min after, the infusion, then every 30 min for about 11 h, then *ad libitum* for a total measurement time of 48 h after the start of the infusion. For the INU input, the model of Fig. 11.4(a) was formulated to fit the response curve; compartments 1 and 2 are the central and peripheral compartments of a standard two-compartment model and compartment 3 is a sink compartment representing excretion via the urine. Urinary recovery was, within experimental accuracy, equivalent to the administered dose, confirming that no additional excretion pathways from compartments 1 and 2 were needed. From Section 3.2 of this book, the transfer function between the input $u_1(t)$ and the state of compartment 1 is

$$G_{11}(s) = \frac{X_1(s)}{U_1(s)} = \frac{s + k_{12}}{s^2 + (k_{12} + k_{31} + k_{21})s + k_{12}k_{31}} \tag{11.8}$$

while that between the input and the state of the observed compartment is

$$G_{31}(s) = \frac{X_3(s)}{U_1(s)} = \frac{k_{31}}{s} G_{11}(s). \tag{11.9}$$

For the INA input, both INA and INU were measured in the urine, and the model of Fig. 11.4(b) was formulated. The recovery was again equivalent,

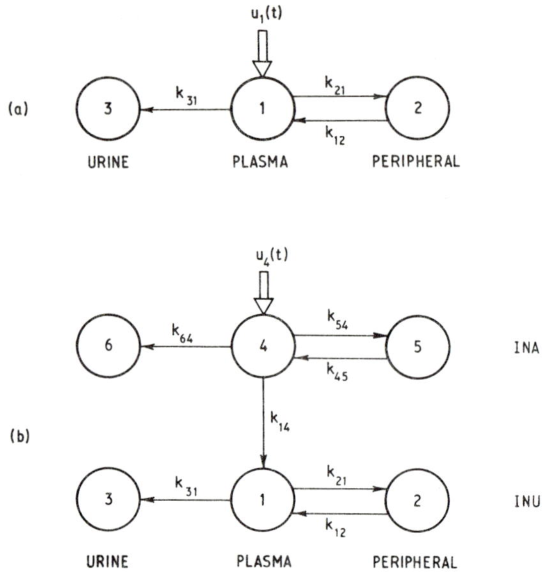

Figure 11.4 Models used by Boxenbaum and Riegelman (1976) (a) to describe the kinetics of INU; (b) to describe the combined kinetics of INA and INU.

to within experimental accuracy, to the quantity administered. In the model, k_{64}, k_{45}, k_{54}, k_{31}, k_{12} and k_{21} are all for pathways representing transport of material from one site to another while k_{14} represents biotransformation (at a first-order rate) from INA to INU, assumed to occur from the central compartment only. The two central compartments (4 and 1) now occupy the same space and represent different species; so do the two peripheral compartments (5 and 2) and the two urine compartments (6 and 3).

By similar analysis to the above, the transfer function between the input and the state of compartment 4 is

$$G_{44}(s) = \frac{X_4(s)}{U_4(s)} = \frac{s + k_{45}}{s^2 + (k_{45} + k_{04} + k_{54})s + k_{45}k_{04}} \tag{11.10}$$

where
$$k_{04} = k_{14} + k_{64}$$

and the transfer function between the input and the observed quantity of INA in the urine is

$$G_{64}(s) = \frac{X_6(s)}{U_4(s)} = \frac{k_{64}}{s} G_{44}(s). \tag{11.11}$$

The transfer function between INA input and INU in the urine is most easily determined by treating the overall six-compartment model as two interconnected three-compartment subsystems (cf. Section 4.2.4). The input rate to compartment 1 of Fig. 11.4(b) is $k_{14}x_4(t)$ so, combining the two transfer functions,

$$G_{14}(s) = \frac{X_1(s)}{U_4(s)} = k_{14}\frac{s + k_{45}}{s^2 + (k_{45} + k_{04} + k_{54})s + k_{45}k_{04}} \times \frac{s + k_{12}}{s^2 + (k_{12} + k_{31} + k_{21})s + k_{12}k_{31}} \tag{11.12}$$

and

$$G_{34}(s) = \frac{X_3(s)}{U_4(s)} = \frac{k_{31}}{s}G_{14}(s). \tag{11.13}$$

In their fitting of the INU response to INA input, Boxenbaum and Riegelman (1976) fixed the value of the ratio $\dfrac{k_{64}}{k_{64} + k_{14}}$ from steady-state values of the urinary excretion data. By doing this, acceptable fits were obtained for all the rate constants in the model of Fig. 11.4(b), but the coefficients of variation of the parameters for the INU part of the model were considerably larger than when fitting the INU input: INU response data. This trend continued as AcINH and then INH responses were modelled and for the 12-compartment model for all four compounds (INH → AcINH → INA → INU) very large coefficients of variation were obtained for some of the rate constants.

This raises the interesting question of whether the parameters for the INU part of the model should be fixed at the values estimated from the INU perturbation experiment, leaving only the parameters for the INA part of the model to be estimated from the INA perturbation experiment. A systematically wrong fit would then indicate incorrect model formulation. In fact, Boxenbaum and Riegelman (1976) found a systematic error in the responses of INA and INU to AcINH perturbation in one of the two subjects. They concluded that this could well result from the INA conjugation being saturable, in which case the metabolism of INA to INU might better have been described by utilizing the Michaelis–Menten equation. The incorporation of nonlinearities into pharmacokinetic models is the subject of Section 11.3.2.

11.3.2 *Nonlinear pharmacokinetic models*

In contrast to the metabolic system kinetics considered previously, perturbations in pharmacokinetics experiments are no longer minute, but are the size required for therapeutic purposes. This means that mathematical models required to describe the kinetics may well require some nonlinearities to be incorporated. It is interesting to note that in a review paper published as long ago as 1973, Wagner (1973b) found over 160 references with evidence of nonlinearities in one or more of drug absorption, distribution, metabolism and excretion or pharmacological response (pharmacodynamics). It is helpful to model these nonlinear systems within a compartmental framework (Wagner, 1975).

We will now consider two Application Examples describing models for drug kinetics alone, one with an absorption nonlinearity and the other with a nonlinear excretion rate.

Application example 5. Limitation on input rate (Colburn, DiSanto and Gibaldi, 1977). When administered orally, many drugs exhibit first-order absorption characteristics, so that the input rate to the compartmental model describing the distribution and elimination of the drug is

$$u(t) = 0, \qquad 0 \leqslant t < \tau \tag{11.14a}$$

$$= FDk_a \exp(-k_a(t-\tau)), \qquad t > \tau \tag{11.14b}$$

where τ is a pure time delay to account for the time taken between the drug being administered and its first appearance in the systematic blood; k_a is the absorption rate constant; D is the dose administered; and F is the bioavailability fraction. Colburn, DiSanto and Gibaldi (1977) note that this equation provides a good description of the absorption phase for single oral administrations of erythromycin, a widely-used antibiotic, and that a one-compartment model (Fig. 11.5) seems adequate to describe the distribution and elimination, at least when the drug is administered orally. In the single-dose case, the quantity of drug in the compartment is given by

$$x_1(t) = 0, \qquad 0 \leqslant t \leqslant \tau \tag{11.15a}$$

$$= \frac{FDk_a}{k_a - k_{el}}[\exp(-k_{el}(t-\tau)) - \exp(-k_a(t-\tau))], \quad t > \tau \tag{11.15b}$$

where k_{el} is the elimination rate constant.

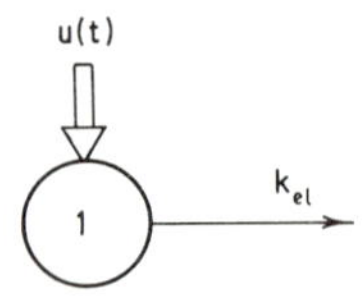

Figure 11.5 One-compartment model used by Colburn, DiSanto and Gibaldi (1977).

Several authors had previously suggested that a more accurate model for the plasma response is obtained using a one-compartment model with zero-order input rate given by

$$u(t) = 0, \qquad 0 \leqslant t < \tau \tag{11.16a}$$

$$= Fk_i, \qquad \tau \leqslant t < T \tag{11.16b}$$

$$= 0, \qquad t \geqslant T. \tag{11.16c}$$

The quantity of drug in the compartment is then:

$$x_1(t) = 0, \qquad 0 \leqslant t \leqslant \tau \tag{11.17a}$$

$$= \frac{Fk_i}{k_{el}}[1 - \exp(-k_{el}(t - \tau))], \qquad \tau \leqslant t < T \tag{11.17b}$$

$$= x_1(T)\exp(-k_{el}(t - T)), \qquad t \geqslant T. \tag{11.17c}$$

It was particularly suggested that these equations applied for multiple dosing (with suitable adjustments for non-zero initial conditions).

Colburn, DiSanto and Gibaldi (1977) investigated this hypothesis, comparing the fits of eqns (11.15) and (11.17), adjusted for multiple dosing, to plasma response data from subjects in three studies, all with administration at regular intervals of 6 h over a 3-day period. In the studies, plasma concentrations were measured at intervals of 1 h or $1\frac{1}{2}$ h for all or part of Day 1 and Day 3. In fitting the model described by eqns (11.15), k_a was allowed to vary from dose to dose, while in fitting the model described by eqns (11.17), k_iF was allowed to vary from dose to dose; in both cases, k_{el} was allowed to vary after every fourth dose (i.e. from day to day). In doing this, the authors were not entirely comparing like with like, and a better first-order fit might well have been obtained by allowing both k_a and F to vary in eqns (11.15). Nevertheless, the results presented in the paper, which are the averaged responses for one of the three studies, clearly indicate a zero-order absorption characteristic, with the discontinuity in slope expected (at time T) from the form of eqns (11.17). Similar results were obtained from the other two studies, which used different formulations of erythromycin. The initial time delay τ was relatively consistent from dose to dose but varied considerably from one formulation to another.

A further interesting finding was that the variability of the absorption rate and the elimination rate was not random. For most subjects, the absorption rate after the second dose on Day 1 or Day 3 was somewhat lower than after the first dose. For all subjects, on both days, the absorption rate after the third dose was considerably lower than after the second dose and the absorption rate after the fourth dose was higher than after the third dose. For all subjects, the elimination rate was lower on Day 3 than on Day 1.

Application example 6. Model with nonlinear elimination kinetics (Gerber and Wagner, 1972). The authors note that, prior to their work, several reports had discussed the dose-dependent decline of diphenylhydantoin concentrations. Gerber and Wagner therefore investigated the use of a one-compartment model with Michaelis–Menten elimination kinetics to fit two different sets of data. The first set consisted of part of the plasma responses of one human subject to three different dose levels of sodium diphenylhydantoin. The doses, of 2.3, 4.7 and 7.9 mg/kg of body weight were administered daily for 3 days, with 14-day intervals between the different dose levels. The second set of data consisted of the average plasma concentration in 6 rats following rapid intravenous doses of 10, 25 and 40 mg/kg of sodium diphenylhydantoin containing ^{14}C diphenylhydantoin.

The model is shown in Fig. 11.6 and, from Section 9.4, the concentration C is described by the differential equation

$$\frac{dC}{dt} = -\frac{V_m C}{K_m + C} + \frac{u_1(t)}{V_1}. \tag{11.18}$$

For the human response data, the first sample used was taken 12 h after the last of the three daily doses; this time was chosen because it was felt that the input rate $u_1(t)$ was then negligible, so that eqn (11.18) can be written

$$\frac{dC}{dt} = -\frac{V_m C}{K_m + C}, \quad t \geqslant 0; \qquad C(0) = C_0 \tag{11.19}$$

where t is set to zero 12 h after the final dose. The solution of eqn (11.19) is then

$$t = \frac{1}{V_m}(C_0 - C) + \frac{K_m}{V_m} \ln \frac{C_0}{C}, \qquad t \geqslant 0. \tag{11.20}$$

For the rat data, the same equation was used with t being set to zero at the time of the first concentration measurement (C_0).

For the human-response data, all three responses were used to estimate single values of K_m (6.77 μg ml^{-1}) and V_m (0.253 μg ml^{-1} h^{-1}). In the paper, the data and curve fits are shown on both linear paper and semilogarithmic paper. The curve fits on semilog paper are repeated here (Fig. 11.7) with

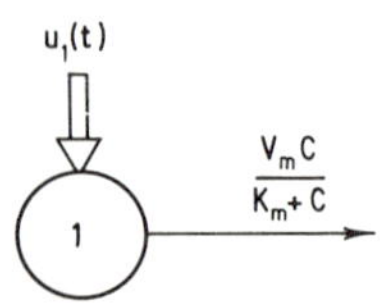

Figure 11.6 One-compartment model with Michaelis–Menten elimination kinetics used by Gerber and Wagner (1972).

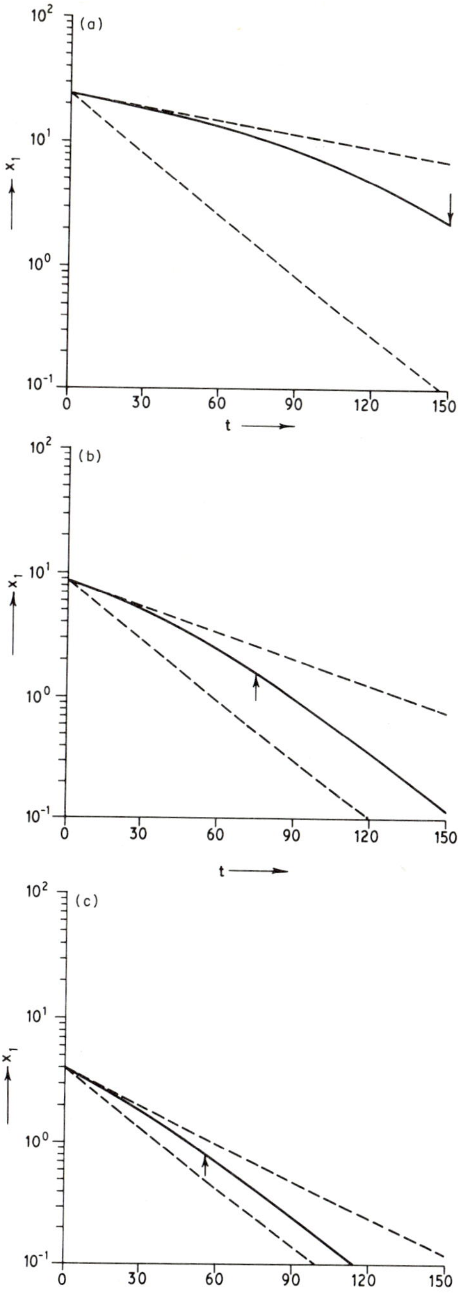

Figure 11.7 Model responses (———) and linear system bounds (– – – –) for the human response data of Gerber and Wagner (1972); $K_m = 6.77\ \mu g\ ml^{-1}$, $V_m = 0.253\ \mu g\ ml^{-1}\ h^{-1}$; (a) $C_0 = 24\ \mu g\ ml^{-1}$; (b) $C_0 = 8.8\ \mu g\ ml^{-1}$; (c) $C_0 = 4\ \mu g\ ml^{-1}$. (Arrows indicate time of final data point.)

upper and lower linear bounds also shown to give an indication of the validity of fitting a nonlinear model. The time of the final data point is arrowed on the three responses. The curve fits provide a very good fit to all three responses; the bounds shown in Fig. 11.7 indicate that it would have been difficult to estimate V_m and K_m with any accuracy from the smallest of the responses, for which a linear model would have given an adequate fit.

For the rat data, it was not possible to fit the three (averaged) responses with same values of K_m and V_m. While the values of V_m decreased as dose increased, there seemed to be no trend in K_m in relation to dose. The curve fits are shown in the paper on linear scales only, and it is found that, while the curve fits for the two smaller doses are good, eqn (11.20) does not provide a very good fit to the largest responses. The curve fits are repeated here with upper and lower linear bounds, and with the time of the final data point indicated, in Fig. 11.8. For the smallest of the responses, an almost equally good fit would have been obtained using a linear model.

Gerber and Wagner (1972) note that the changes in V_m and K_m values with dose for the rat data may well be a reflection of parallel metabolic pathways, since there is more than one metabolite of diphenylhydantoin. This may also be the reason for the comparatively poor fit of the largest response. The excellence of the fits for the human data, with single values of V_m and K_m for all three responses, may be an indication that, at the dose levels considered, elimination is predominantly via a single pathway.

11.3.3 *The relationship between pharmacokinetics and pharmacological response (phamacodynamics)*

The pharmacological effect E of a drug is traditionally described by the Hill equation (Wagner, 1968):

$$\frac{E}{E_{\max}} = \frac{C^\alpha}{K + C^\alpha} \tag{11.21}$$

where $E_{\max}$ is the maximum effect possible; K is a constant; C is the drug concentration in the vicinity of the receptor; and α is a positive constant. A plot of E v. $\log C$ produces a sigmoidal curve which is very nearly linear over the range $0.2E_{\max} < E < 0.8E_{\max}$, so that the simpler relationship

$$E = m \log C + e \tag{11.22}$$

Figure 11.8 Model responses (———) and linear system bounds (----) for the rat response data of Gerber and Wagner (1972); (a) $C_0 = 30\ \mu g\ ml^{-1}$, $K_m = 15.1\ \mu g\ ml^{-1}$, $V_m = 7.86\ \mu g\ ml^{-1}\ h^{-1}$; (b) $C_0 = 16.5\ \mu g\ ml^{-1}$, $K_m = 9.0\ \mu g\ ml^{-1}$, $V_m = 12.2\ \mu g\ ml^{-1}\ h^{-1}$; (c) $C_0 = 7.4\ \mu g\ ml^{-1}$, $K_m = 15.0\ \mu g\ ml^{-1}$, $V_m = 20.0\ \mu g\ ml^{-1}\ h^{-1}$. (Arrows indicate time of final data point.)

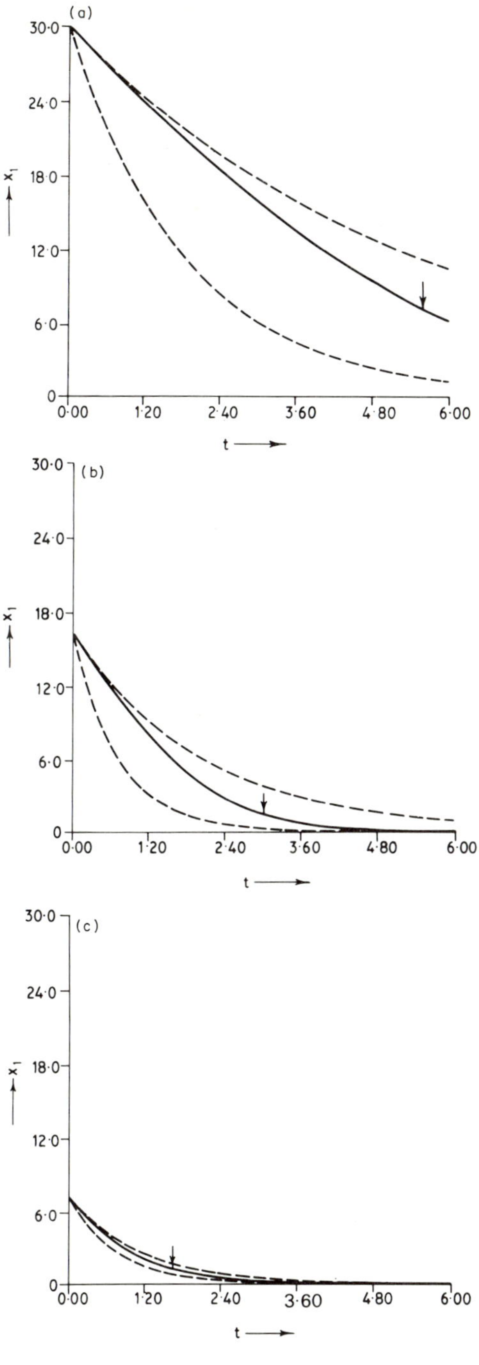
(a)
30·0
24·0
18·0
12·0
6·0
0
x1
0·00
1·20
2·40
3·60
4·80
6·00
t
(b)
30·0
24·0
18·0
12·0
6·0
0
x1
0·00
1·20
2·40
3·60
4·80
6·00
t
(c)
30·0
24·0
18·0
12·0
6·0
0
x1
0·00
1·20
2·40
3·60
4·80
6·00
t

where m and e are constants, is appropriate over a large proportion of the therapeutic range.

The effect–time curve is thus obtained by applying eqn (11.22) (or eqn (11.21) if a wide range is covered) to the appropriate concentration–time curve $C(t)$, and it is interesting to note that the more nearly $C(t)$ approximates to a single exponential, the more nearly $E(t)$ approximates to a straight line. Thus, for a one-compartment model, $E(t)$ is a straight line (over much of the range) for a single rapid intravascular administration and tends towards a straight line for large values of t for extravascular administration.

For multi-compartment disposition models, there is a problem in knowing how to find the appropriate curve for $C(t)$. Taking the two-compartment model as an example, it is often found that insertion of either $C_1(t)$ or $C_2(t)$ into eqns (11.21) or (11.22) gives a poor approximation to $E(t)$. Two methods have been proposed to overcome this.

The first method is to extend the classical pharmacokinetic model by introducing an effect compartment E, as shown in Fig. 11.9, in which the new compartment has a negligible effect on the pharmacokinetic model (k_{E1} is very small in comparison with any of k_{01}, k_{12} and k_{21}). The effect compartment response is different from those of either of the other compartments. The idea of an effect compartment has been put forward in the anaesthetics literature by Hull *et al.* (1978) and in the pharmacokinetics literature by Sheiner *et al.* (1979). The differential equation for the quantity, x_E, in the effect compartment is

$$\frac{dx_E}{dt} = -k_{0E}x_E + k_{E1}x_1 \tag{11.23}$$

so that

$$X_E(s) = \frac{k_{E1}}{s + k_{0E}} X_1(s). \tag{11.24}$$

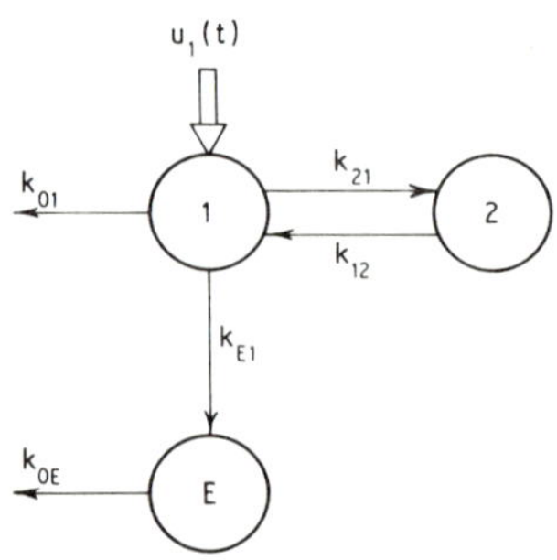

Figure 11.9 Two-compartment pharmacokinetic model with effect compartment (E) to model pharmacodynamics.

Since x_1 is unchanged by the addition of the extra compartment, we can substitute for $X_1(s)$ from the standard expressions given in Section 3.2. For example, if $u_1(t) = D_1\,\delta(t)$, and $b_1 = 1$,

$$X_E(s) = \frac{k_{E1}}{s + k_{0E}} \frac{D_1(s + k_{12})}{s^2 + (k_{01} + k_{12} + k_{21})s + k_{01}k_{12}}. \qquad (11.25)$$

The response of the effect compartment is thus third-order, and the freedom given by the extra order in the response enables the pharmacodynamic data to be fitted much more closely.

The second method is to model the effect as a sum of contributions from the compartments of the pharmacokinetic model:

$$E(t) = \sum_{i=1}^{p} f_i(x_i(t)) \qquad (11.26)$$

where $f_i(x_i(t))$ is a function, to be determined empirically, of the time response $x_i(t)$ of the quantity in compartment i (Whiting and Kelman, 1980). A comparison of the two approaches has been given by Kelman and Whiting (1980).

We will now illustrate these methods by considering two recently published examples of the simultaneous fitting of pharmacokinetics and pharmacodynamics which illustrate the need to span the full range of a nonlinearity in order to identify the parameters describing the nonlinearity with any meaning.

Application example 7. Spanning the range of a nonlinearity (Sheiner *et al.*, 1979; Whiting, Holford and Sheiner, 1980; Kelman and Whiting, 1980). Sheiner *et al.* (1979) measured the pharmacokinetics of the drug D-tubocurarine (DTC), and its pharmacological effect in the form of force of thumb adduction following electrical stimulus of the ulnar nerve. The principal findings are based on two groups of subjects. The first group consisted of seven healthy patients undergoing elective surgery. The drug was administered in the form of intravenous infusions of 16.8 μg (kg body weight)$^{-1}$ min^{-1} for 10 min, then 1.2 μg (kg body weight)$^{-1}$ (min)$^{-1}$, until the end of surgery (70 min to 240 min). Blood samples were taken at 1 min intervals up to 10 min then at 2 min intervals up to 25 min, then at 15 min intervals until the end of surgery, making a total of between 20 and 30 samples depending on the length of surgery. The second group consisted of 10 normal subjects undergoing elective surgery and 10 subjects with chronic end-stage kidney failure, undergoing renal transplantation. This group received a single bolus intravenous injection of DTC, half receiving 0.3 mg (kg body weight)$^{-1}$ and the other half 0.5 mg (kg body weight)$^{-1}$. Blood samples were taken at 3, 15, 30, 45, 60, 90 and 120 min in all subjects and additionally at 150 and 180 min for those who had received a 0.5 mg (kg body weight)$^{-1}$ dose.

Effect was measured by a force displacement transducer and recorded on a polygraph, the effect being quantified between 0 (no effect) and 1 (100% effect). For the first group, the range covered was between 0 and 0.9, while for the second group, the effect at $t = 0^+$ was 1 in all subjects. For many subjects in the second group, the range of effect covered was rather small, particularly for those with renal failure who had been administered 0.5 mg (kg body weight)$^{-1}$, where the effect was still high at the end of surgery. Even for the normal subjects in this group, the range of effect following a 0.5 mg (kg body weight)$^{-1}$ dose was not great, in two cases dropping only to 0.7 at $t = 180$ min.

It was found that the plasma concentration–time curves could be fitted by a two-compartment model, but that the effect kinetics did not follow the kinetics of either compartment. An effect compartment was therefore added, as in Fig. 11.9, and the whole set of data (plasma concentration and effect, with equal weighting) used to estimate the parameters of the model and to estimate α in the Hill eqn (11.21) relating pharmacological effect to concentration in the effect compartment. The pharmacokinetic model parameters are not given in the paper, but the values of the two parameters describing the pharmacodynamics were, for the first group (expressed as (mean $\pm$ standard deviation)) $k_{0E} = (0.13 \pm 0.015)\ \text{min}^{-1}$; $\alpha = 2.53 \pm 0.16$ and for the second group $k_{0E} = (0.16 \pm 0.14)\ \text{min}^{-1}$; $\alpha = 2.30 \pm 0.01$. For both groups, the plasma concentration giving 50% of full effect was about 37% of that required to give full effect. For the first group, results are also presented for fitting with $k_{0E} = \infty$ so that concentration in the effect compartment is the same as that in the plasma compartment, and the sum of squared deviations of the effect data was significantly worse for all subjects. The fitting was also significantly worse when α was fixed at 1.

It is interesting to note the much greater variability in k_{0E} in the second group due to the smaller range of the nonlinearity being spanned and to the smaller number of observations (7 or 9, depending on the dose, compared with 20 to 30 for the first group). There was no statistically significant difference between the pharmacokinetic parameters for the two groups, and in neither case was k_{0E} approximately equal to k_{12}. If it had been, then, from eqn (11.25) the effect dynamic response would have been similar in shape to the response of the peripheral compartment (no. 2).

An interesting illustration of how an effect compartment may become unnecessary when only a small part of a nonlinear range is spanned is given by Whiting, Holford and Sheiner (1980) who examined the pharmacokinetics and pharmacodynamics of disopyramide. The subjects were eight healthy volunteers who were given three different doses well separated in time, two oral preparations of 300 mg and an intravenous infusion over a period of 2 min, of 2 mg (kg body weight)$^{-1}$. Plasma samples were collected at frequent intervals up to between 8 h and 10 h, and the effect, measured at the same times, was shortening of the QT interval on an electrocardiogram.

The pharmacokinetics following the intravenous administration were modelled by a two-compartment model, but for the oral administration, a one-compartment model with delayed first-order input proved adequate. Comparing the concentration and effect responses, it was clear that, for the i.v. data, the effect did not follow the time course of the response of either the central (plasma) compartment or the peripheral compartment, so an effect compartment was added with exchange from the central compartment. As a first attempt, a linear model for the effect was tried:

$$E(t) = mC_E(t) + i \tag{11.27}$$

where $C_E(t)$ is the response in the effect compartment and m and i are constants.

Attempts to fit the more traditional effect–concentration relationships (11.21) or (11.22) did not result in significantly better fits. For the oral data, the effect followed the time course of the (single) compartment closely, making an effect compartment unnecessary, and a linear relationship (of the form of eqn (11.27)) applied to the plasma concentration gave a good fit to the effect data.

As pointed out by the authors, the simplicity of the model used, employing a linear relationship rather than a logarithmic one, and in the case of the oral data not even needing an effect compartment, is a result of the limited part of the nonlinear range which is spanned. Such a simple model would not necessarily be valid for an experiment with a greater range of plasma concentrations and changes in QT interval.

The second method of fitting effect data, using eqn (11.26), has been applied by Kelman and Whiting (1980) to the intravenous response data of Whiting, Holman and Sheiner (1980), specifically with linear functions of the compartmental quantities:

$$E(t) = A_1 x_1(t) + A_2 x_2(t) + B \tag{11.28}$$

where A_1, A_2 and B are constants and subscripts 1 and 2 refer to the central (plasma) and peripheral compartments, respectively. Note that the approach dispenses with the notion of an effect compartment. With this form of relationship, it also dispenses with the notion of a nonlinear relationship, so for the oral data, where a one-compartment model proved adequate to describe the pharmacokinetics and no effect compartment was needed to describe the pharmacodynamics, the two approaches are the same.

The values of A_1 and A_2 for the eight subjects (mean $\pm$ standard deviation) were (0.64 ± 0.35) ms mg^{-1} and (0.35 ± 0.23) ms mg^{-1}, and for seven of the eight subjects, $A_1 > A_2$. The goodness of fit of the effect compartment model (eqn (11.27)) was very similar to that of the functional relationship model (eqn (11.28)). Kelman and Whiting (1980) also report on the application of the latter approach to digoxin data, for which a three-compartment model was necessary to describe the pharmacokinetics. The effect was thus fitted as

$$E(t) = A_1 x_1(t) + A_2 x_2(t) + A_3 x_3(t) + B \qquad (11.29)$$

and it is interesting to note that, in five of six subjects, A_1 was not significantly different from zero.

The two approaches give considerable flexibility in describing combined pharmacokinetic/pharmacodynamic data. If the pharmacodynamic data alone were available, it would not be possible to estimate as many parameters as there are in the two models with any meaning. Indeed, a single response may be adequately described by a linear model with a small number of compartments; the nonlinear effects would only become apparent if responses to doses of different amplitudes were recorded.

11.4 Ecosystem modelling

A recent text (Matis, Patten and White (eds.), 1979) presented sixteen papers from a conference selected to show the state of the art of compartmental modelling of ecosystems. Some applications have been to tracer experiments, and so are similar to those already considered in this chapter, but many compartment-oriented models of ecosystems describe the steady state, with appropriate inputs from and outputs to the environment. Extensive perturbation tests are often carried out on the simulated system. These types of applications differ markedly from those discussed earlier in the chapter in that much use is made of prior knowledge, and to illustrate this we will consider below a linear model of a forested watershed ecosystem. Although some authors (e.g. Bledsoe, 1976) have criticized the use of linear models, a recent review paper (O'Neill, 1979) of linear compartmental models in ecology contains over a hundred references on the application of such models, and the author notes that the considerable mathematical tools which can be brought to bear on a linear problem mean that linear compartmental models will continue to be used either as preliminary models or as final models subject to constraints on size of perturbation.

The substantial body of theory on stochastic compartmental modelling has been developed with ecosystem modelling very much in mind.

Application example 8. Incorporation of prior knowledge into a compart-

mental model (Waide *et al.*, 1974). The way in which prior knowledge is incorporated in ecosystem modelling is particularly clearly illustrated in a well-documented study of the calcium cycle in a temperate forest ecosystem (Waide *et al.*, 1974). The vertical limits of the system were taken to be the height of the vegetation and the depth to which roots and other organisms penetrate the soil. The particular deciduous forest ecosystem considered (the Hubbard Brook Experimental Forest) had been the subject of a comprehensive examination of the nutrient cycling relationships, so that information was available on the structure and function of natural and disturbed ecosystems.

Since calcium does not pass through a gaseous phase as it cycles through terrestrial ecosystems, the model developed did not include an atmosphere compartment. On the other hand, nutrients behave differently in living and decaying organic matter, so that the single organic compartment considered in some previous studies was split into two separate compartments, vegetation and litter. The model thus had four compartments: vegetation (x_1), litter (x_2), available nutrients (x_3), and soil and rock minerals (x_4). Calcium inputs to the model occur via precipitation, partly to vegetation and partly to litter, and new rock supplies which become available for weathering reactions. Calcium outputs from the system are loss of dissolved and particulate organic matter from the litter, dissolved inorganic matter from the available nutrients and particulate inorganic matter from the soil and rock minerals. Inter-compartment transfers are from vegetation to litter by litterfall and leaching, from litter to available nutrients by decomposition and leaching, from available nutrients to vegetation through plant uptake and from soil and rock minerals to available nutrients by weathering and vice versa via mineralization. The structure of the compartmental model, with the external inputs, is thus as shown in Fig. 11.10.

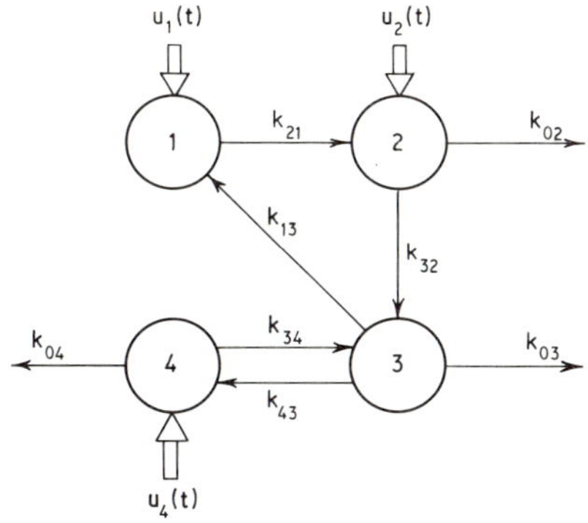

Figure 11.10 Four-compartment linear model used by Waide *et al.* (1974) to describe a forested watershed ecosystem.

Using the previously published data on the Hubbard Brook system, Waide *et al.* (1974) were able to assign values to the compartmental masses, flows, inputs and outputs, and these are shown on their Fig. 2. From these values, it is possible to determine the rate constants of the model, in units of yr^{-1}:

$$k_{21} = (7.814) \times 10^{-2} \qquad k_{02} = (0.005) \times 10^{-2}$$

$$k_{32} = (2.966) \times 10^{-2} \qquad k_{03} = (1.696) \times 10^{-2}$$

$$k_{13} = (7.145) \times 10^{-2} \qquad k_{04} = (0.077) \times 10^{-4}$$

$$k_{34} = (3.370) \times 10^{-4} \qquad (k_{43} \text{ negligible}).$$

The external inputs were $u_1(t) = 2.34$, $u_2(t) = 0.26$, $u_4(t) = 9.0$, all in units of kg $(\text{hectare})^{-1}$ yr^{-1}.

The authors discuss at some length many of the assumptions used in the model, such as the linearity, the donor-control formulation and the time-invariance, in particular with respect to the stated objective of using the model to characterize behaviour at or near steady-state conditions. When the system had been simulated on a computer, a number of experiments were conducted on the simulation. Firstly, washout curves were generated and then small-perturbation experiments were carried out on each input in turn, with results presented in both the time domain and the frequency domain. Sensitivity analyses were also carried out to see the effects on the system states of small changes in the rate constants. The results of these studies were interpreted in ecological terms, so giving further insight to the system.

Linear compartmental models incorporating much prior knowledge have also been used in other areas of ecology. For example, Walter (1979) has described a compartmental model of a marine ecosystem, modelling the interactions between various finfish species in a traditional fishing area of the North West Atlantic where production declined sharply in the 1970s. The flowrates of the model were estimated using stomach content data, biomass estimates and energy consumption estimates for eleven fish species plus two other species (squid and other invertebrates). Walter notes that a linear compartmental model can form a useful first approximation, for it is often possible to infer certain properties of the ecosystem from the structure of the model without exact knowledge of the flow rates. In his example, Walter compares the effects of eliminating mackerel and herring.

11.5 Modelling of engineering systems

The role of compartmental models to describe engineering systems has been discussed at the end of Chapter 1 of this book, where it is noted that, while many equations are of compartmental form, many systems have some equations which are not, so that the benefits of a compartmental approach are hard to define. Nevertheless, quite substantial parts of an overall system model may be compartmental—for example the mass balance equations of a grinding process (Kortela and Niemi, 1978) or the energy balance equations of a distillation column (Espana and Landau, 1978). The latter example brings out some interesting points concerning model-order reduction, and it will now be discussed in more detail as the final Application Example.

Application example 9. Model-order reduction (Espana and Landau, 1978). A distillation column is a device for separating more volatile components from less volatile components in a feed. In the body of the column, there are a number of plates; at the top of the column there is a condenser and at the bottom of the column there is a reboiler. The vapours moving up the column from the reboiler are increasingly enriched in more volatile components, and the liquid stream moving down the column is increasingly diminished in these components. The bottom product is taken off from the reboiler while the distillate, or top product is taken off from underneath the condenser. Part of the condenser output, the reflux, forms the liquid flow back down the column. The feed is introduced directly onto one of the plates in the column (the feed plate).

Control is exercised through the heat, Q_B, supplied to the reboiler, or the reflux flow rate (L_0), and the system outputs are the distillate flow rate, D, and the bottom product flow rate, L_B. The reflux ratio is defined as L_0/D.

Espana and Landau (1978) formulated their model of a column in terms of one output (the concentration of one particular component of the top product) and four inputs (the feed flow rate, L_F, the molal concentration vector $\mathbf{x}_F$ of the feed, and L_0 and Q_B as defined above), and they discuss various hypotheses used in the formulation of models. A full model can be constructed by considering each plate separately. Numbering the plates from the top, the equation for plate j for a particular component is

$$\frac{\mathrm{d}}{\mathrm{d}t}(c_j(t)x_j) = -L_j x_j - V_j y_j + L_{j-1}x_{j-1} + V_{j+1}y_{j+1} + \alpha L_F x_F \quad (11.30)$$

where c_j is the molal liquid holdup on plate j (vapour holdup assumed negligible by comparison); L_j is the liquid volumetric flow rate from plate j; x_j is the molal liquid concentration at plate j; V_j is the vapour volumetric flow rate from plate j; y_j is the molal vapour concentration at plate j; subscript F denotes the feed plate; and $\alpha = 1$ at the feed plate and zero at other plates.

The authors show that the state equation for $\mathbf{x}$ can be written in bilinear form, with cross-products between the inputs and the states. The full model, with one equation per plate, is not easy to solve and there is strong motivation to develop a reduced model.

The reduced model considered by Espana and Landau (1978) has only three states, corresponding to the condenser, the rectifying section (those plates above the feed plate up to the condenser) and the stripping section (from the feed plate downwards, including the reboiler). The assumptions in the reduced formulation are discussed; the principal ones are that the vapour and liquid enthalpies are constant in each section. The liquid flow rate is taken as L_0 in the rectifying section and $(L_0 + L_F)$ in the stripping section. For the reduced model, shown in Fig. 11.11, the compartmental equations are:

$$c_1\dot{x}_1 = -Vx_1 + Vy_2 \tag{11.31a}$$

$$c_2\dot{x}_2 = L_0x_1 - L_0x_2 - Vy_2 + Vy_3 \tag{11.31b}$$

$$c_3\dot{x}_3 = L_0x_2 - Vy_3 - (L_0 + L_F - V)x_3 + L_Fx_F. \tag{11.31c}$$

(It is assumed that the condenser does its job perfectly so that $y_1 = 0$.)

These can be written in the form of a state equation, with the origin at any of the equilibrium points in the input and state space. Denoting the new origin by superscript e, deviations δu_k and δx_j are defined as

$$\delta u_k = u_k - u_k^e, \qquad k = 1, 2, 3 \tag{11.32a}$$

$$\delta x_j = x_j - x_j^e, \qquad j = 1, 2, 3 \tag{11.32b}$$

Under hypotheses used in deriving the model, V and Q_B are related through a proportionality constant so that $\mathbf{u} = [L_F \quad L_0 \quad V]^T$; x_F is maintained constant.

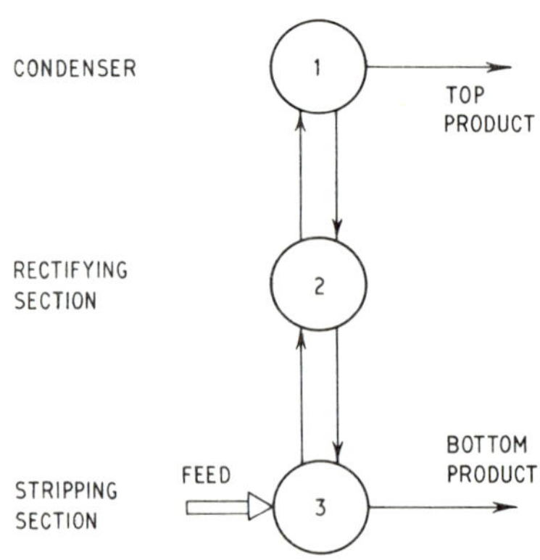

Figure 11.11 Structure of the reduced compartmental model used by Espana and Landau (1978) to describe a distillation column.

Further,

$$y_2 = k_2 x_2 + b_2 \tag{11.33a}$$

and

$$y_3 = k_3 x_3 + b_3. \tag{11.33b}$$

The incremental state equation is then given by

$$\mathbf{c}\,\delta\dot{\mathbf{x}} = \left\{ \begin{bmatrix} 0 & 0 & 0 \\ 0 & 0 & 0 \\ 0 & 0 & -1 \end{bmatrix} (u_1^e + \delta u_1) + \begin{bmatrix} 0 & 0 & 0 \\ 1 & -1 & 0 \\ 0 & 0 & -1 \end{bmatrix} (u_2^e + \delta u_2) + \begin{bmatrix} -1 & k_2 & 0 \\ 0 & -k_2 & k_3 \\ 0 & 0 & 1 - k_3 \end{bmatrix} (u_3^e + \delta u_3) \right\} \delta\mathbf{x} + \mathbf{B}^*\,\delta\mathbf{u} \tag{11.34}$$

where

$$\mathbf{B}^*\,\delta\mathbf{u} = \begin{bmatrix} 0 \\ 0 \\ x_F - x_3^e \end{bmatrix} \delta u_1 + \begin{bmatrix} 0 \\ x_1^e - x_2^e \\ x_2^e - x_3^e \end{bmatrix} \delta u_2 + \begin{bmatrix} 0 \\ y_3^e - y_2^e \\ x_3^e - y_3^e \end{bmatrix} \delta u_3 \tag{11.35}$$

and $\mathbf{c} = \operatorname{diag}\,[c_1 \quad c_2 \quad c_3]$.

The column vectors of $\mathbf{B}^*$ are not independent, however, since the equilibrium point in the $\mathbf{x}$-space is associated with a direction in the $\mathbf{u}$-space so that

$$\mathbf{B}^*\mathbf{u}^e = 0. \tag{11.36}$$

This gives three equality constraints on $(\mathbf{x}^e, \mathbf{u}^e)$ jointly so that if any three of the six are chosen, the other three follow. The reduced model is defined in terms of only eight parameters, the three independent coefficients in $\mathbf{B}^*$ together with c_1, c_2, c_3, k_2 and k_3. Espana and Landau report experiments which show that the responses of the reduced model to perturbations of L_F, L_0 and Q_B agree closely with the responses of a full model, which has $(N + 2)$ states, where N is the number of plates.

12 Some Unsolved Problems in Compartmental Modelling

In the ten years following the publication of the book by Jacquez (1972), the most substantial contribution to the understanding of compartment systems has been in the area of deterministic identifiability (the inverse problem) of linear, time-invariant models. Laplace transform identifiability analysis, considered in Chapter 6, while conceptually simple, may be straightforward or surprisingly difficult, even for a system of low order. An alternative method, based on normal-mode analysis and designed to overcome some of the drawbacks of the Laplace transform approach (Norton, 1980; Norton, Brown and Godfrey, 1980) is more systematic and is to be preferred to the Laplace transform approach when examining several different cases with the same number of compartments, but in individual cases, it is difficult to forecast which of the two methods will involve less work.

Many other methods for analyzing deterministic identifiability of linear, time-invariant systems continue to appear in the literature, but while they may prove easier in particular cases, they do not appear to offer any great advantage in general. The magnitude of the problem and the wide variety of behaviour have been highlighted by Norton (1982) in a detailed deterministic identifiability analysis of three-compartment systems. A general and comprehensive analysis of identifiability is considerably more difficult than expected.

While deterministic identifiability is now much better understood, at least for linear, time-invariant systems, several other problems concerned with compartmental modelling remain unsolved or relatively unexplored, and in the remainder of this chapter, some of these will be discussed.

Continuing for the moment with linear, time-invariant compartmental systems, some aspects of analysis remain rather poorly understood. We do know that the non-negativity of the rate constants results in constraints on system responses (see Section 2.4). For example, responses to positive perturbations and/or positive initial conditions cannot go negative (Bellman, 1960), the washout curves decrease monotonically (Hearon, 1968) and, if the eigenvalues are real, the impulse response of the perturbed compartment decreases monotonically (Hearon, 1979). However, with the notable exception of the work of Rubinow and Winzer (1971), very little has appeared concerning the constraints which a compartmental structure imposes on amplitudes and exponents of a multi-exponential fit to an observation. The problem, which can be formulated as "given the structure of a compartmental model, determine the constraints on the amplitudes and exponents of the response of compartment j to perturbation of compartment i" is an important one in model selection, for if the model originally proposed to describe a particular experimental situation cannot possibly fit the available data, it would be useful to be able to check the structure of any other models which do fit the data, to see whether they provide acceptable alternatives. An example of inadmissible responses, for a three-compartment model including a source compartment (Chau, 1977), was given in Section 6.3.5.

Further work seems desirable on the bounds on the parameters of an unidentifiable compartmental system, due to the compartmental structure of the equations. In Chapter 7, such bounds are given for three models of practical interest, but it would be interesting to determine bounds for other models and to investigate how the widths of the bounds are related to the amplitudes and exponents of the curves fitted to the observations.

There is still comparatively little known about linear, time-invariant models with complex eigenvalues (possible when three or more compartments are connected in a cycle). In Sections 4.1.6 and 4.2.5 of this book, the behaviour of the closed unilateral circulation model with equal rate constants has been discussed and while it seems likely that all other models have less oscillatory responses, no global proof exists, as far as the author is aware.

While deterministic identifiability has been discussed extensively, little has appeared in the literature on numerical identifiability of compartmental models and the problems posed by real data seem less well appreciated. As we have seen in Chapter 8, it is often necessary to illustrate the problems which arise by considering specific examples.

One interesting and hitherto unresolved numerical identifiability problem of considerable practical interest concerns the estimation of parameters when more than one measurement is made in a model consisting of interlinked subsystems, in particular when the links describe a change of species. For an experiment consisting of perturbation with species A and observation of the responses of both species A and B (for example, a drug and its metabolite), the first response gives parameters in that part of the model describing species A, while if the model is correct, the response of species B will, in theory, often have more parameters than could realistically be identified if it were the only response available. However, as we have seen in Section 4.2.4, some of the exponents are the same as those already identified from the species A response. The problem is to determine under what circumstances the fact that some of the exponents are already known makes the remaining parameters identifiable.

Turning now to nonlinear systems, much less material has been published, and the field is wide open for new developments. Few analytical results appear possible, and it will usually be necessary to compute solutions numerically. At the simplest possible level, the effects on compartmental responses of incorporating some well-known nonlinearities into various links of simple compartmental models should be examined. In the biomedical area, nonlinearities of interest include Michaelis–Menten saturation kinetics (eqn (1.3)), including zero-order kinetics as a limiting case, and the Langmuir nonlinearity considered in Section 9.2 Example 3, while if pharmacological response is being considered, the Hill equation (Section 11.3.3) is important. Numerical identifiability is extremely important because the nonlinear system must be distinguishable from a linear system. The approach given in Chapter 9 of determining bounds for the responses of corresponding linear systems would seem quite widely applicable.

For linear but time-varying systems, as with nonlinear systems, few analytical solutions appear possible, an exception being when all the rate coefficients are varying in the same manner (i.e. same proportion of variation and, if the variation is cyclic, same frequency and phase). For other cases, numerical solutions are still possible.

The problem of deterministic identifiability for nonlinear systems and time-varying systems is tricky. A test for unique identifiability, based on Taylor series expansion of the observations, is available, but as we have seen in Chapters 9 and 10, it can rapidly get out of hand even for systems of low order. Also, as we have seen from comparison of the two 2-compartment models considered in Section 9.4.3, it is currently far from obvious what are the features in model and experiment which make a Taylor series identifiability analysis difficult. Any methods of easing these problems would be valuable.

Stochastic compartmental modelling is currently a very active area of research. One interesting point which has been raised is that, for the semi-Markov models, responses can depart in the limit from the sums of exponentials of linear, deterministic models (Purdue, 1979). Responses of nonlinear models and of time-varying models also depart from sums of exponentials and there seems a need for a study of the most realistic way of modelling non-exponential responses within a compartmental framework. A specific example which can be modelled in two different ways is the calcium metabolism example of Staub *et al.* (1981) considered in Application Example 3 in Chapter 11. The authors used a nonlinear model to describe the responses which were highly oscillatory, but a linear but time-varying model could also have produced similar responses. The question then is to determine which approach is more plausible in terms of the physiology of the system. Other factors which govern the choice between models include the parsimony of each alternative, the extent to which the alternatives can be validated and the relative difficulty of designing experiments to fill in gaps in knowledge of structure or parameter values. The use to which models will be put is also important, and the choice if the main objective of the model were to discriminate between different conditions may well be different from that if the main objective were predictive.

References

Aamodt, R. L., Rumble, W. F., Johnston, G. S., Foster, D. M. and Henkin, R. I. (1979). Zinc metabolism in humans after oral and intravenous administration of Zn-69m. *Am. Jnl. Clin. Nutrition*, **32**, 559–569.

Aamodt, R. L., Rumble, W. F., Babcock, A. K., Foster, D. M. and Henkin, R. I. (1982). Effects of oral zinc loading on zinc metabolism in humans. I: experimental studies. *Metabolism*, **31**, 326–334.

Atkins, G. L. (1969). "Multicompartment Models for Biological Systems". Methuen, London.

Atkins, G. L. and Nimmo, I. A. (1975). A comparison of seven methods for fitting the Michaelis–Menten equation. *Biochem. J.*, **149**, 775–777.

Babcock, A. K., Henkin, R. I., Aamodt, R. L., Foster, D. M. and Berman, M. (1982). Effects of oral zinc loading on zinc metabolism in humans. II: in vivo kinetics. *Metabolism*, **31**, 335–347.

Bache, R. A., Gray, W. M. and Murray-Smith, D. J. (1981). Time-domain system identification applied to non-invasive estimation of cardiopulmonary quantities. *IEE Proc.*, **128 (Part D)**, 56–64.

Bajzer, Z. and Nosil, J. (1977). A simple mathematical lung model for quantitative regional ventilation measurement using $^{81}Kr^{m}$. *Phys. Med. Biol.*, **22**, 975–980.

Bellman, R. (1960). "Introduction to Matrix Analysis". McGraw Hill, New York.

Bellman, R. and Åström, K. J. (1970). On structural identifiability. *Math. Biosci.*, **7**, 329–339.

Bergner, P. E. E. and Lushbaugh, C. C. (eds) (1967). "Compartments, Pools and Spaces in Medical Physiology". U.S.A.E.C. Symposium Series No. 11.

Berk, P. D., Howe, R. B., Bloomer, J. R. and Berlin, N. I. (1969). Studies of bilirubin kinetics in normal adults. *J. Clin. Inv.*, **48**, 2176–2190.

Berk, P. D., Bloomer, J. R., Howe, R. B. and Berlin, N. I. (1970). Constitutional hepatic dysfunction (Gilbert's syndrome). *Am. Jnl. Med.*, **49**, 296–305.

Berman, M. and Schoenfeld, R. (1956). Invariants in experimental data on linear kinetics and formulation of models. *J. Appl. Phys.*, **27**, 1361–1370.

Bledsoe, L. L. (1976). Linear and nonlinear approaches for ecosystem dynamic modeling. *In* "Systems Analysis and Simulation in Ecology", Vol. 4 (B. C. Patten, ed.), pp. 283–298. Academic Press, New York and London.

Boxenbaum, H. G. and Riegelman, S. (1976). Pharmacokinetics of isoniazid and some metabolities in man. *Jnl. Pharmacokin. Biopharm.*, **4**, 287–325.

Campello, L. and Cobelli, C. (1978). Parameter estimation of biological stochastic compartmental models—an application. *IEEE Trans. Biomed. Eng.*, **BME-25**, 139–146.

Carson, E. R., Cobelli, C. and Finkelstein, L. (1981). The identification of metabolic systems—a review. *Am. J. Physiol.*, **240**, R120–R129.

Carson, E. R., Godfrey, K. R. and Reeve, J. (1982). A review of modelling and the role of dynamic tracer studies in metabolism. *In* "A Quantitative Approach to Metabolism: the Role of Tracers and Models in Clinical Medicine" (D. G. Cramp, ed.), Chapter 1. John Wiley, New York.

Chau, N. P. (1977). Linear pharmacokinetic models: geometric construction to determine transfer and elimination rate constants. *Jnl. Pharmacokin. Biopharm.*, **5**, 147–159.

Cobelli, C. and DiStefano III, J. J. (1980). Parameter and structural identifiability concepts and ambiguities: a critical review and analysis. *Am. J. Physiol.*, **239**, R7–R24.

Cobelli, C. and Morato, L. M. (1978). On the identification by filtering techniques of a biological *n*-compartment model in which the transport rate parameters are assumed to be stochastic processes. *Bull. Math. Biol.*, **40**, 651–660.

Cobelli, C., Frezza, M. and Tiribelli, C. (1975). Modeling, identification and parameter estimation of bilirubin kinetics in normal, hemolytic and Gilbert's states. *Comput. Biomed. Res.*, **8**, 522–537.

Cobelli, C., Lepschy, A. and Romanin-Jacur, G. (1978a). Identifiability of time-variant compartmental models of biological processes. *In* "A Link between Science and Applications of Automatic Control, 7th. IFAC World Congress" (A. Niemi, B. Wahlström and J. Virkkunen, eds), Vol. 1, pp. 509–512.

Cobelli, C., Lepschy, A. and Romanin-Jacur, G. (1978b). Comments on "On the relationships between structural identifiability and the controllability, observability properties". *IEEE Trans. Autom. Contr.*, **AC-23**, 965.

Cobelli, C., Lepschy, A. and Romanin-Jacur, G. (1979a). Identifiability of compartmental systems and related structural properties. *Math. Biosci.*, **44**, 1–18.

Cobelli, C., Lepschy, A. and Romanin-Jacur, G. (1979b). Identifiability results on some constrained compartmental systems. *Math. Biosci.*, **47**, 173–195.

Colburn, W. A., DiSanto, A. R. and Gibaldi, M. (1977). Pharmacokinetics of erythromycin on repetitive dosing. *J. Clin. Pharmacol.*, **17**, 592–600.

Cramér, H. (1946). "Mathematical Methods of Statistics". Princeton University Press, Princeton, New Jersey.

D'Argenio, D. Z. (1981). Optimal sampling times for pharmacokinetic experiments. *Jnl. Pharmacokin. Biopharm.*, **9**, 739–756.

Davison, E. J. (1977). Connectability and structural controllability of composite systems. *Automatica*, **13**, 109–123.

Delforge, J. (1980). New results on the problem of identifiability of a linear system. *Math. Biosci.*, **52**, 73–96.

Delforge, J. (1981). Necessary and sufficient structural condition for local identifiability of a system with linear compartments. *Math. Biosci.*, **54**, 159–180.

DiStefano III, J. J. (1976). Tracer experiment design for unique identification of nonlinear physiological systems. *Am. J. Physiol.*, **230**, 476–485.

DiStefano III, J. J. (1977). On the relationships between structural identifiability and the controllability, observability properties. *IEEE Trans. Autom. Contr.*, **AC-22**, 652.

DiStefano III, J. J. (1981). Optimized blood sampling protocols and sequential design of kinetic experiments. *Am. J. Physiol.*, **240**, R259–R265.

DiStefano III, J. J. and Cobelli, C. (1980). On parameter and structural identifiability: nonunique observability/reconstructability for identifiable systems, other ambiguities and new definitions. *IEEE Trans. Autom. Contr.* **AC-25**, 830–833.

DiStefano III, J. J. and Mori, F. (1977). Parameter identifiability and experiment design: thyroid hormone parameters. *Am. J. Physiol.*, **233**, R134–R144.

DiStefano III, J. J., Wilson, K. C., Jang, M. and Mak, P. H. (1975). Identification of the dynamics of thyroid hormone metabolism. *Automatica*, **11**, 149–159.

DiStefano III, J. J., Jang, M., Malone, T. K. and Broutman, M. (1982). Comprehensive kinetics of triiodothyronine (T_3) production, distribution and metabolism in blood and tissue pools of the rat using optimized blood sampling protocols. *Endocrinology*, **110**, 198–213.

Eisenthal, R. and Cornish-Bowden, A. (1974). The direct linear plot: a new graphical method for estimating enzyme kinetic parameters. *Biochem. J.*, **139**, 715–720.

Espana, M. and Landau, I. D. (1978). Reduced order bilinear models for distillation columns. *Automatica*, **14**, 345–355.

Eykhoff, P. (1974). "System Identification: Parameter and State Estimation". John Wiley, New York.

Finkelstein, L. and Carson, E. R. (1979). "Mathematical Modelling of Dynamic Biological Systems". Research Studies Press, Forest Grove, Oregon.

Foster, D. M. and Jacquez, J. A. (1975). Multiple zeros for eigenvalues and the multiplicity of traps of a linear compartmental system. *Math. Biosci.*, **26**, 89–97.

Foster, D. M., Aamodt, R. L., Henkin, R. I. and Berman, M. (1979). Zinc metabolism in humans: a kinetic model. *Am. J. Physiol.*, **237**, R340–R349.

Garrettson, L. K. and Jusko, W. J. (1975). Diphenylhydantoin elimination kinetics in overdosed children. *Clin. Pharmacol. Ther.*, **17**, 481–491.

Gavalas, G. R. (1968). "Nonlinear Differential Equations of Chemically Reacting Systems". Springer, New York.

Gerber, N. and Wagner, J. G. (1972). Explanation of dose-dependent decline of diphenylhydantoin plasma levels by fitting to the integrated form of the Michaelis–Menten equation. *Res. Commun. Chem. Pathol. Pharmacol.*, **3**, 455–466.

Gibaldi, M. and Perrier, D. (1975). "Pharmacokinetics". Marcel Dekker, New York.

Glover, K. and Willems, J. C. (1974). Parametrization of linear dynamical systems: canonical forms and identifiability. *IEEE Trans. Autom. Contr.*, **AC-19**, 640–645.

Goldberg, K. (1956). A matrix with real characteristic roots. *Jnl. Res. National Bureau of Standards*, **56**, 87.

Grewal, M. S. and Glover, K. (1976). Identifiability of linear and nonlinear dynamical systems. *IEEE Trans. Autom. Contr.*, **AC-21**, 833–837.

Hart, H. E. (ed.) (1963). Multi-compartment analysis of tracer experiments. *Ann. N.Y. Acad. Sci.*, **108**, 1–338.

Hearon, J. Z. (1968). The washout curve in tracer kinetics. *Math. Biosci.*, **3**, 31–39.

Hearon, J. Z. (1979). A monotonicity theorem for compartmental systems. *Math. Biosci.*, **46**, 293–300.

Henri, V. (1902). Théorie générale de l'action de quelques diastases. *Compt. Rend. Hebd. Séanc Acad. Sci. (Paris)*, **135**, 916–919.

Himmelstein, K. and Lutz, R. J. (1979). A review of physiologically-based pharmacokinetic modelling. *Jnl. Pharmacokin. Biopharm.*, **7**, 127–145.

Hull, C. J. (1982). Pharmacodynamics of non-depolarizing neuromuscular blocking agents. *Br. J. Anaesth.*, **54**, 169–182.

Hull, C. J., Van Beem, H. B. H., McLeod, K., Sibbald, A. and Watson, M. J. (1978). A pharmacodynamic model for pancuronium. *Br. J. Anaesth.*, **50**, 1113–1122.

Jacquez, J. A. (1972). "Compartmental Analysis in Biology and Medicine". Elsevier, Amsterdam.

Jennrich, R. I. (1969). Asymptotic properties of nonlinear least squares estimators. *Ann. Math. Stat.*, **40**, 633–643.

Kelman, A. W. and Whiting, B. (1980). Modeling of drug response in individual subjects. *Jnl. Pharmacokin. Biopharm.*, **8**, 115–130.

Kortela, U. K. J. and Niemi, T. J. (1978). Modelling and prediction of copper concentration of a grinding process. *Automatica*, **14**, 547–556.

Krüger-Thiemer, E. V., Diller, W., Dettli, L., Bünger, P. and Seydel, J. (1964). Demonstration des einflusses der eiweissbindung und der ionisation auf die pharmacokinetic am kombinierten gaskinetischen modell nach van't Hoff und Langmuir. *Antibiotica et Chemotherapia*, **12**, 171–193.

Ladde, G. S. (1976a). Cellular systems. I—Stability of chemical systems. *Math. Biosci.*, **29**, 309–330.

Ladde, G. S. (1976b). Cellular systems. II—Stability of compartmental systems. *Math. Biosci.*, **30**, 1–21.

Lanczos, C. (1957). "Applied Analysis". Pitman, London.

Landahl, H. (1972). Note on oscillations in a closed chain of compartments with linear, unidirectional transport. *Bull. Math. Biophys.*, **34**, 293–296.

LeCourtier, Y. and Walter, E. (1981). Comments on "On parameter and structural identifiability: nonunique observability/reconstructability for identifiable systems, other ambiguities and new definitions". *IEEE Trans. Autom. Contr.*, **AC-26**, 800–801.

Lewis, R. M. and Anderson, B. D. O. (1980). Insensitivity of a class of nonlinear compartmental systems to the introduction of arbitrary time delays. *IEEE Trans. Circuits Syst.*, **CAS-27**, 604–612.

Lineweaver, H. and Burk, D. (1934). The determination of enzyme dissociation constants. *J. Amer. Chem. Soc.*, **56**, 658–666.

Llaurado, J. G. (1973). Compartmental approaches to water and electrolyte distribution. *In* "Engineering Principles in Physiology, Vol. II" (J. H. U. Brown and D. S. Gann, eds.), Chapter 25. Academic Press, New York and London.

Lundquist, F. and Wolthers, H. (1958). The kinetics of alcohol elimination in man. *Acta Pharmacol. Toxicol.*, **14**, 265–289.

McCollum, P. A. and Brown, B. F. (1965). "Laplace Transform Tables and Theorems". Holt, Rinehart and Winston, New York.

Maeda, H., Kodama, S. and Ohta, Y. (1978). Asymptotic behaviour of nonlinear compartmental systems: non-oscillation and stability. *IEEE Trans. Circuits Syst.*, **CAS-25**, 372–378.

Mandel, J. (1964). "The Statistical Analysis of Experimental Data". Interscience, New York.

Marcus, A. H. (1979). Semi-Markov compartmental models in ecology and environmental health. *In* "Compartmental Analysis of Ecosystem Models" (J. H. Matis, B. C. Patten and G. C. White, eds.) pp. 261–278. International Co-operative Publishing House, Fairland, Maryland.

Martin, J. F., Vierling, J. M., Wolkoff, A. W., Scharschmidt, B. P., Vergalla, J., Waggoner, J. G. and Berk, P. D. (1976). Abnormal hepatic transport of indocyanine green in Gilbert's syndrome. *Gastroenterology*, **70**, 385–391.

Matis, J. H. and Hartley, H. O. (1971). Stochastic compartmental analysis: model and least-squares estimation from time series data. *Biometrics*, **27**, 77–102.

Matis, J. H. and Tolley, H. D. (1979). Compartmental models with multiple sources of stochastic variability: the one-compartment, time-invariant hazard rate case. *Bull. Math. Biol.*, **41**, 491–515.

Matis, J. H. and Wehrly, T. E. (1979). An approach to a compartmental model with multiple sources of stochasticity for modeling ecological systems. *In* "Compartmental Analysis of Ecosystem Models" (J. H. Matis, B. C. Patten and G. C. White, eds) pp. 195–222. International Co-operative Publishing House, Fairland, Maryland.

Matis, J. H., Kodell, R. L. and Cardenas, M. (1976). A note on the use of a stochastic mammillary compartmental model as an environmental safety model. *Bull. Math. Biol.*, **38**, 467–478.

Matis, J. H., Patten, B. C. and White, G. C. (eds) (1979). "Compartmental Analysis of Ecosystem Models". International Co-operative Publishing House, Fairland, Maryland.

Mazanov, A. (1976). Stability of multi-pool models with lags. *J. Theor. Biol.*, **59**, 429–442.

Michaelis, L. and Menten, M. L. (1913). Der Kinetic der Invertinwirkung. *Biochem. Z.*, **49**, 333–369.

Milanese, M. and Molino, G. P. (1975). Structural identifiability of compartmental models and pathophysiological information from the kinetics of drugs. *Math. Biosci.*, **26**, 175–190.

Miller, R. D. (1982). Pharmacokinetics of competitive muscle relaxants. *Br. J. Anaesth.*, **54**, 161–167.

Mori, F. and DiStefano III, J. J. (1979). Optimal nonuniform sampling interval and test-input design for identification of physiological systems from very limited data. *IEEE Trans. Autom. Contr.*, **AC-24**, 893–900.

Mulholland, R. S. and Keener, M. S. (1974). Analysis of linear compartmental models for ecosystems. *J. Theor. Biol.*, **44**, 105–116.

Norton, J. P. (1980a). Normal-mode identifiability analysis of linear compartmental systems in linear stages. *Math. Biosci.*, **50**, 95–115.

Norton, J. P. (1980b). Structural zeros in the modal matrix and its inverse. *IEEE Trans. Autom. Contr.*, **AC-25**, 980–981.

Norton, J. P. (1982). An investigation of the sources of non-uniqueness in deterministic identifiability. *Math. Biosci.*, **60**, 89–108.

Norton, J. P., Brown, R. F. and Godfrey, K. R. (1980). Modal analysis of identifiability of linear compartmental models. *IEE Proc.*, **127** (**Part D**), 83–92.

O'Neill, R. V. (1979). A review of linear compartmental analysis in ecosystem science. *In* "Compartmental Analysis of Ecosystem Models" (J. H. Matis, B. C. Patten and G. C. White, eds) pp. 3–27. International Co-operative Publishing House, Fairland, Maryland.

Ottaway, J. H. and Mowbray, J. (1973). The computation of parameters in isotopic tracer systems not in a steady state. *Eur. J. Biochem.*, **36**, 380–384.

Owens, D., Jones, E. A., and Carson, E. R. (1977). Studies on the kinetics of unconjugated ^{14}C bilirubin metabolism in normal subjects and patients with compensated cirrhosis. *Clin. Sci. Mol. Med.*, **52**, 555–570.

Owens, D. H. (1981). "Multivariable and Optimal Systems". Academic Press, New York and London.

Paumgartner, G., Probst, P., Kraines, R. and Leevy, C. M. (1970). Kinetics of indocyanine green removal from the blood. *Ann. N.Y. Acad. Sci.*, **170**, 134–147.

Pohjanpalo, H. (1978). System identifiability based on the power series expansion of the solution. *Math. Biosci.*, **41**, 21–33.

Power, H. M. and Simpson, R. J. (1978). "Introduction to Dynamics and Control". McGraw Hill, London and New York.

Purdue, P. (1979). Stochastic compartmental models: a review of the mathematical theory with ecological applications. *In* "Compartmental Analysis of Ecosystem Models" (J. H. Matis, B. C. Patten and G. C. White, eds) pp. 223–260. International Co-operative Publishing House, Fairland, Maryland.

Reid, J. G. (1977). Structural identifiability in linear time-invariant systems. *IEEE Trans. Autom. Contr.*, **AC-22**, 242–246.

Rescigno, A. and Beck, J. S. (1972). Compartments. *In* "Foundations of Mathematical Biology, Vol. II—Cellular Systems" (R. Rosen, ed.), Chapter 5. Academic Press, New York and London.

Richards, R. J. (1979). "An Introduction to Dynamics and Control". Longman, London and New York.

Riggs, D. S. (1963). "The Mathematical Approach to Physiological Problems". M.I.T. Press, Cambridge, Massachusetts and London.

Rogus, E. and Zierler, K. L. (1973). Sodium and water contents of sarcoplasm and sarcoplasmic reticulum in rat skeletal muscle: effects of anisotonic media, ouabain and external sodium. *J. Physiol.*, **233**, 227–270.

Rubinow, S. I. and Winzer, A. (1971). Compartment analysis: an inverse problem. *Math. Biosci.*, **11**, 203–247.

Sandberg, I. W. (1978). On the mathematical foundations of compartmental analysis in biology, medicine and ecology. *IEEE Trans. Circuits Syst.*, **CAS-25**, 273–279.

Schoenfeld, R. L. (1963). Linear network theory and tracer analysis. *Ann. N.Y. Acad. Sci.*, **108**, 69–91.

Sedman, A. J. and Wagner, J. G. (1974). Importance of the use of the appropriate pharmacokinetic model to analyse in vivo enzyme constants. *Jnl. Pharmacokin. Biopharm.*, **2**, 161–173.

Sel'Kov, E. (1968). Self-oscillation in glycolysis. 1—a simple kinetic model. *Eur. J. Biochem.*, **4**, 79–86.

Sheiner, L. B., Stanski, D. R., Vozeh, S., Miller, R. D. and Ham, J. (1979). Simultaneous modelling of pharmacokinetics and pharmacodynamics: application to d-tubocurarine. *Clin. Pharmacol. Ther.* **25**, 358–371.

Sheppard, C. W. (1948). The theory of transfers within a multi-compartment system using isotopic tracers. *J. Appl. Phys.*, **19**, 70–76.

Sheppard, C. W. and Householder, A. S. (1951). The mathematical basis of the interpretation of tracer experiments in closed steady-state systems. *J. Appl. Phys.*, **22**, 510–520.

Shipley, R. A. and Clark, R. E. (1972). "Tracer Methods for in-vivo Kinetics". Academic Press, New York and London.

Skinner, S. M., Clark, R. E., Baker, N. and Shipley, R. A. (1959). Complete solution of the three-compartment model in steady-state after single injection of radioactive tracer. *Am. J. Physiol.*, **196**, 238–244.

Smith, G. A. and Llaurado, J. G. (1974). Computer modeling of nonsteady state sodium kinetics in liver. *IEEE Trans. Biomed. Eng.*, **BME-21**, 433–443.

Soong, T. T. and Dowdee, J. W. (1974). Pharmacokinetics with uncertainties in rate constants—III: the inverse problem. *Math. Biosci.*, **19**, 343–353.

Sprinson, D. B. and Rittenberg, D. (1949). The rate of interaction of the amino acids of the diet with the tissue proteins. *J. Biol. Chem.*, **180**, 715–726.

Staub, J-F., Brezillon, P., Perault-Staub, A. M. and Milhaud, G. (1981). Nonlinear modelling of calcium metabolism—first attempt in the calcium-deficient rat. *Trans. Inst. Meas. Contr.*, **3**, 89–97.

Teorell, T. (1937a). Kinetics of distribution of substances administered to the body. I—the extravascular routes of administration. *Arch. Internat. Pharmacodynamie Thérapie*, **57**, 205–225.

Teorell, T. (1937b). Kinetics of distribution of substances administered to the body. II—the intravascular routes of administration. *Arch. Internat. Pharmacodynamie Thérapie*, **57**, 226–240.

Thron, C. D. (1972). Structure and kinetic behaviour of linear multicompartment systems. *Bull. Math. Biophys.*, **34**, 277–291.

Tolley, H. D., Burdick, D., Manton, K. G. and Stallard, E. (1978). A compartmental model approach to the estimation of tumor incidence and growth: investigation of a model of cancer latency. *Biometrics*, **34**, 377–389.

Tong, D. D. M. and Metzler, C. M. (1980). Mathematical properties of compartment models with Michaelis–Menten type elimination. *Math. Biosci.*, **48**, 293–306.

Travis, C. C. and Haddock, G. (1981). On structural identification. *Math. Biosci.*, **56**, 157–173.

Van den Bos, A. (1980). A class of small sample nonlinear least squares problems. *Automatica*, **16**, 487–490.

Van den Bos, A. (1981). Degeneracy in nonlinear least squares. *IEE Proc.*, **128 (Part D)**, 109–116.

Wagner, J. G. (1968). Kinetics of pharmacological response. *J. Theor. Biol.*, **20**, 173–201.

Wagner, J. G. (1973a). Properties of the Michaelis–Menten equation and its integrated form which are useful in pharmacokinetics. *Jnl. Pharmacokin. Biopharm.*, **1**, 103–121 and 337–338.

Wagner, J. G. (1973b). A modern view of pharmacokinetics. *Jnl. Pharmacokin. Biopharm.*, **1**, 363–401.

Wagner, J. G. (1975). Do you need a pharmacokinetic model and, if so, which one? *Jnl. Pharmacokin. Biopharm.*, **3**, 457–478.

Wagner, J. G. (1978). Time to reach steady state and prediction of steady-state concentrations for drugs obeying Michaelis–Menten elimination kinetics. *Jnl. Pharmacokin. Biopharm.*, **6**, 209–225.

Waide, J. B., Krebs, J. E., Clarkson, S. P. and Setzler, E. M. (1974). A linear systems analysis of the calcium cycle in a forested watershed ecosystem. *Prog. Theor. Biol.*, **3**, 261–345.

Walter, E., Le Cardinal, G. and Bertrand, P. (1976). On the identifiability of linear state systems. *Math. Biosci.*, **31**, 131–141.

Walter, E., Bertrand, P. and Le Cardinal, G. (1979). Comments on "Identifiability of linear and nonlinear dynamical systems". *IEEE Trans. Autom. Contr.*, **AC-24**, 329–331.

Walter, G. G. (1979). A compartmental model of a marine ecosystem. *In* "Compartmental Analysis of Ecosystem Models" (J. H. Matis, B. C. Patten and G. C. White, eds), pp. 29–42. International Co-operative Publishing House, Fairland, Maryland.

Weiner, D. and Purdue, P. (1977). A semi-Markov approach to stochastic compartmental models. *Commun. Stat., Part A—Theory and Methods*, **A6**, 1231–1243.

Whiting, B. and Kelman, A. W. (1980). The modelling of drug response. *Clin. Sci.*, **59**, 311–315.

Whiting, B., Holford, N. H. G. and Sheiner, L. B. (1980). Quantitative analysis of disopyramide concentration–effect relationship. *Br. J. Clin. Pharmacol.*, **9**, 67–75.

Wilson, K. C., DiStefano III, J. J., Fisher, D. A. and Sack, J. (1977). Systems analysis and estimation of key parameters of thyroid hormone metabolism in sheep. *Annals Biomed. Eng.*, **5**, 70–84.

Zilversmit, D. B., Entenman, C. and Fishler, M. C. (1943). On the calculation of turnover time and turnover rate from experiments involving the use of labeling agents. *J. Gen. Physiol.*, **26**, 325–331.

Appendix 1 Laplace Transforms

The Laplace transform is used for solving ordinary linear differential equations. Essentially, it replaces the time domain by the complex domain of the Laplace operator s, so allowing differential equations to be manipulated easily by conventional algebraic techniques. As we will see below, the transformed expressions (which include values for initial conditions) may be rearranged in a form which can generally be found in a table of Laplace transforms so that, on transformation back into the time domain, the complete solution of the differential equations is found.

1 Definition and some simple examples

The Laplace transform of a function of time $f(t)$ is defined as:

$$\mathscr{L}[f(t)] \equiv F(s) \triangleq \int_{0^-}^{\infty} f(t)\,\mathrm{e}^{-st}\,\mathrm{d}t \tag{A1.1}$$

where s is the complex frequency variable $\sigma + j\omega$. Note that the lower limit of integration is 0^-, i.e. immediately before $t = 0$. This is different from that used in many other texts and takes into account that $f(t)$ may be an impulse or one of its higher derivatives.

For a Laplace transform to exist, it is sufficient that

$$\int_{0^-}^{\infty} |f(t)|\,\mathrm{e}^{-\sigma_1 t}\,\mathrm{d}t < \infty \tag{A1.2}$$

for some real, positive σ_1. This is true for almost all time functions of practical significance, but it does mean that certain functions mainly of mathematical interest, such as e^{t^2}, do not possess a Laplace transform.

We will now find the Laplace transforms of some familiar functions.

(i). $f(t) = 0,\ t < 0$

$\qquad = k_i,\ t > 0 \qquad (k_i \text{ a constant})$

From the definition,

$$F(s) = \int_{0^-}^{\infty} k_i \, e^{-st} \, dt$$

$$= -\left[\frac{k_i}{s} e^{-st}\right]_{0^-}^{\infty}$$

$$= \frac{k_i}{s}.$$

(ii). $f(t) = 0, \; t < 0$

$= k_a e^{-k_a t}, \; t > 0 \qquad (k_a \text{ a constant})$

$$F(s) = k_a \int_{0^-}^{\infty} e^{-(s+k_a)t} \, dt$$

$$= -\frac{k_a}{s + k_a} [e^{-(s+k_a)t}]_{0^-}^{\infty}$$

$$= \frac{k_a}{s + k_a}.$$

(iii). $f(t) = D\,\delta(t) \qquad (D \text{ a constant})$

$$F(s) = \int_{0^-}^{\infty} D\,\delta(t)\, e^{-st} \, dt$$

$$= D\, e^{-st}|_{t=0}$$

$$= D.$$

(Note here that if the value of 0^+, which appears in older text-books, had been used as the lower limit, $F(s)$ would have been zero.)

The Laplace transform of a derivative is of particular interest. From the definition,

$$\mathscr{L}\left(\frac{df}{dt}\right) = \int_{0^-}^{\infty} \frac{df}{dt} e^{-st} \, dt$$

so that, integrating by parts,

$$\mathscr{L}\left(\frac{df}{dt}\right) = [f(t)\, e^{-st}]_{0^-}^{\infty} + \int_{0^-}^{\infty} sf(t)\, e^{-st} \, dt$$

$$= -f(0^-) + sF(s). \qquad \text{(A1.3)}$$

Note that it has been assumed that $\lim_{t \to \infty} [f(t)\, e^{-st}] = 0$.

Similarly,
$$\mathscr{L}\left(\frac{d^2 f}{dt^2}\right) = s^2 F(s) - sf(0^-) - \frac{df}{dt}(0^-) \tag{A1.4}$$

and
$$\mathscr{L}\left(\frac{d^r f}{dt^r}\right) = s^r F(s) - s^{r-1} f(0^-) - s^{r-2}\frac{df}{dt}(0^-) - \ldots - \frac{d^{r-1} f}{dt^{r-1}}(0^-). \tag{A1.5}$$

A short list of important Laplace transform pairs is given in Table A.1. A much more extensive table, containing 520 pairs, is given by McCollum and Brown (1965).

Having found a Laplace transform $F(s)$, the time solution $f(t)$ is obtained by inverse Laplace transformation. This is given by the contour integral:

$$f(t) \equiv \mathscr{L}^{-1}[F(s)] = \frac{1}{2\pi j} \oint_{\sigma - j\infty}^{\sigma + j\infty} F(s)\, e^{st}\, ds \tag{A1.6}$$

where σ is a positive real quantity greater than the σ_1 convergence factor in eqn (A1.2). For most cases of interest, rather than having to use this expression, it is possible to use the Laplace transform table in the reverse direction to obtain $f(t)$ from $F(s)$. This often requires some manipulation of $F(s)$ to get it into a form which appears in the table, and this will be considered in Section 2. Before that, however, let us consider the simple constant-coefficient first-order ordinary differential equation relating the input $u(t)$ and output $x(t)$ of a linear system:

$$\frac{dx}{dt} = ax + bu \tag{A1.7}$$

where a and b are constants. Suppose the system is in a steady state with $u(t) = 0$ for $t < 0$. Taking Laplace transforms,

$$sX(s) - x(0^-) = aX(s) + bU(s)$$

so that
$$(s - a)X(s) = x(0^-) + bU(s)$$

giving
$$X(s) = \frac{x(0^-)}{s - a} + \frac{bU(s)}{s - a}. \tag{A1.8}$$

If $u(t) = \delta(t)$, for example, so that $U(s) = 1$,

$$X(s) = \frac{x(0^-)}{s - a} + \frac{b}{s - a}$$

with solution $x(t) = x(0^-)\, e^{at} + b\, e^{at}$.

If $u(t)$ were a different function of time, it would usually be necessary to use partial fractions to obtain the time solution from $X(s)$.

TABLE A.1. Important Laplace transform pairs

$f(t)$, $t > 0$	$F(s)$
Unit impulse, $\delta(t)$	1
Unit step, 1	$\frac{1}{s}$
t	$\frac{1}{s^2}$
t^2	$\frac{2}{s^3}$
t^N (N a positive integer)	$\frac{N!}{s^{N+1}}$
$e^{-\alpha t}$	$\frac{1}{s+\alpha}$
$t\,e^{-\alpha t}$	$\frac{1}{(s+\alpha)^2}$
$(1-\alpha t)\,e^{-\alpha t}$	$\frac{s}{(s+\alpha)^2}$
$\frac{\sin \omega t}{\omega}$	$\frac{1}{s^2+\omega^2}$
$\cos \omega t$	$\frac{s}{s^2+\omega^2}$
$e^{-\alpha t}\frac{\sin \omega t}{\omega}$	$\frac{1}{(s+\alpha)^2+\omega^2}$
$e^{-at}\cos \omega t$	$\frac{s+\alpha}{(s+\alpha)^2+\omega^2}$
$\frac{\sinh \omega t}{\omega}$	$\frac{1}{s^2-\omega^2}$
$\cosh \omega t$	$\frac{s}{s^2-\omega^2}$
$e^{-\alpha t}f(t)$	$F(s+\alpha)$
$f(t-a)$, with $f(t-a)=0$, $t<a$	$e^{-as}F(s)$
$\int_0^t f(u)\,du$	$\frac{1}{s}F(s)$
Convolution, $\int_0^t f_1(t-\lambda)f_2(\lambda)\,d\lambda$	$F_1(s)F_2(s)$

In all cases, $f(t) = 0$, $t < 0$.

2 Partial fractions and completing the square

Suppose that we wish to find the inverse Laplace transform of

$$F(s) \equiv \frac{P(s)}{Q(s)}$$

but that the form of $Q(s)$ does not correspond with any appearing in Table A.1. The denominator can often be factorized, and if $(s - \lambda_j)$ is an unrepeated factor of $Q(s)$, then a term $\dfrac{k_j}{s - \lambda_j}$ appears in $F(s)$ where the constant k_j is given by

$$k_j = \left.\frac{(s - \lambda_j)P(s)}{Q(s)}\right|_{s=\lambda_j}. \tag{A1.9}$$

To demonstrate this, let us return to eqn (A1.8) and consider the case where $u(t)$ is a unit step, so that $U(s) = \dfrac{1}{s}$. Then $X(s) = \dfrac{x(0^-)}{s-a} + \dfrac{b}{s(s-a)}$.

Taking the second term on the right-hand side,

$$\frac{b}{s(s-a)} = \frac{k_1}{s} + \frac{k_2}{s-a}, \qquad \text{where, from eqn (A1.9),}$$

$$k_1 = \left.\frac{sb}{s(s-a)}\right|_{s=0} = -\frac{b}{a}$$

and

$$k_2 = \left.\frac{(s-a)b}{s(s-a)}\right|_{s=a} = \frac{b}{a}.$$

Hence

$$X(s) = \frac{x(0^-)}{s-a} + \frac{b}{a}\left(\frac{1}{s-a} - \frac{1}{s}\right)$$

and

$$x(t) = x(0^-)\,\mathrm{e}^{at} + \frac{b}{a}(\mathrm{e}^{at} - 1). \tag{A1.10}$$

(Remember that a will be negative for most practical systems).

If $Q(s)$ cannot be split into distinct factors $(s - \lambda_1)$, $(s - \lambda_2)$ ... it can still often be factorized into a combination of first-order and second-order factors, the latter of the form $s^2 + \alpha s + \beta$ (including either or both of $\alpha = 0$, $\beta = 0$) which give rise to terms $\dfrac{As + B}{s^2 + \alpha s + \beta}$ in $F(s)$. The constants A and B are then evaluated by comparing the numerator of the partial-fractioned expression with that of the original expression. This will now be illustrated with two examples drawn from the main body of the book.

Example 1. (*Repeated eigenvalue*). In Section 3.2.2, we saw that, for a two-compartment model with $k_{12} = 0$ and $(k_{01} + k_{21}) = k_{02}$, the response, $y_2(t)$, to a constant continuous infusion to compartment 1 ($u_1(t) = k_i$, $t > 0$) has a Laplace transform

$$Y_2(s) = c_2 b_1 k_i k_{21}\left[\frac{1}{s(s + k_{02})^2}\right]. \tag{A1.11}$$

Considering the part in the square bracket only,

$$\frac{1}{s(s + k_{02})^2} = \frac{k_1}{s} + \frac{As + B}{(s + k_{02})^2}$$

where, from eqn (A1.9),

$$k_1 = \left.\frac{s}{s(s + k_{02})^2}\right|_{s=0} = \frac{1}{k_{02}^2}.$$

The other constants A and B are found by comparing numerators:

$$\frac{1}{s(s + k_{02})^2} = \frac{1}{k_{02}^2}\frac{1}{s} + \frac{As + B}{(s + k_{02})^2}$$

$$= \frac{\frac{1}{k_{02}^2}(s + k_{02})^2 + s(As + B)}{s(s + k_{02})^2}$$

so that $$1 = \frac{1}{k_{02}^2}s^2 + \frac{2}{k_{02}}s + 1 + As^2 + Bs.$$

Comparing terms in s^2,

$$0 = \frac{1}{k_{02}^2} + A, \qquad \text{whence} \quad A = -\frac{1}{k_{02}^2}.$$

Comparing terms in s,

$$0 = \frac{2}{k_{02}} + B, \qquad \text{whence} \quad B = -\frac{2}{k_{02}}.$$

Thus $$\frac{1}{s(s + k_{02})^2} = \frac{1}{k_{02}^2}\frac{1}{s} - \frac{\frac{1}{k_{02}^2}s + \frac{2}{k_{02}}}{(s + k_{02})^2}$$

$$= \frac{1}{k_{02}^2}\left(\frac{1}{s} - \frac{(s + k_{02}) + k_{02}}{(s + k_{02})^2}\right).$$

From the Laplace transform table,

$$\mathscr{L}^{-1}\left(\frac{1}{s(s+k_{02})^2}\right) = \frac{1}{k_{02}^2}(1 - \mathrm{e}^{-k_{02}t}) - \frac{1}{\mathrm{k}_{02}}\, t\, \mathrm{e}^{-k_{02}t}$$

and eqn (3.35b) follows on multiplying by the original numerator constant $c_2 b_1 k_i k_{21}$ in eqn (A1.11).

Example 2. In Section 4.1.6, we found that the Laplace transforms of the responses of the three compartments of the model of Fig. 4.13 to a unit impulse perturbation of compartment 1 were given by

$$X_1(s) = \frac{(s+k)^2}{s(s^2+3ks+3k^2)}$$

$$X_2(s) = \frac{k(s+k)}{s(s^2+3ks+3k^2)}$$

$$X_3(s) = \frac{k^2}{s(s^2+3ks+3k^2)}.$$

The quadratic term in the denominators does not factorize. Taking $X_1(s)$ first,

$$X_1(s) = \frac{k_1}{s} + \frac{As+B}{s^2+3ks+3k^2}$$

where, from eqn (A1.9),

$$k_1 = \left.\frac{s(s+k)^2}{s(s^2+3ks+3k^2)}\right|_{s=0} = \frac{1}{3}.$$

Equating numerators,

$$s^2 + 2ks + k^2 = \tfrac{1}{3}(s^2 + 3ks + 3k^2) + As^2 + Bs.$$

Comparing the terms in s^2, $1 = \frac{1}{3} + A$, whence $A = \frac{2}{3}$.

Comparing the terms in s, $2k = k + B$, whence $B = k$.

Thus $$X_1(s) = \frac{1}{3s} + \frac{1}{3}\,\frac{2s+3k}{s^2+3ks+3k^2}.$$

The denominator of the second term on the right-hand side does not appear in this form in Table A.1, and it is necessary to complete the square by writing

$$s^2 + 3ks + 3k^2 \equiv (s + \tfrac{3}{2}k)^2 + (3k^2 - \tfrac{9}{4}k^2) = (s + 3k/2)^2 + \tfrac{3}{4}k^2,$$

so that $$X_1(s) = \frac{1}{3s} + \frac{1}{3}\,\frac{2(s+3k/2)}{(s+3k/2)^2 + 3k^2/4}.$$

From the entry $\mathscr{L}(\mathrm{e}^{-\alpha t}\cos\omega t) = \dfrac{s+\alpha}{(s+\alpha)^2+\omega^2}$ in the Table, it may be seen that for our example, $\alpha = \dfrac{3k}{2}$ and $\omega = \dfrac{k\sqrt{3}}{2}$, so

$$x_1(t) = \frac{1}{3}\left(1 + 2\,\mathrm{e}^{-3kt/2}\cos\frac{k\sqrt{3}}{2}t\right)$$

as given in eqn (4.22b).

Taking partial fractions for $X_2(s)$ gives

$$X_2(s) = \frac{1}{3}\left[\frac{1}{s} - \frac{s}{s^2+3ks+3k^2}\right]$$

so that, on completing the square in the denominator of the second term

$$X_2(s) = \frac{1}{3}\left[\frac{1}{s} - \frac{s}{(s+3k/2)^2+3k^2/4}\right].$$

To make use of the entries in Table A.1 that

$$\mathscr{L}(\mathrm{e}^{-\alpha t}\cos\omega t) = \frac{s+\alpha}{(s+\alpha)^2+\omega^2}$$

and

$$\mathscr{L}\left(\mathrm{e}^{-\alpha t}\frac{\sin\omega t}{\omega}\right) = \frac{1}{(s+\alpha)^2+\omega^2},$$

the numerator of the second term must be written in the form $(s+3k/2) - 3k/2$, so that, on inverse Laplace transformation,

$$x_2(t) = \frac{1}{3}\left[1 - \mathrm{e}^{-3kt/2}\left(\cos\frac{k\sqrt{3}}{2}t - \frac{3k}{2}\frac{\sin(k\sqrt{3}/2)t}{k\sqrt{3}/2}\right)\right]$$

as given in eqn (4.22c). The response $x_3(t)$ follows in a similar manner.

As an exercise, the reader is invited to confirm the inverse Laplace transformation of $\dfrac{(s+2k)^2}{(s+k)(s^2+5ks+7k^2)}$, given in Example 4.5.

3 Transfer functions

The transfer function, $G(s)$, of a system is the ratio of the Laplace transform of the output of the system to that of the input. The initial conditions are assumed zero. Thus

$$G(s) = \frac{X(s)}{U(s)}. \tag{A1.12}$$

For eqn (A1.6),

$$G(s) = \frac{b}{s - a}.$$

As an example of its use, the unit step response for this system (with zero initial conditions) is obtained from

$$X(s) = G(s)U(s) = \frac{b}{s - a}\,\frac{1}{s}$$

giving $x(t) = \frac{b}{a}(e^{at} - 1)$, which is the same as the solution (A1.10) with $x(0^-)$ zero.

4 Impulsive inputs

In the definition of the Laplace transform, we have deliberately taken the lower limit as $t = 0^-$ so that $\mathscr{L}[\delta(t)] = 1$. On some occasions, as in the Taylor series identifiability analysis given in Chapters 9 and 10, it is necessary to consider the behaviour of a system from $t = 0^+$ onwards. Consider again eqn (A1.7):

$$\frac{dx}{dt} = ax + bu$$

with $x(0^-) = 0$. Assume $u(t) = \delta(t)$. We have seen above that

$$X(s) = \frac{b}{s - a}$$

so that
$$x(t) = b\,e^{at}.$$

If we now consider the solution for $t > 0$, $u(t)$ within the time range 0^+ to ∞ is zero, so that the equation to be solved is

$$\frac{dx}{dt} = ax, \qquad t > 0$$

with $x(0^+) = b$ which has the same solution as above.

The solution of a differential equation with impulsive input at $t = 0$ can therefore be thought of either in terms of the initial condition at $t = 0^-$ (immediately before the impulse is applied) with the impulsive input included in the differential equation or in terms of the initial condition at $t = 0^+$ (immediately after the impulse) with the impulsive input excluded from the differential equation.

5 Initial-value and final-value theorems

Two theorems which often prove useful (for example, in checking Laplace transforms) are the initial-value theorem:

$$\lim_{t \to 0^+} f(t) = \lim_{s \to 0} sF(s). \tag{A1.13}$$

and the final-value theorem:

$$\lim_{t \to \infty} f(t) = \lim_{s \to 0} sF(s). \tag{A1.14}$$

(In the initial-value theorem, it is necessary that the degree of the denominator polynomial of $F(s)$ should be greater than that of the numerator polynomial of $F(s)$).

We have made use of the initial-value theorem in Chapters 3 and 4. If the numerator degree of $F(s)$ is two less than the denominator degree, then, from eqn (A1.13), the unit impulse response $f(t)$ has initial value $f(0^+)$ zero. For example, if

$$F(s) = \frac{K}{s^2 + \alpha s + \beta}$$

then
$$f(0^+) = \lim_{s \to \infty} sF(s) = \lim_{s \to \infty} \frac{Ks}{s^2 + \alpha s + \beta} = 0.$$

Note that the initial value of the slope, $\frac{\mathrm{d}f}{\mathrm{d}t}(0^+)$ is given by

$$\lim_{s \to \infty} s(sF(s)) = \lim_{s \to \infty} \frac{Ks^2}{s^2 + \alpha s + \beta} = K.$$

The initial value of the slope would have been zero if the degree of the numerator of $F(s)$ had been at least three less than that of the denominator.

Appendix 2 Elementary Matrix Algebra

A matrix is an array of real (or possibly complex) numbers written in the form

$$\mathbf{A} = \begin{bmatrix} a_{11} & a_{12} & \cdots & a_{1n} \\ a_{21} & a_{22} & \cdots & a_{2n} \\ \vdots & & & \vdots \\ a_{m1} & a_{m2} & \cdots & a_{mn} \end{bmatrix}.$$

The numbers that constitute **A** are the elements of **A**. In the above example, there are m rows and n columns and **A** is referred to as an $(m \times n)$ matrix.

Two matrices of the same size can be added together, element by element. Multiplication of a matrix by a scalar (i.e. a number) is accomplished by multiplying every element by the scalar.

The transpose, $\mathbf{A}^{\mathrm{T}}$, of a matrix is the $(n \times m)$ matrix obtained by interchanging rows and columns of **A**:

$$\mathbf{A}^{\mathrm{T}} = \begin{bmatrix} a_{11} & a_{21} & \cdots & a_{m1} \\ a_{12} & a_{22} & \cdots & a_{m2} \\ \vdots & & & \vdots \\ a_{1n} & a_{2n} & \cdots & a_{mn} \end{bmatrix}.$$

A square matrix has $m = n$. If a square matrix has $\mathbf{A}^{\mathrm{T}} = \mathbf{A}$, i.e. $a_{ij} = a_{ji}$, then **A** is a symmetric matrix.

A vector is conventionally taken to be a column:

$$\mathbf{x} = \begin{bmatrix} x_1 \\ x_2 \\ \vdots \\ x_n \end{bmatrix}$$

i.e. an $(n \times 1)$ matrix. Thus $\mathbf{x}^{\mathrm{T}} = [x_1 x_2 \ldots x_n]$, a row vector, is a $(1 \times n)$ matrix.

Two matrices **A** and **B** can be multiplied together provided the number of columns in **A** equals the number of rows in **B**. If **A** is $(m \times n)$ and **B** is $(n \times p)$, then the product **Q** is $(m \times p)$ and has elements given by

$$Q_{ij} = \sum_{k=1}^{n} a_{ik} b_{kj}.$$

Thus
$$\begin{bmatrix} a & b & c \\ d & e & f \end{bmatrix} \times \begin{bmatrix} u & v \\ w & x \\ y & z \end{bmatrix} = \begin{bmatrix} (au + bw + cy) & (av + bx + cz) \\ (du + ew + fy) & (dv + ex + fz) \end{bmatrix}.$$

For two square matrices, **AB** and **BA** both exist, but, in general are unequal. If **AB** = **BA**, then **A** and **B** are said to commute.

A diagonal matrix is square and has all elements except those on the main diagonal (the diagonal from the top left-hand corner to the bottom right-hand corner) zero. In particular, the unit matrix

$$\mathbf{I} \triangleq \begin{bmatrix} 1 & 0 & 0 & \ldots & 0 \\ 0 & 1 & 0 & \ldots & 0 \\ 0 & 0 & 1 & \ldots & 0 \\ . & . & . & \ldots & . \\ 0 & 0 & 0 & \ldots & 1 \end{bmatrix}$$

is a diagonal matrix.

The determinant $|\mathbf{A}|$ of a matrix can only exist if **A** is square; $|\mathbf{A}|$ is a scalar quantity. For a 2×2 matrix,

$$|\mathbf{A}| = \begin{vmatrix} a_{11} & a_{12} \\ a_{21} & a_{22} \end{vmatrix} = a_{11}a_{22} - a_{12}a_{21}$$

while for a 3×3 matrix,

$$|\mathbf{A}| = \begin{vmatrix} a_{11} & a_{12} & a_{13} \\ a_{21} & a_{22} & a_{23} \\ a_{31} & a_{32} & a_{33} \end{vmatrix} = a_{11} \begin{vmatrix} a_{22} & a_{23} \\ a_{32} & a_{33} \end{vmatrix} - a_{12} \begin{vmatrix} a_{21} & a_{23} \\ a_{31} & a_{33} \end{vmatrix} + a_{13} \begin{vmatrix} a_{21} & a_{22} \\ a_{31} & a_{32} \end{vmatrix}$$

$$= a_{11}(a_{22}a_{33} - a_{23}a_{32}) - a_{12}(a_{21}a_{33} - a_{23}a_{31}) + a_{13}(a_{21}a_{32} - a_{22}a_{31}).$$

Higher-order determinants follow readily; note the alternation in sign of the successive terms.

Solutions of simultaneous algebraic equations can easily be written in terms of determinants using Cramer's rule. If

$$a_{11}x + a_{12}y + a_{13}z = b_1$$
$$a_{21}x + a_{22}y + a_{23}z = b_2$$
$$a_{31}x + a_{32}y + a_{33}z = b_3$$

then

$$x = \frac{\begin{vmatrix} b_1 & a_{12} & a_{13} \\ b_2 & a_{22} & a_{23} \\ b_3 & a_{32} & a_{33} \end{vmatrix}}{|\mathbf{A}|} \quad y = \frac{\begin{vmatrix} a_{11} & b_1 & a_{13} \\ a_{21} & b_2 & a_{23} \\ a_{31} & b_3 & a_{33} \end{vmatrix}}{|\mathbf{A}|} \quad \text{and} \quad z = \frac{\begin{vmatrix} a_{11} & a_{12} & b_1 \\ a_{21} & a_{22} & b_2 \\ a_{31} & a_{32} & b_3 \end{vmatrix}}{|\mathbf{A}|}$$

The minor of element a_{ij} of a square matrix $\mathbf{A}$ is the determinant of the matrix $\mathbf{E}_{ij}$ obtained by deletion of the ith row and jth column of $\mathbf{A}$. Thus if

$$\mathbf{A} = \begin{bmatrix} a_{11} & a_{12} & \dots & a_{1n} \\ a_{21} & a_{22} & \dots & a_{2n} \\ \vdots & \vdots & & \\ a_{n1} & a_{n2} & \dots & a_{nn} \end{bmatrix}$$

then

$$\mathbf{E}_{12} = \begin{bmatrix} a_{21} & a_{23} & \dots & a_{2n} \\ a_{31} & a_{33} & \dots & a_{3n} \\ \vdots & & & \\ a_{n1} & a_{n3} & \dots & a_{nn} \end{bmatrix}.$$

The cofactor, c_{ij}, of a_{ij} is defined as $(-1)^{i+j}|\mathbf{E}_{ij}|$. The transposed matrix of cofactors, $\mathbf{C}^{\mathrm{T}}$, is called the adjugate of $\mathbf{A}$ (adj $\mathbf{A}$ for short) and is useful in finding the inverse $\mathbf{A}$.

The inverse, $\mathbf{A}^{-1}$, of a square matrix $\mathbf{A}$ is defined from

$$\mathbf{A}\mathbf{A}^{-1} = \mathbf{A}^{-1}\mathbf{A} = \mathbf{I}.$$

For $|\mathbf{A}| \neq 0$,

$$\mathbf{A}^{-1} = \frac{\text{adj } \mathbf{A}}{|\mathbf{A}|}.$$

By way of example, consider

$$\mathbf{A} = \begin{bmatrix} 1 & 1 & 1 \\ 1 & 2 & 1 \\ 0 & 1 & -1 \end{bmatrix}.$$

From the above,

$$|\mathbf{A}| = \begin{vmatrix} 2 & 1 \\ 1 & -1 \end{vmatrix} - \begin{vmatrix} 1 & 1 \\ 0 & -1 \end{vmatrix} + \begin{vmatrix} 1 & 2 \\ 0 & 1 \end{vmatrix}$$

$$= -3 + 1 + 1 = -1.$$

The minor of $a_{11} = \begin{vmatrix} 2 & 1 \\ 1 & -1 \end{vmatrix} = -3$

while the minor of $a_{12} = \begin{vmatrix} 1 & 1 \\ 0 & -1 \end{vmatrix} = -1$ and so on.

Thus the co-factor of $a_{11} = (-1)^2(-3) = -3$

while that of $a_{12} = (-1)^3(-1) = 1$,

so that the matrix of co-factors is

$$\mathbf{C} = \begin{bmatrix} -3 & 1 & 1 \\ 2 & -1 & -1 \\ -1 & 0 & 1 \end{bmatrix} \quad \text{and} \quad \operatorname{adj} \mathbf{A} \triangleq \mathbf{C}^{\mathrm{T}} = \begin{bmatrix} -3 & 2 & -1 \\ 1 & -1 & 0 \\ 1 & -1 & 1 \end{bmatrix}.$$

Hence $$\mathbf{A}^{-1} = -\begin{bmatrix} -3 & 2 & -1 \\ 1 & -1 & 0 \\ 1 & -1 & 1 \end{bmatrix} = \begin{bmatrix} 3 & -2 & 1 \\ -1 & 1 & 0 \\ -1 & 1 & -1 \end{bmatrix}.$$

Checking

$$\mathbf{A}\mathbf{A}^{-1} = \begin{bmatrix} 1 & 1 & 1 \\ 1 & 2 & 1 \\ 0 & 1 & -1 \end{bmatrix} \begin{bmatrix} 3 & -2 & 1 \\ -1 & 1 & 0 \\ -1 & 1 & -1 \end{bmatrix} = \begin{bmatrix} 1 & 0 & 0 \\ 0 & 1 & 0 \\ 0 & 0 & 1 \end{bmatrix} = \mathbf{I}.$$

Index

H

I

K

L

M

N

O

P

R